School of Athens, from the Stanza della Segnatura, 1510-11 (fresco), Raphael (Raffaello Sanzio of Urbino) (1483-1520)/Vatican Museums and Galleries, Vatican City, Italy/Giraudon/The Bridgeman Art Library. Legend has it that over the door to Plato's Academy in Athens there was an inscription "Let no man ignorant of geometry enter here." Words to live by, in antiquity and today.

ISOGEOMETRIC ANALYSIS

ISOGEOMETRIC ANALYSIS
TOWARD INTEGRATION OF CAD AND FEA

J. Austin Cottrell

Systematic Options Trading
Citigroup Inc, USA

Thomas J. R. Hughes

Institute for Computational Engineering and Sciences
The University of Texas at Austin, USA

Yuri Bazilevs

Department of Structural Engineering
University of California, San Diego, USA

A John Wiley and Sons, Ltd., Publication

Registered office
John Wiley & Sons Ltd, The Atrium, Southern Gate, Chichester, West Sussex, PO19 8SQ, United Kingdom

For details of our global editorial offices, for customer services and for information about how to apply for permission to reuse the copyright material in this book please see our website at www.wiley.com.

Library of Congress Cataloguing-in-Publication Data

Cottrell, J. Austin.
   Isogeometric analysis : toward integration of CAD and FEA / J. Austin Cottrell, Thomas J. R. Hughes, Yuri Bazilevs.
      p. cm.
   Includes bibliographical references and index.
   ISBN 978-0-470-74873-2 (cloth)
   1. Finite element method–Data processing.   2. Spline theory–Data processing.   3. Isogeometric analysis–Data processing.   4. Computer-aided design.   I. Thomas J. R. Hughes.   II. Bazilevs, Yuri.   III. Title.
   TA347.F5C685 2009
   620.001'51825–dc22

                                                                                  2009019919

A catalogue record for this book is available from the British Library.

ISBN: 978-0-470-74873-2

Typeset in 10/12pt Times by Aptara Inc., New Delhi, India

Contents

Preface

The initial work in isogeometric analysis was motivated by the existing gap between the worlds of finite element analysis (FEA) and computer-aided design (CAD); see Hughes *et al.*, 2005. As the number of people involved with isogeometric analysis from both the FEA and the CAD communities has grown, this gap has become increasingly apparent to all involved. It is not only a shortcoming of the current technology but of the entire engineering process. Indeed, technological barriers are often easier to overcome than the inertia of the status quo. At this early stage, one of the most important contributions of the research in isogeometric analysis has been to initiate conversation between the design and analysis camps, and to begin to make each side aware of the hurdles that the other faces, as well as what each has to offer. This book is meant to be part of that dialogue.

What are we providing and for whom?

Isogeometric analysis seeks to unify the fields of CAD and FEA. In pursuing this end we have found, with very few exceptions, that FEA people know very little about computational geometry, and computational geometry people know very little about FEA. Our background is in FEA. We have attempted to cross the divide and learn from and work with computational geometers in order to orchestrate changes in CAD and FEA that will result in an agreed upon isogeometric technology satisfactory to both constituencies. That being said, we are neophytes in computational geometry so nothing fundamentally new on that topic will be found herein. Our most immediate goals are to encourage computational analysts to learn about isogeometric analysis and to begin to take advantage of it in their work. Specifically, we have attempted to build upon a knowledge of finite element analysis and to indicate what is new and different about isogeometric analysis. A background in finite element analysis at the level of Hughes, 2000 is ideal preparation for understanding this book. Most of the book, however, is sufficiently self-contained as to not require that much finite element background. We wrote this book so that the reader could learn how to *do* isogeometric analysis.

We wanted this book to be accessible, in fact, easy to read and learn from, but we did not want to superficially gloss over important details to achieve simplicity. Although computational mechanics has become a sophisticated and complex discipline, the essence of the finite element method is quite simple and straightforward. The same may be said of isogeometric analysis, and we endeavored to express this viewpoint in this book. Nevertheless, certain basics of computational geometry need to be learned, and these are not part of the traditional training and repertoire of finite element analysts. We have tried to present them in a clear and direct manner. We at least hope the book is simple enough that most motivated readers will be able to learn the essential ideas.

We assumed that many readers would want to add isogeometric capabilities to existing finite element computer programs, so we developed this theme right from the start. The early chapters deal with the basic concepts, how to implement them, and how to handcraft isogeometric models. The latter chapters attempt to demonstrate convincingly why one might want to do so. By explaining the details of Non-Uniform Rational B-Spline (NURBS) basis functions and showing how their unique properties come to bear on a wide variety of applications, we hope to motivate others to consider how their own research might benefit from these powerful functions.

There are many computational geometry technologies that could serve as a basis for isogeometric analysis. The reason for selecting NURBS as the initial basis is compelling: It is the most widely used computational geometry technology in engineering design. Unfortunately, at this stage of the game, an isogeometric modeling toolset is not available. We hope that this void will be filled in the not-too-distant future and be made available to the community. Research projects are already underway with this as one of the goals.

Although we present some applications of isogeometric analysis that have appeared previously in research papers, a conscious effort has been made to present material *not* in research papers, in particular, detailed examples and data sets are presented that one needs to thoroughly understand to gain a working knowledge of the material. Another theme has been to only show examples and applications that exhibit some unique feature of isogeometric analysis not available in traditional finite element analysis. One might consider isogeometric analysis as simply an expansion and powerful generalization of traditional finite element analysis.

Channeling developments in order to make them more relevant to downstream engineering

We would like to help people on each side of the CAD/FEA divide to further the state of their respective arts. By being aware of the their own place in the idea-to-product process, both the geometer and analyst might strive to design technologies that are integrative and avoid creating bottlenecks at any stage of the engineering process. We have no doubt that the futures of CAD and FEA lie much closer together than do their pasts. The reader is invited to participate in the effort to unify these fields.

Organization of the text

This book begins in Chapter 1 with an historical perspective on the fields of finite element analysis and computer aided design. This provides a context from which the ideas throughout the book have emerged. Additionally, we briefly point out some of the issues of isogeometric analysis that seem to cause some confusion for researchers coming from a classical FEA background. Each of these issues is discussed in detail within the body of the text, but it may prove useful for the reader to be aware of them before embarking. We then introduce Non-Uniform Rational B-Splines (NURBS) in Chapter 2, with an initial focus on geometric design and the particular features that make this technology unique. A brief tutorial on the construction of a NURBS geometry is included. Chapter 3 describes how computer aided design technology can be used within an analysis framework. Particular attention is given to a detailed explanation of the Galerkin finite element method as this is the framework within which the bulk of isogeometric analysis has been performed. Chapter 3 also includes a discussion of how classical finite element software might be modified to create isogeometric analysis software.

The bulk of the remainder of this book contains examples of the many different applications to which isogeometric analysis has been applied. The specific choice of material is meant to emphasize the interesting properties of NURBS basis functions and to display the unique capabilities of an analysis framework built upon them. The examples increase in complexity as the book progresses, loosely chronicling the evolution of the technology. For the most part, linear problems are discussed before nonlinear problems, and static problems precede time-dependent problems. Chapters 6 and 7 provide general discussions of time-dependent problems and nonlinear problems, respectively. The reader unfamiliar with these topics may want to review these chapters before proceeding to chapters on such applications. We attempt to be quite thorough on the simpler examples, providing everything needed for an individual just getting started to be able to perform a calculation. Contrastingly, there is a bias towards brevity for the more complex problems. The treatment of examples from the forefront of research is meant to highlight the specific features of isogeometric analysis upon which these applications rely. Whenever details necessary to replicate the work are omitted, references to the literature where those details may be found are included. Still, every effort is made to tie the implementation used and the results obtained to the features of isogeometric analysis that differ from classical finite elements.

Chapter 4 discusses linear elasticity, with a particular emphasis on the analysis of thin-walled structures. Chapter 5 covers vibrations and wave propagation. Whereas the examples considered in Chapter 4 particularly benefit from the geometrical accuracy of isogeometric analysis, the examples in Chapter 5 demonstrate the accuracy advantages NURBS have over classical finite elements due to their increased smoothness. In Chapter 6 we move from static to dynamic problems and discuss various time-integration techniques that are in common usage. Chapter 7 discusses the solution of nonlinear equations, and it expands on the discussion of Chapter 6 to address solving nonlinear, time-dependent problems by means of the generalized-α method. Chapter 8 discusses one approach to addressing the locking phenomenon common in the analysis of both linear and nonlinear nearly incompressible solids. Chapter 9 features many examples from the field of computational fluid dynamics, ranging from the linear advection-diffusion equation to turbulence. In all cases, smooth NURBS basis functions are shown to achieve superior accuracy per degree-of-freedom than the classical FEA basis functions of the same order. Chapter 9 also presents the variational multiscale method. Fluid-structure interaction and fluids problems posed on moving domains are discussed in Chapter 10. Each of these problems requires care in tracking the motion of the mesh and correctly formulating the equations on the moving domain. Chapter 11 discusses partial differential equations in which the highest order derivative is greater than two. A traditional variational treatment of such problems requires the use of basis functions that are smoother than C^0. This is frequently difficult or impossible in a classical FEA setting, but is quite easy within isogeometric analysis. Chapter 12 discusses polar forms, which offer an alternative mathematical description of splines. The use of polar forms has been instrumental in the development of efficient algorithms for the manipulation of spline objects. Lastly, Chapter 13 discusses the current state-of-the-art in isogeometric analysis, as well as many promising directions for future work in the subject.

Additional resources

There are many places for the interested reader to seek more information about the topics discussed in this book. Though an effort has been made to make this book as self-contained as possible, it is not possible to address every topic in the full generality that it deserves. For a more thorough discussion of NURBS we suggest starting with Rogers, 2001 and then going on to Piegl and Tiller, 1997. The former is quite readable and features many historical perspectives on NURBS and those whose work has led to their development; the latter is quite comprehensive and served as an indispensable guide when we were developing our initial software. Here is a list of geometry books we have found helpful, including the two already mentioned. It is by no means complete, and we are still learning from them.

- *Geometric Modeling with Splines: An Introduction*, E. Cohen *et al.*, 2001
- *Curves and Surfaces for CAGD, A Practical Guide, Fifth Edition*, G.E. Farin, 1999a
- *NURBS Curves and Surfaces: from Projective Geometry to Practical Use, Second Edition*, G.E. Farin, 1999b
- *Computational Conformal Geometry, Theory and Algorithms*, X.D. Gu and S.-T. Yau, 2008
- *The NURBS Book*, L. Piegl and W. Tiller, 1997
- *Bézier and B-Spline Techniques*, H. Prautzsch, W. Boehm and M. Paluszny, 2002
- *An Introduction to NURBS: With Historical Perspective*, D.F. Rogers, 2001
- *Spline Functions: Basic Theory (third edition)*, L.L. Schumaker, 2007

For an introductory but thorough treatise on the finite element method, see Hughes, 2000. We attempt as far as possible to be consistent with the notation of Hughes, 2000, which we will make reference to many times throughout this book. The best source for information on the many applications contained herein is in the research papers upon which much of the content is based. Each chapter provides references to original journal articles, which frequently discuss the topics in a great deal more depth, and with many more examples, than is possible here.

Notation

A word of caution is in order. Notational conventions that are very illustrative in simple settings, particularly when introducing a concept for the first time, frequently become unwieldy as things become more complex. For this reason, we attempt to use the notation that provides the most clarity in a given situation, though this choice is sometimes at odds with other usage. Whenever there is the potential for confusion, the issue is addressed directly herein.

How work on isogeometric analysis began

Isogeometric analysis began when Tom Hughes was privy to a conversation concerning the creation of finite element models from CAD representations. The gist of the conversation expressed the theme that despite years of research into mesh generation, the model creation problem was a significant bottleneck to the effective use of FEA and, for complex engineering designs, the problem seemed to be getting worse. It appeared to Tom that if the situation was that bad, the problem must either be very difficult *or* the research community was pursuing a solution from the wrong perspective. After some study he concluded that the problem *as posed* was indeed very difficult, but not only was the research community pursuing it from the wrong perspective, it was pursuing the wrong problem.

CAD and FEA grew up independently. Despite dealing with the *same* objects, engineering designs, they represent them with entirely different geometrical constructs. This seemed to be the fundamental problem. Tom hoped to replace this situation with a single, agreed upon, geometrical description. He thought that he might be able to reconstitute analysis within the geometric framework of CAD technologies. This seemed doable, but it also became apparent that CAD representations would have to be enhanced. He was surprised to find that newer technologies emanating from the computational geometry research literature were actually moving in that direction and that some of these technologies were finding their way into commercial products. The final piece of the puzzle, developing analysis suitable trivariate parameterizations from surface representations, is an open problem but one that is beginning to be addressed by the computational geometry community with new and promising mathematical approaches. The confluence of all these activities is ***Isogeometric Analysis***. Through the combined efforts of the computational geometry and computational analysis communities, we believe the potential of isogeometric analysis can be realized.

Work on isogeometric analysis began in earnest in 2003 almost a year after Tom Hughes joined the University of Texas at Austin. He had received a research grant to pursue the topic, but did not have a PhD student assigned to it. A then first-year graduate student, Austin Cottrell, in the Computational and Applied Mathematics program in the Institute for Computational Engineering and Sciences came to talk to Tom about research topics and possibly doing his PhD under Tom's supervision. Among other topics, Tom described to Austin his vision of this as yet nameless new approach to analysis and geometry. After thinking things over, Austin said he would like to pursue it and could get started in the summer of 2003. Tom gave Austin a copy of Rogers' book on NURBS.

As Austin was making progress with NURBS technology, another of Tom's graduate students, Yuri Bazilevs, started to interact with him on it, and the two of them implemented the first NURBS based finite element codes. By the fall of 2003, linear problems were being solved and good results were being obtained. It was around that time that the name "isogeometric analysis" was coined. Rapid progress was being made developing the technology. Before long, isogeometric analysis became an integral part of all work in Tom's group. After completing his PhD, Victor Calo also joined the effort, as did a number of other students, post-docs, and visitors to the Institute, including Ido Akkerman, Laurenco Beirão da Veiga, David Benson, Thomas Elguedj, John Evans, Héctor Gómez, Scott Lipton, Alessandro Reali, Giancarlo Sangalli, Mike Scott, and Jessica Zhang. The effectiveness of the procedures and the richness of the subject exceeded everyone's expectations.

How this book was written

Discussions about writing a book occurred frequently during the course of the work. It was decided that a good time to start would be after Austin and Yuri completed their PhDs. The project began in earnest in September of 2007. The plan was to release Austin from all other obligations and have him rough draft as much as possible, as quickly as possible, and then he and Tom would begin to iterate on the drafts. Austin and Tom put together an outline and set as a goal to be finished, or at least declare they were finished, by the end of July, 2008, when Austin was scheduled to leave for New York City. Realizing this schedule might be a bit optimistic, it was intended that Yuri, who provided numerous results and helped in a variety of ways throughout the project, would step in after Austin left and that he and Tom would complete the project. Things more or less unfolded as planned.

Acknowledgments

We would like to thank our collaborators on the work contained in this volume. In particular, the efforts of Ido Akkerman, Laurenco Beirão da Veiga, Victor Calo, Thomas Elguedj, John Evans, Héctor Gómez, Scott Lipton, Alessandro Reali, Mike Scott, Guglielmo Scovazzi, and Jessica Zhang have all led directly to the examples in this book. Your efforts are all greatly appreciated, and we look forward to many fruitful collaborations again in the future. We would also like to thank Tom Sederberg for making geometry "analysis-suitable," Omar Ghattas for providing insights into current trends in supercomputing and describing how isogeometric analysis may be a key enabling technology for taking advantage of modern computer hardware, and Ted Belytschko, Elaine Cohen, Tom Lyche, Rich Riesenfeld, and Tom Sederberg for helpful comments and suggestions concerning an initial draft of this book.

Much of the funding for the research which led to this book was provided by the Office of Naval Research Computational Mechanics Program, under the direction of Dr. Luise Couchman. This support is gratefully acknowledged.

J. Austin Cottrell
Thomas J. R. Hughes
Yuri Bazilevs

1

From CAD and FEA to Isogeometric Analysis: An Historical Perspective

1.1 Introduction

1.1.1 The need for isogeometric analysis

It may seem inconceivable to young engineers, but it was not long ago that computers were nowhere to be seen in design offices. Designers worked at drawing boards and designs were drawn with pencils on vellum or Mylar[1]. The design drawings were passed to stress analysts and the interaction between designer and analyst was simple and direct. Times have changed. Designers now generate CAD (Computer Aided Design) files and these must be translated into analysis-suitable geometries, meshed and input to large-scale finite element analysis (FEA) codes. This task is far from trivial and for complex engineering designs is now estimated to take over 80% of the overall analysis time, and engineering designs are becoming increasingly more complex; see Figure 1.1. For example, presently, a typical automobile consists of about 3,000 parts, a fighter jet over 30,000, the Boeing 777 over 100,000, and a modern nuclear submarine over 1,000,000. Engineering design and analysis are not separate endeavors. Design of sophisticated engineering systems is based on a wide range of computational analysis and simulation methods, such as structural mechanics, fluid dynamics, acoustics, electromagnetics, heat transfer, etc. Design speaks to analysis, and analysis speaks to design. However, analysis-suitable models are not automatically created or readily meshed from CAD geometry. Although not always appreciated in the academic analysis community, model generation is much more involved than simply generating a mesh. There are many time consuming, preparatory steps involved. And one mesh is no longer enough. According to Steve Gordon, Principal Engineer, General Dynamics / Electric Boat Corporation, "We find that today's bottleneck in CAD-CAE integration is not only automated mesh generation, it lies with efficient creation of appropriate 'simulation-specific' geometry." (In the commercial sector analysis is usually referred to as CAE, which stands for Computer Aided Engineering.) The anatomy of the process has been studied by Ted Blacker, Manager of Simulation Sciences, Sandia National Laboratories. At

Isogeometric Analysis: Toward Integration of CAD and FEA by J. A. Cottrell, T. J. R. Hughes, Y. Bazilevs
© 2009, John Wiley & Sons, Ltd

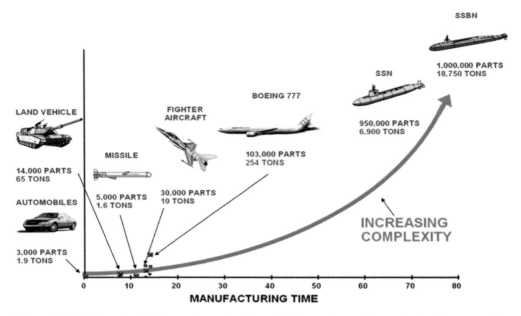

Figure 1.1 Engineering designs are becoming increasingly complex, making analysis a time consuming and expensive endeavor. (Courtesy of General Dynamics / Electric Boat Corporation).

Sandia, mesh generation accounts for about 20% of overall analysis time, whereas creation of the analysis-suitable geometry requires about 60%, and only 20% of overall time is actually devoted to analysis per se; see Figure 1.2. The 80/20 modeling/analysis ratio seems to be a very common industrial experience, and there is a strong desire to reverse it, but so far little progress has been made, despite enormous effort to do so. The integration of CAD and FEA has proven a formidable problem. It seems that fundamental changes must take place to fully integrate engineering design and analysis processes.

Recent trends taking place in engineering analysis and high-performance computing are also demanding greater precision and tighter integration of the overall modeling-analysis process. We note that a finite element mesh is only an approximation of the CAD geometry, which we view as "exact." This approximation can in many situations create errors in analytical results. The following examples may be mentioned: Shell buckling analysis is very sensitive to geometric imperfections (see Figure 1.3), boundary layer phenomena are sensitive to the precise geometry of aerodynamic and hydrodynamic configurations (see Figures 1.4 and 1.5), and sliding contact between bodies cannot be accurately represented without precise geometric descriptions (see Figure 1.6). The Babuška paradox (see Birkhoff and Lynch, 1987) is another example of the pitfalls of polygonal approximations to curved boundaries. Automatic adaptive mesh refinement has not been as widely adopted in industry as one might assume from the extensive academic literature, because mesh refinement requires access to the exact geometry and thus seamless and automatic communication with CAD, which simply does not exist. Without accurate geometry and mesh adaptivity, convergence and high-precision results are impossible.

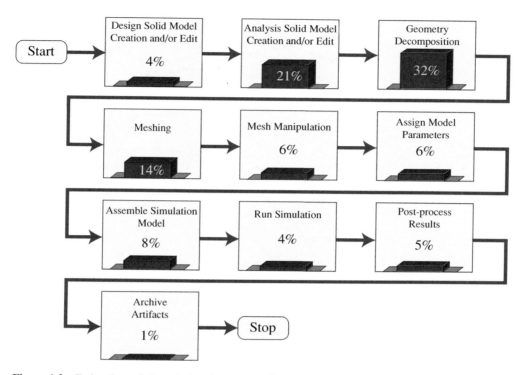

Figure 1.2 Estimation of the relative time costs of each component of the model generation and analysis process at Sandia National Laboratories. Note that the process of building the model completely dominates the time spent performing analysis. (Courtesy of Michael Hardwick and Robert Clay, Sandia National Laboratories.).

Deficiencies in current engineering analysis procedures also preclude successful application of important pace-setting technologies, such as design optimization, verification and validation (V&V), uncertainty quantification (UQ), and petascale computing.

The benefits of design optimization have been largely unavailable to industry. The bottleneck is that to do shape optimization the CAD geometry-to-mesh mapping needs to be automatic, differentiable, and tightly integrated with the solver and optimizer. This is simply not the case as meshes are disconnected from the CAD geometries from which they were generated.

V&V requires error estimation and adaptivity, which in turn requires tight integration of CAD, geometry, meshing, and analysis. UQ requires simulations with numerous samples of models needed to characterize probability distributions. Sampling puts a premium on the ability to rapidly generate geometry models, meshes, and analyses, which again leads to the need for tightly integrated geometry, meshing, and analysis.

The era of petaflop computing is on the horizon. Parallelism keeps increasing, but the largest unstructured mesh simulations have stalled, because no one truly knows how to generate and adapt massive meshes that keep up with increasing concurrency. To be able to capitalize on the era of $O(100,000)$ core parallel systems, CAD, geometry, meshing, analysis, adaptivity, and visualization all have to run in a tightly integrated way, in parallel, and in a scalable fashion.

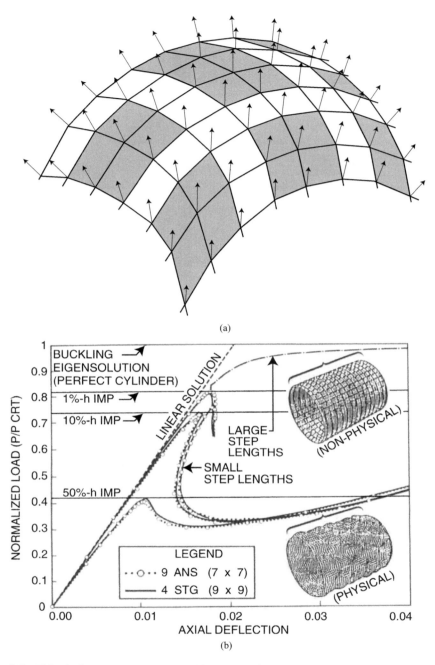

(a)

(b)

Figure 1.3   Thin shell structures exhibit significant imperfection sensitivity. (a) Faceted geometry of typical finite element meshes introduces geometric imperfections (adapted from Gee *et al.*, 2005). (b) Buckling of cylindrical shell with random geometric imperfections. The buckling load depends significantly upon the magnitude of the imperfections (from Stanley, 1985).

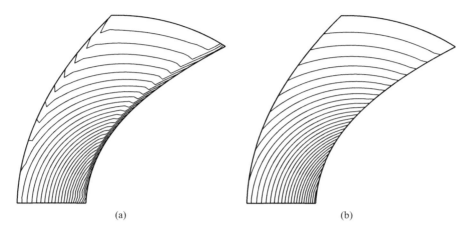

(a) (b)

Figure 1.4 Isodensity contours of Galerkin/least-squares (GLS) discretization of Ringleb flow. (a) Isoparametric linear triangular element approximation: both solution and geometry are represented by piecewise linear functions. (b) Super-isoparametric element approximation: solution is piecewise linear, while geometry is piecewise quadratic. Smooth geometry avoids spurious entropy layers associated with piecewise-linear geometric approximations (from Barth, 1998).

It is apparent that the way to break down the barriers between engineering design and analysis is to reconstitute the entire process, but at the same time maintain compatibility with existing practices. A fundamental step is to focus on one, and only one, geometric model, which can be utilized directly as an analysis model, or from which geometrically precise analysis models can be automatically built. This will require a change from classical finite

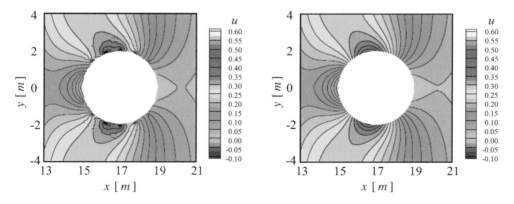

Figure 1.5 The two-dimensional Boussinesq equations. The x-component of velocity obtained using 552 triangles with fifth order polynomials on each triangle. On the left, the elements are straight-sided. The spurious oscillations in the solution on the left are due to the use of straight-sided elements for the geometric approximation. On the right, the cylinder is approximated by elements with curved edges, and the oscillations are eliminated (from Eskilsson and Sherwin, 2006).

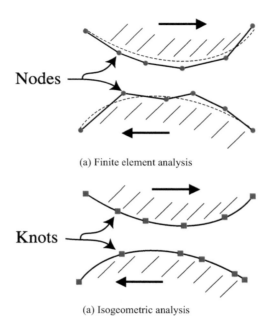

(a) Finite element analysis

(a) Isogeometric analysis

Figure 1.6   Sliding contact. (a) Faceted polynomial finite elements create problems in sliding contact (see Laursen, 2002 and Wriggers, 2002). (b) NURBS geometries can attain the smoothness of real bodies.

element analysis to an analysis procedure based on CAD representations. This concept is referred to as *Isogeometric Analysis*, and it was introduced in Hughes *et al.*, 2005. Since then a number of additional papers have appeared (Bazilevs *et al.*, 2006a, 2006b; Cottrell *et al.*, 2006, 2007; Zhang *et al.*, 2007; Gomez *et al.*, 2008).

Here are the reasons why the time may be right to transform design and analysis technologies: Initiatory investigations of the isogeometric concept have proven very successful. Backward compatibility with existing design and analysis technologies is attainable. There is interest in both the computational geometry and analysis communities to embark on isogeometric research. Several mini-symposia and workshops at international meetings have been held and several very large multi-institutional research projects have begun in Europe. In particular, EXCITING – exact geometry simulation for optimized design of vehicles and vessels – is a three year, six million dollar project focused on developing computational tools for the optimized design of functional free-form surfaces, and the Integrated Computer Aided Design and Analysis (ICADA) project is a five year, five million dollar initiative focused on bridging the gap between design and analysis in industry through isogeometric analysis.

There is an inexorable march toward higher precision and greater reality. New technologies are being introduced and adopted rapidly in design software to gain competitive advantage. New and better analysis technologies can be built upon and influence these new CAD technologies. Engineering analysis can leverage these developments as a basis for the isogeometric concept.

Anyone who has lived the last 60 years is acutely aware of the profound changes that have occurred due to the emergence of new technologies. History has demonstrated repeatedly that statements to the effect that "people will not change" are false. An interesting example of a paradigm shift concerns the slide rule, a mechanical device that dominated computing for approximately 350 years. In the 20th century alone nearly 40 million slide rules were produced throughout the world. The first transistorized electronic calculators emerged in the early 1960s, with portable four-function models available by the end of the decade. The first hand-held scientific calculator, Hewlett-Packard's HP35, became commercially available in 1972. Keuffel and Esser Co., the world's largest producer of slide rules, manufactured its last slide rule in 1975, just 3 years later (see Stoll, 2006).

1.1.2 Computational geometry

There are a number of candidate computational geometry technologies that may be used in isogeometric analysis. The most widely used in engineering design are NURBS (non-uniform rational B-splines), the industry standard (see, Piegl and Tiller, 1997; Farin, 1999a, 1999b; Cohen et al., 2001; Rogers, 2001). The major strengths of NURBS are that they are convenient for free-form surface modeling, can exactly represent all conic sections, and therefore circles, cylinders, spheres, ellipsoids, etc., and that there exist many efficient and numerically stable algorithms to generate NURBS objects. They also possess useful mathematical properties, such as the ability to be refined through knot insertion, C^{p-1}-continuity for pth-order NURBS, and the variation diminishing and convex hull properties. NURBS are ubiquitous in CAD systems, representing billions of dollars in development investment. One may argue the merits of NURBS versus other computational geometry technologies, but their preeminence in engineering design is indisputable. As such, they were the natural starting point for isogeometric analysis and their use in an analysis setting is the focus of this book.

T-splines (Sederberg et al., 2003; Sederberg et al., 2004) are a recently developed forward and backward generalization of NURBS technology. T-splines extend NURBS to permit local refinement and coarsening, and are very robust in their ability to efficiently sew together adjacent patches. Commercial T-spline plug-ins have been introduced in Maya and Rhino, two NURBS-based design systems (see references T-Splines, Inc., 2008a and T-Splines, Inc., 2008b). Initiatory investigations of T-splines in an isogeometric analysis context have been undertaken by Bazilevs et al., 2009 and Dorfel et al., 2008. These works point to a promising future for T-splines as an isogeometric technology.

There are other computational geometry technologies that also warrant investigation as a basis of isogeometric analysis. One is subdivision surfaces which use a limiting process to define a smooth surface from a mesh of triangles or quadrilaterals (see, e.g., Warren and Weimer, 2002; Peters and Reif, 2008). They have already been used in analysis of shell structures by Cirak et al., 2000; Cirak and Ortiz, 2001, 2002. The appeal of subdivision surfaces is there is no restriction on the topology of the control grid. Like T-splines, they also create gap-free models. Most of the characters in Pixar animations are modeled using subdivision surfaces. The CAD industry has not adopted subdivision surfaces very widely because they are not compatible with NURBS. With billions of dollars of infrastructure invested in NURBS, the financial cost would be prohibitive. Nevertheless, subdivision surfaces should play an important role in isogeometric technology. Subdivision solids have been studied by Bajaj *et al.*, 2002.

Other geometric technologies that may play a role in the future of isogeometric analysis include Gordon patches (Gordon, 1969), Gregory patches (Gregory, 1983), S-patches (Loop and DeRose, 1989), and A-patches (Bajaj *et al.*, 1995). Provatidis has recently solved a number of problems using Coons patches (see Provatidis, 2009, and references therein). Others may be invented specifically with the intent of fostering the isogeometric concept, namely, to use the surface design model directly in analysis. This would only suffice if analysis only requires the surface geometry, such as in the stress or buckling analysis of a shell. In many cases, the surface will enclose a volume and an analysis model will need to be created for the volume. The basic problem is to develop a three-dimensional (trivariate) representation of the solid in such a way that the surface representation is preserved. This is far from a trivial problem. Surface differential and computational geometry and topology are now fairly well understood, but the three-dimensional problem is still open (the Thurston conjecture characterizing its solution remains to be proven, see Thurston, 1982, 1997). The hope is that through the use of new technologies, such as, for example, Ricci flows and polycube splines (see Gu and Yau, 2008), progress will be forthcoming.

1.2 The evolution of FEA basis functions

Solution of partial differential equations by the finite element method consists, roughly speaking, of a variational formulation and trial and weighting function spaces defined by their respective basis functions. These basis functions are defined in turn by finite elements, local representations of the spaces. The elements are a non-overlapping decomposition of the problem domain into simple shapes (*e.g.*, triangles, quadrilaterals, tetrahedra, hexahedra, etc.). In the most widely used variational methods, the trial and weighting functions are essentially the same. Specifically, the same elements are used in their construction. There are three ways to improve a finite element method:

1 Improve the variational method. Sometimes this can be done in such a way as to correct a shortcoming in the finite elements for the problem under consideration, such as, for example, through the use of selective integration (see Hughes, 2000). Another way is to use an alternative variational formulation with improved properties, an example being "stabilized methods." See Brooks and Hughes, 1982; Hughes *et al.*, 2004.
2 Improve the finite element spaces, that is, the elements themselves.
3 Improve both, that is, the variational method *and* the elements.

Our focus here is on finite element spaces and ultimately how they perform in comparison to spaces of functions built from NURBS, T-splines, etc. Consequently, we will give a brief review of the historical milestones in finite elements.

Typically, finite elements are defined in terms of interpolatory polynomials. The classical families of polynomials, especially the Lagrange and Hermite polynomials, are widely utilized (see Hughes, 2000). These may be considered the historical antecedents of finite elements.

Figure 1.7 Finite element picture gallery.

J. Tinsley Oden   D. R. J. Owen   Pierre Arnaud Raviart

Robert L. Taylor   Olgierd C. Zienkiewicz

Figure 1.7   *(continued)*

Early publications in the engineering literature describing what is now known as the finite element method were Argyris and Kelsey, 1960, which is a collection of articles by those authors dating from 1954 and 1955, and Turner *et al.*, 1956. The term "finite elements" was coined by Clough, 1960. However, the first finite element, the linear triangle, can be traced all the way back to Courant, 1943. It is perhaps the simplest element and is still widely used today. It is interesting to note that the engineering finite element literature was unaware of this reference until sometime in the late 1960s by which time the essential features of the finite element method were well established. The linear tetrahedron appeared in Gallagher *et al.*, 1962. Through the use of triangular and tetrahedral coordinates (*i.e.*, barycentric coordinates) and the Pascal triangle and tetrahedron, it became a simple matter to generate C^0-continuous finite elements for straight-edged triangles and flat-surfaced tetrahedra. The bilinear quadrilateral was developed by Taig, 1961, and it presaged the development of *isoparametric elements* (Irons, 1966; Zienkiewicz and Cheung, 1968), perhaps the most important concept in the history of element technology.

The idea of isoparametric elements immediately generalized elements which could be developed on a regular parent domain, such as a square, or a cube, to an element which could take on a smoothly curved shape in physical space. Furthermore, it was applicable to any element topology, including triangles, tetrahedra, etc. An essential feature was that the spaces so constructed satisfied basic mathematical convergence criteria, as well as physical attributes in problems of mechanics, namely, the ability to represent all affine motions (*i.e.*, rigid translations and rotations, uniform stretchings and shearings) exactly. Curved quadrilateral and hexahedral elements became popular in structural and solid mechanics applications. The classical isoparametric elements were developed using tensor-product constructs. Subsequently, procedures were developed to circumvent the necessity of the tensor-product format. The eight-node serendipity quadrilateral was an early noteworthy example (Zienkiewicz *et al.*, 1971). This eventually led to the variable-number-of-nodes concept (Zienkiewicz *et al.*, 1970; Taylor, 1972; see Hughes, 2000, chapter 3, for examples).

In practical applications computational efficiency is critical. In nonlinear dynamic applications low-order elements have played a dominant role. The constant pressure bilinear quadrilateral element (Hughes and Allik, 1969; Nagtegaal *et al.*, 1974; Hughes, 1977, 1980; Malkus and Hughes, 1978) and its three-dimensional generalization, the constant pressure trilinear hexahedral element, have dominated nonlinear solid mechanical calculations. Effective one-point integration quadrilateral bending elements (Hughes et al., 1977, 1978; Hughes and Liu, 1981a, 1981b; Flanagan and Belytschko, 1981; Belytschko and Tsay, 1983; Belytschko *et al.*, 1984) with scaled lumped rotatory inertia mass matrices (Hughes *et al.*, 1978; Hughes, 2000) enabled automobile crash analysis to become a standard design tool. The cost of analysis prior to these developments precluded its practical use.

A limitation of the isoparametric concept was that while it worked for C^0-continuous interpolation, it did not for C^1 or higher. There was a strong interest in the development of C^1-continuous interpolation schemes primarily because of the desire to construct thin plate and shell elements for structural analysis. Thin bending elements require square-integrability of generalized second derivatives and so C^1-continuous elements constitute a suitable subspace. Many researchers sought solutions to this problem. Noteworthy successes were due to Clough and Tocher, 1965; Argyris *et al.*, 1968; Cowper *et al.*, 1968; de Veubeke, 1968; Bell, 1969. However, these elements were complicated to use and expensive, and interest moved to different variational formulations to circumvent the need for C^1-continuous basis functions. This is an example where it was more convenient to adopt a different variational formulation than construct appropriate discrete approximation subspaces for the original one. It should be said, however, that the development of effective Reissner-Mindlin bending elements, requiring only C^0-continuity, was not without its own difficulties.

Mathematicians have played a prominent role in devising discrete approximation spaces for certain classes of variational formulations. Noteworthy examples are due to Raviart and Thomas, 1977, and Brezzi *et al.*, 1985; see also Brezzi and Fortin, 1991, for Darcy flow (these are referred to as $H(\text{div})$ elements) and Nedelec, 1980, Demkowicz, 2007, and Demkowicz *et al.*, 2008 for Maxwell's equations (these are referred to as $H(\text{curl})$ and $H(\text{div}) \oplus H(\text{curl})$ elements). The engineering and mathematics literatures are also replete with various alternative variational formulations that enhance the performance of simple elements.

Another recent trend in basis function construction has been away from the classical concept of an element decomposition. These approaches have come to be known as meshless methods (Nayroles *et al.*, 1992) and they have generated considerable interest. Noteworthy contributions to meshless methods are the element-free Galerkin method of Belytschko *et al.*, 1994, the reproducing kernel particle method of Liu *et al.*, 1995, the partition of unity method of Melenk and Babuska, 1996, and the *hp*-clouds of Duarte and Oden, 1996. This is another subject entirely, but we note that, as in the case of the finite element method, the link to CAD geometry, at best, is tenuous (see, *e.g.*, Sakurai, 2006). A timeline of FEA and meshless basis function development is presented in Table 1.1

Table 1.1   Timeline: Milestones in FEA and meshless basis function development

1779	Lagrange polynomials
1864	Hermite polynomials
1943	Linear triangle
1960	Clough coins the name "finite elements"
1961	Bilinear quadrilateral
1962	Linear tetrahedron
1965–1968	C^1-continuous triangles and quadrilaterals
1966	Isoparametric elements
1968–1971	Variable-number-of-nodes elements
1977–1986	$H(\mathrm{div})$, $H(\mathrm{curl})$, and $H(\mathrm{div}) \oplus H(\mathrm{curl})$ elements
1992–1996	Meshless methods

Another class of meshless methods that has enjoyed recent popularity is that of particle methods. An early variant is so-called smoothed particle hydrodynamics (Gingold and Monaghan, 1977). The particle finite element method of Oñate *et al.*, 1996 utilizes geometric reconstruction from particles combined with finite element remeshing strategies and thus has features in common with meshless methods and classical finite element discretizations. The discrete element method of Cundall and Strack, 1979 (see also Munjiza *et al.*, 1995) likewise combines ideas of particles and finite elements. These procedures have opened the way to solution of very complex engineering problems that are beyond the scope of classical finite element procedures.

It needs to be mentioned again that finite elements never faithfully replicate the CAD geometry. It is always a piecewise polynomial approximation. In most cases involving complex engineering designs, it has now become a much more formidable task to generate a finite element model from the CAD geometry than to perform the analysis. This is the primary motivation behind the development of the isogeometric concept.

1.3 The evolution of CAD representations

It is generally agreed that present day CAD had its origins in the work of two French automotive engineers, Pierre Bézier of Renault and Paul de Faget de Casteljau of Citroën. Bézier, 1966, 1967, and 1972 utilized the Bernstein polynomial basis (Bernstein, 1912) to generate curves and surfaces. De Casteljau, 1959, developed similar ideas, but his work was never published in the open literature. Although there seem to be earlier instances of work utilizing splines, the term "spline" was introduced in the mathematical literature by Schoenberg, 1946, whose work drew attention to the possibilities of spline approximations, but the subject did not become active until the 1960s (see Curry and Schoenberg, 1966). During the early years, the role of the Coons patch (Coons, 1967), based on the idea of generalized Hermite interpolation (http://en.wikipedia.org/wiki/Hermite_interpolation), predominated but its influence faded subsequently in favor of the methods of Bézier and de Casteljau. A number of fundamental contributions occurred during the 1970s beginning with Reisenfeld's Ph.D. dissertation on B-splines (Riesenfeld, 1972). This was followed shortly thereafter by Versprille's Ph.D. dissertation on rational B-splines, which have become known as NURBS (Versprille, 1975).

Figure 1.8　Computational Geometry Picture Gallery.

Rich Riesenfeld   Malcolm Sabin   Isaac Schoenberg

Peter Schröder   Larry Schumaker   Tom Sederberg

Figure 1.8   (*continued*)

There are many efficient and numerically stable algorithms that have been developed to manipulate B-splines, for example, the Cox–de Boor recursion (Cox, 1971; de Boor, 1972), the de Boor algorithm (de Boor, 1978), the Oslo algorithm (Cohen *et al.*, 1980), polar forms and blossoms (Ramshaw, 1987a; Ramshaw, 1989), etc.

Another major development in the 1970s was the pioneering work on subdivision surfaces (Catmull and Clark, 1978; Doo and Sabin, 1978). Ed Catmull is the CEO of Pixar and Walt Disney Animation Studios and Jim Clark was the founder of Silicon Graphics and Netscape. The seminal ideas of subdivision are generally attributed to de Rham, 1956 and Chaikin, 1974. Other works of note are Lane and Riesenfeld, 1980, which is intimately linked to Bézier and B-spline surfaces, and Loop, 1987, which is box spline based. Subdivision surfaces have become popular in the field of animation. They generate smooth surfaces from quadrilateral or triangular (Loop, 1987) surface meshes. For engineering design, NURBS are still the dominant technology. Recent generalizations of NURBS-based technology that allow some unstructuredness are T-splines (Sederberg *et al.*, 2003, 2004). T-splines constitute a superset of NURBS (*i.e.*, every NURBS is a T-spline) and the local refinement properties of T-splines facilitate solution of the gap/overlap problem of intersecting NURBS surfaces. A recent work

Table 1.2 Timeline: Milestones in CAD representations

1912	Bernstein polynomials
1946	Schoenberg coins the name "spline"
1959	de Casteljau algorithm
1966–1972	Bézier curves and surfaces
1971, 1972	Cox-de Boor recursion
1972	B-splines
1975	NURBS
1978	Catmull–Clark and Doo–Sabin subdivision surfaces
1980	Oslo knot insertion algorithm
1987	Loop subdivision
1987, 1989	Polar forms, blossoms
1996–present	Triangular and tetrahedral B-splines
2003	T-splines

shows how to replace trimmed NURBS surfaces with untrimmed T-splines (Sederberg *et al.*, 2008). Table 1.2 presents a timeline of important developments in CAD.

Other technologies of note include triangular and tetrahedral generalizations of B-splines (see Lai and Schumaker, 2007).

Splines have also been used as a basis for solving variational problems (see, *e.g.*, Schultz, 1973; Prenter 1975; Höllig 2003; Kwok *et al.*, 2001), but these efforts have been dwarfed by activity in finite element analysis. Spline finite elements were also developed in the (second) Ph.D. thesis of Malcolm Sabin (Sabin, 1997).

It is interesting to note that isoparametric elements developed in the 1960s are still the most widely utilized elements in commercial FEA codes, and even in research activities in FEA. This is in contrast to CAD in which fundamentally new technologies, such as T-splines, have only recently been introduced. It seems very likely that this trend may continue, presenting new opportunities to unify CAD and FEA.

Earlier attempts to integrate finite element analysis and computational geometry were referred to as "physically-based modeling." Several researchers developed tools for free-form geometric design based on mechanical principles (see, *e.g.*, Celniker and Gossard, 1991; Terzopoulos and Qin, 1994; Kagan *et al.*, 1998; Volpin *et al.*, 1999; Bronstein *et al.*, 2008). For example, rather than manipulating a B-spline surface by explicitly moving the control points, the material properties of a thin metal shell are ascribed to the surface so that the geometry may be deformed by applying fictitious forces wherever desired by the designer to "mold" the surface into the desired configuration. This mechanical approach to geometrical modeling is appealing in that the geometries respond in very intuitive ways. The difficulty is that it requires solving differential equations, frequently using an FEA-based approach, each time the designer modifies its shape. Many approaches to such modeling are inherently isogeometric. Those who develop physically-based design systems and those who develop isogeometric analysis capabilities have many goals in common. The futures of these technologies are probably linked.

1.4 Things you need to get used to in order to understand NURBS-based isogeometric analysis

In FEA there is one notion of a mesh and one notion of an element, but an element has two representations, one in the parent domain and one in the physical space. Elements are usually defined by their nodal coordinates and the degrees-of-freedom are usually the values of the basis functions at the nodes. Finite element basis functions are typically interpolatory and may take on positive and negative values. Finite element basis functions are often referred to as "interpolation functions," of "shape functions." See Hughes, 2000 for a discussion of the basic concepts.

In NURBS, the basis functions are usually not interpolatory. There are two notions of meshes, the control mesh[2] and the physical mesh. The control points define the control mesh, and the control mesh interpolates the control points. The control mesh consists of multilinear elements, in two dimensions they are bilinear quadrilateral elements, and in three dimensions they are trilinear hexahedra. The control mesh does not conform to the actual geometry. Rather, it is like a scaffold that controls the geometry. The control mesh has the look of a typical finite element mesh of multilinear elements. The control variables are the degrees-of-freedom and they are located at the control points. They may be thought of as "generalized coordinates." Control elements may be degenerated to more primitive shapes, such as triangles and tetrahedra. The control mesh may also be severely distorted and even inverted to an extent, while at the same time, for sufficiently smooth NURBS, the physical geometry may still remain valid (in contrast with finite elements).

The physical mesh is a decomposition of the actual geometry. There are two notions of elements in the physical mesh, the patch and the knot span. The patch may be thought of as a macro-element or subdomain. Most geometries utilized for academic test cases can be modeled with a single patch. Each patch has two representations, one in a parent domain and one in physical space. In two-dimensional topologies, a patch is a rectangle in the parent domain representation. In three dimensions it is a cuboid.

Each patch can be decomposed into knot spans. Knots are points, lines, and surfaces in one-, two-, and three-dimensional topologies, respectively. Knot spans are bounded by knots. These define element domains where basis functions are smooth (*i.e.*, C^∞). Across knots, basis functions will be C^{p-m} where p is the degree[3] of the polynomial and m is the multiplicity of the knot in question. Knot spans are convenient for numerical quadrature. They may be thought of as micro-elements because they are the smallest entities we deal with. They also have representations in both a parent domain and physical space. When we speak of "elements" without further description, we usually mean knot spans.

There is one other very important notion that is a key to understanding NURBS, the *index space* of a patch. It uniquely identifies each knot and discriminates among knots having multiplicity greater than one.

See Table 1.3 for a summary of NURBS paraphernalia employed in isogeometric analysis. A schematic illustration of the ideas is presented in Figure 1.9 for a NURBS surface in $\mathbb{R}^3$. Detailed examples will be provided in subsequent chapters.

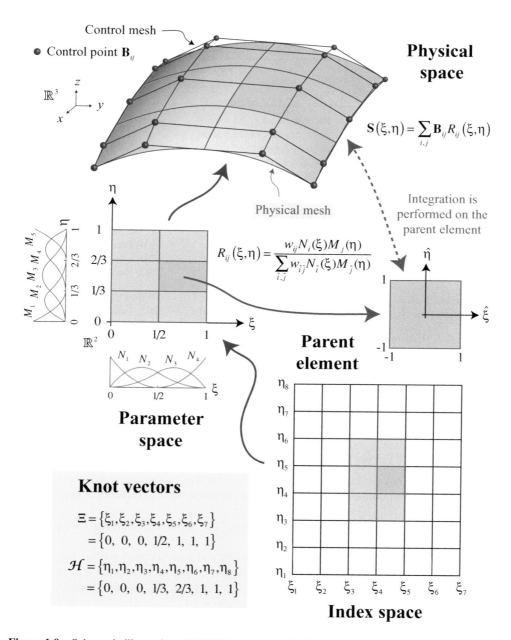

Figure 1.9 Schematic illustration of NURBS paraphernalia for a one-patch surface model. Open knot vectors and quadratic C^1-continuous basis functions are used. Complex multi-patch geometries may be constructed by assembling control meshes as in standard finite element analysis. Also depicted are C^1-quadratic ($p = 2$) basis functions determined by the knot vectors. Basis functions are multiplied by control points and summed to construct geometrical objects, in this case a surface in $\mathbb{R}^3$. The procedure used to define basis functions from knot vectors will be described in detail in Chapter 2.

Table 1.3   NURBS paraphernalia in isogeometric analysis

Index Space		
Control Mesh	Physical Mesh	
Multilinear Control Elements	Patches	Knot Spans
Topology:	Patches: Images of rectangular meshes in the parent domain mapped into the actual geometry. Patches may be thought of as macro-elements or subdomains.	Topology of knots in the parent domain:
1D: Straight lines defined by two consecutive control points		**1D:** Points
		2D: Lines
		3D: Planes
2D: Bilinear quadrilaterals defined by four control points	Topology:	Topology of knots in the physical space:
	1D: Curves	**1D:** Points
	2D: Surfaces	**2D:** Curves
	3D: Volumes	**3D:** Surfaces
3D: Trilinear hexahedra defined by eight control points	Patches are decomposed into knot spans, the smallest notion of an element.	Topology of knots spans, *i.e.*, "elements":
		1D: Curved segments connecting consecutive knots
		2D: Curved quadrilaterals bounded by four curves
		3D: Curved hexahedra bounded by six curved surfaces

Notes

1. Young engineers may not know what vellum and Mylar are. Vellum is a translucent drafting material made from cotton fiber. Mylar is the trade name of a translucent polyester film used for drafting.
2. The control mesh is also known as the "control net," the "control lattice," and curiously the "control polygon" in the univariate case.
3. There is a terminology conflict between the geometry and analysis communities. Geometers will say a cubic polynomial has degree 3 and order 4. In geometry, order equals degree plus one. Analysts will say a cubic polynomial is order three, and use the terms order and degree synonymously. This is the convention we adhere to.

2

NURBS as a Pre-analysis Tool: Geometric Design and Mesh Generation

2.1 B-splines

NURBS are built from B-splines and so a discussion of B-splines is a natural starting point for their investigation. Unlike in standard finite element analysis, the B-spline *parameter space* is local to *patches* rather than elements. That is, the parameter space in FEA (tellingly dubbed the "reference element" or "parent element") is mapped into a single element in the physical space, and each element has its own such mapping, as in Figure 2.1. Alternatively, the B-spline mapping takes a patch of multiple elements in the parameter space into the physical space, as seen in Figure 2.2. Each element in the physical space is the image of a corresponding element in the parameter space, but the mapping itself is global to the whole patch, rather than to the elements themselves. Patches play the role of *subdomains* within which element types and material models are assumed to be uniform. Many simple domains can be represented by a single patch.

2.1.1 Knot vectors

A *knot vector* in one dimension is a non-decreasing set of coordinates in the parameter space, written $\Xi = \{\xi_1, \xi_2, \ldots, \xi_{n+p+1}\}$, where $\xi_i \in \mathbb{R}$ is the ith *knot*, i is the knot index, $i = 1, 2, \ldots, n + p + 1$, p is the polynomial order, and n is the number of basis functions used to construct the B-spline curve. The knots partition the parameter space into elements. Element boundaries in the physical space are simply the images of knot lines under the B-spline mapping. See, again, Figure 2.2.

Note that the distinction between "elements" and "patches" may be thought of in two different ways. In Kagan *et al.*, 1998 and 2003, the patches themselves are referred to as elements. This is not unreasonable as the parameter space is local to patches and a finite element code must include a loop over the patches during assembly. As is clear from the discussion thus far, we take the alternate view that patches are subdomains comprised of many

Isogeometric Analysis: Toward Integration of CAD and FEA   by J. A. Cottrell, T. J. R. Hughes, Y. Bazilevs
© 2009, John Wiley & Sons, Ltd

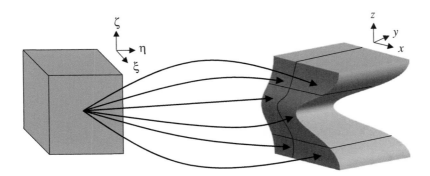

Figure 2.1   In classical finite element analysis, the parameter space is local to individual elements. Each element has its own mapping from the reference element.

elements, namely the "knot spans." This latter view seems more appropriate as, in our work, numerical quadrature is usually carried out at the knot span level. Furthermore, in the case of B-splines, the functions are piecewise polynomials where the different "pieces" join along knot lines. In this way the functions are C^∞ within an element. Lastly, surprisingly complicated domains can be described by a single patch (*e.g.*, all of the numerical examples in Hughes *et al.*, 2005). Describing such domains as being comprised of one element seems inconsistent with the traditional notion of what an element is.

Knot vectors may be ***uniform*** if the knots are equally spaced in the parameter space. If they are unequally spaced, the knot vector is ***non-uniform***. Knot values may be repeated, that is, more than one knot may take on the same value. The multiplicities of knot values have important implications for the properties of the basis. A knot vector is said to be ***open*** if its first and last knot values appear $p + 1$ times. Open knot vectors are the standard in the CAD literature. In one dimension, basis functions formed from open knot vectors are interpolatory at the ends of the parameter space interval, $[\xi_1, \xi_{n+p+1}]$, and at the corners of patches in multiple dimensions, but they are not, in general, interpolatory at interior knots. This is a distinguishing feature between knots and "nodes" in finite element analysis. A further consequence of the use

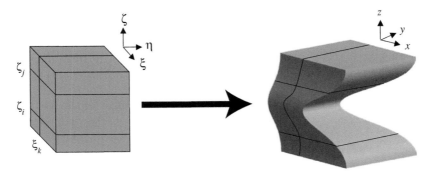

Figure 2.2   The B-spline parameter space is local to the entire patch. Internal knots partition the patch into elements. A single B-spline map takes the patch from the parameter space to the physical space.

of open knot vectors in multiple dimensions is that the boundary of a B-spline object with d parametric dimensions is itself a B-spline object of dimension $d - 1$. For example, each edge of a B-spline surface is itself a B-spline curve.

2.1.2 Basis functions

With a knot vector in hand, the B-spline basis functions are defined recursively starting with piecewise constants ($p = 0$) :

$$N_{i,0}(\xi) = \begin{cases} 1 & \text{if } \xi_i \leq \xi < \xi_{i+1}, \\ 0 & \text{otherwise.} \end{cases} \tag{2.1}$$

For $p = 1, 2, 3, \ldots$, they are defined by

$$N_{i,p}(\xi) = \frac{\xi - \xi_i}{\xi_{i+p} - \xi_i} N_{i,p-1}(\xi) + \frac{\xi_{i+p+1} - \xi}{\xi_{i+p+1} - \xi_{i+1}} N_{i+1,p-1}(\xi). \tag{2.2}$$

This is referred to as the **Cox–de Boor recursion formula** (Cox, 1971; de Boor, 1972). The results of applying (2.1) and (2.2) to a uniform knot vector are presented in Figure 2.3. For B-spline functions with $p = 0$ and $p = 1$, we have the same result as for standard piecewise constant and linear finite element functions, respectively. Quadratic B-spline basis functions,

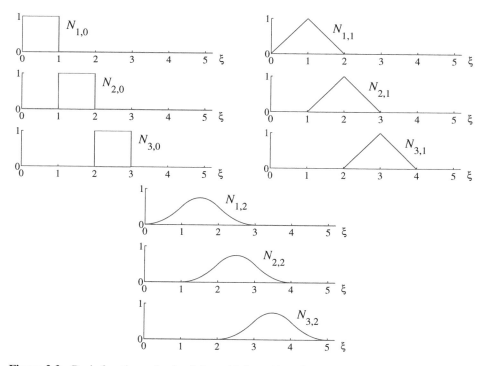

Figure 2.3 Basis functions of order 0, 1, and 2 for uniform knot vector $\Xi = \{0, 1, 2, 3, 4, \ldots\}$.

however, differ from their FEA counterparts. They are each identical but shifted relative to each other, whereas the shape of a quadratic finite element function depends on whether it corresponds to an internal node or an end node. This "homogeneous" pattern continues for the B-splines as we continue to higher-orders.

There are several important features to observe in Figure 2.3 in addition to the homogeneity of the functions. The first is that the basis constitutes a partition of unity, that is, $\forall \xi$,

$$\sum_{i=1}^{n} N_{i,p}(\xi) = 1. \tag{2.3}$$

Also observe that each basis function is pointwise nonnegative over the entire domain, that is, $N_{i,p}(\xi) \geq 0, \forall \xi$. This means that all of the entries of a mass matrix will be positive (see Chapter 6), which has implications for developing lumped mass schemes. The third major feature to note is that each pth order function has $p - 1$ continuous derivatives across the element boundaries (*i.e.*, across the knots). This feature has many extremely important implications for the use of splines as a basis for analysis and is one of the most distinctive features of isogeometric analysis. Lastly, the support of the B-spline functions of order p is always $p + 1$ knot spans. Thus higher-order functions have support over much larger portions of the domain than do classical FEA functions. It is a common misconception that this increasing support of the functions leads to increased bandwidth in a numerical method. This simply is not the case. As we see in Figure 2.4 for cubics, the total number of functions that any given function shares support with (including itself) is $2p + 1$ regardless of whether we are using an FEA basis or B-splines.

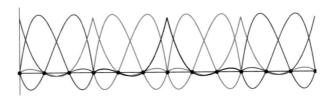

(a) Standard cubic finite element basis functions with equally spaced nodes

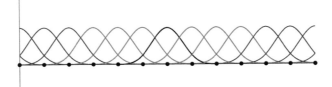

(b) Cubic B-spline basis functions with equally spaced knots

Figure 2.4   Bandwidth comparison for FEA and B-spline functions. Regardless of whether we use the C^0 FEA cubics or the C^2 B-spline cubics, the bandwidth of the resulting matrices will be $2p + 1 = 7$. In each case, the function in black has overlapping support with each of the functions in red, as well as with itself.

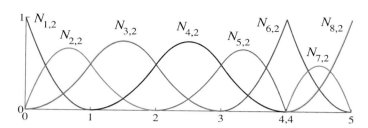

Figure 2.5 Quadratic basis functions for open, non-uniform knot vector $\Xi = \{0, 0, 0, 1, 2, 3, 4, 4, 5, 5, 5\}$.

The use of a non-uniform knot vector allows us to obtain much richer behavior than is possible with a simple uniform one. A quadratic example is presented in Figure 2.5 for the open, non-uniform knot vector $\Xi = \{\xi_1, \xi_2, \xi_3, \xi_4, \xi_5, \xi_6, \xi_7, \xi_8, \xi_9, \xi_{10}, \xi_{11}\} = \{0, 0, 0, 1, 2, 3, 4, 4, 5, 5, 5\}$. Note that the basis functions are interpolatory at the ends of the interval and also at $\xi = 4$, the location of a repeated knot. At this repeated knot, only C^0-continuity is attained. Elsewhere, the functions are C^1-continuous. In general, basis functions of order p have $p - m_i$ continuous derivatives across knot ξ_i, where m_i is the multiplicity of the value of ξ_i in the knot vector. When the multiplicity of a knot value is exactly p, the basis is interpolatory at that knot. When the multiplicity is $p + 1$, the basis becomes discontinuous and the patch boundary is formed.

This relationship between continuity and the multiplicity of the knots is even more apparent in Figure 2.6, in which we have a fourth order curve with differing levels of continuity at every element boundary. At the first internal element boundary, $\xi = 1$, the knot value appears only once in the knot vector, and so we have the maximum level of continuity possible: $C^{p-1} = C^3$. At each subsequent internal knot value, the multiplicity is increased by one, and so the number of continuous derivatives is decreased by one. Note, as before, that when a knot

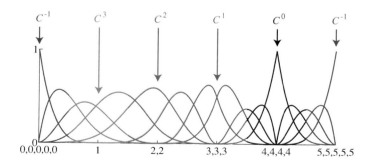

Figure 2.6 Quartic $(p = 4)$ basis functions for an open, non-uniform knot vector $\Xi = \{0, 0, 0, 0, 0, 1, 2, 2, 3, 3, 3, 4, 4, 4, 4, 5, 5, 5, 5, 5\}$. The continuity across an interior element boundary is a direct result of the polynomial order and the multiplicity of the corresponding knot value.

value is repeated p times, in this case at $\xi = 4$, the C^0 basis is interpolatory. The basis is also interpolatory at the boundary of the domain, where the open knot vector demands that the first and last knot value be repeated $p + 1$ times. The result is "C^{-1}"-continuity, that is, the basis is fully discontinuous, naturally terminating the domain.

Observe that increasing the multiplicities of the knot values seems to have decreased the support of some of the functions. This is not a contradiction with the trend we observed previously as the support of each function $N_{i,p}$ still begins at knot ξ_i and ends at ξ_{i+p+1}. That is, the support of each function is still $p + 1$ knot spans, but some of those knot spans have zero measure due to the repetition of knot values. Surprisingly, none of this has any effect on the bandwidth.

2.1.2.1 Building functions from non-uniform knot vectors

To see how the repeated knot values come into play in the definition of the basis functions, let us explicitly build the quadratic functions corresponding to knot vector $\Xi = \{\xi_1, \xi_2, \xi_3, \xi_4, \xi_5, \xi_6, \xi_7\} = \{0, 0, 0, 1, 2, 2, 2\}$ by carefully applying (2.1) and (2.2) within each knot span. Beginning with $i = 1$, we have that

$$N_{1,0}(\xi) = \begin{cases} 1 & \xi_1 \le \xi < \xi_2, \\ 0 & \text{otherwise.} \end{cases} \tag{2.4}$$

As $\xi_1 = \xi_2 = 0$, we observe that there exists no value of ξ such that $0 \le \xi$ *and* $\xi < 0$, and therefore $N_{1,0}(\xi) \equiv 0$. There is no ambiguity in the definition; we need only interpret (2.1) literally. Applying the same logic to the remaining indices, we arrive at the following piecewise constant functions:

$$N_{1,0}(\xi) = 0, \tag{2.5a}$$

$$N_{2,0}(\xi) = 0, \tag{2.5b}$$

$$N_{3,0}(\xi) = \begin{cases} 1 & 0 \le \xi < 1, \\ 0 & \text{otherwise,} \end{cases} \tag{2.5c}$$

$$N_{4,0}(\xi) = \begin{cases} 1 & 1 \le \xi < 2, \\ 0 & \text{otherwise.} \end{cases} \tag{2.5d}$$

We could proceed to $i = 5$ and $i = 6$, but the corresponding functions, $N_{5,p}$ and $N_{6,p}$, would be identically zero for all polynomial orders.

In Figure 2.7, we plot the functions of (2.5a)–(2.5d) in the **index space**, meaning that we equally space all of the knots independent of their actual values. The constant functions corresponding to "trivial" knot spans are always identically zero.

We now build the linear functions, $N_{i,1}(\xi)$, from these constant functions using (2.2). For $i = 1$, now with $p = 1$, we have

$$N_{1,1}(\xi) = \frac{\xi - 0}{0 - 0} N_{1,0}(\xi) + \frac{0 - \xi}{0 - 0} N_{2,0}(\xi) \tag{2.6}$$

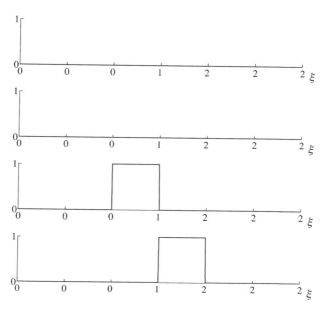

Figure 2.7 Constant basis functions corresponding to $\Xi = \{0, 0, 0, 1, 2, 2, 2\}$ plotted in the index space.

Recognizing immediately that $N_{1,0}(\xi) = N_{2,0}(\xi) = 0$, we have only to augment (2.1) and (2.2) with the *definition*

$$\frac{0}{0} \doteq 0. \tag{2.7}$$

Thus, we have that $N_{1,1}(\xi) \equiv 0$.

As an additional example, consider $i = 3$. We have

$$N_{3,1}(\xi) = \frac{\xi - 0}{1 - 0} N_{3,0}(\xi) + \frac{2 - \xi}{2 - 1} N_{4,0}(\xi)$$

$$= \xi \begin{cases} 1 & 0 \leq \xi < 1, \\ 0 & \text{otherwise,} \end{cases}$$

$$+ (2 - \xi) \begin{cases} 1 & 1 \leq \xi < 2, \\ 0 & \text{otherwise,} \end{cases}$$

$$= \begin{cases} \xi & 0 \leq \xi < 1, \\ 2 - \xi & 1 \leq \xi < 2, \\ 0 & \text{otherwise.} \end{cases} \tag{2.8}$$

Performing the same steps for the remaining indices results in piecewise linear functions

$$N_{1,1}(\xi) = 0, \tag{2.9a}$$

$$N_{2,1}(\xi) = \begin{cases} 1 - \xi & 0 \leq \xi < 1, \\ 0 & \text{otherwise,} \end{cases} \tag{2.9b}$$

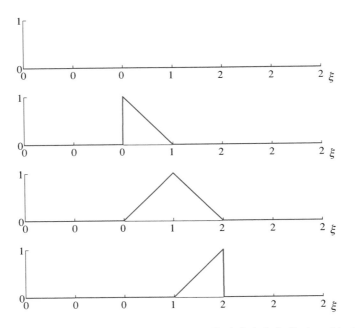

Figure 2.8   Linear basis functions corresponding to $\Xi = \{0, 0, 0, 1, 2, 2, 2\}$ plotted in the index space.

$$N_{3,1}(\xi) = \begin{cases} \xi & 0 \le \xi < 1, \\ 2 - \xi & 1 \le \xi < 2, \\ 0 & \text{otherwise.} \end{cases} \tag{2.9c}$$

$$N_{4,1}(\xi) = \begin{cases} \xi - 1 & 1 \le \xi < 2, \\ 0 & \text{otherwise.} \end{cases} \tag{2.9d}$$

shown in Figure 2.8.

Finally, we may build the piecewise quadratic functions from these piecewise linears. Taking, for example, $i = 2$, we have

$$N_{2,2}(\xi) = \frac{\xi - 0}{1 - 0} N_{2,1}(\xi) + \frac{2 - \xi}{2 - 0} N_{3,1}(\xi)$$

$$= \xi \begin{cases} (1 - \xi) & 0 \le \xi < 1, \\ 0 & \text{otherwise,} \end{cases}$$

$$+ \frac{1}{2}(2 - \xi) \begin{cases} \xi & 0 \le \xi < 1, \\ (2 - \xi) & 1 \le \xi < 2, \\ 0 & \text{otherwise,} \end{cases}$$

$$= \begin{cases} \xi(1 - \xi) + \frac{1}{2}(2 - \xi)\xi & 0 \le \xi < 1, \\ \frac{1}{2}(2 - \xi)^2 & 1 \le \xi < 2, \\ 0 & \text{otherwise.} \end{cases} \tag{2.10}$$

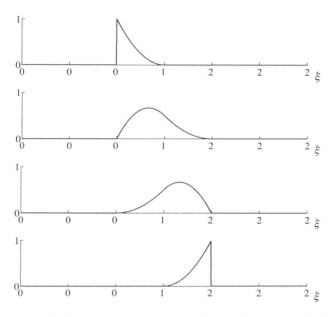

Figure 2.9 Quadratic basis functions corresponding to $\Xi = \{0, 0, 0, 1, 2, 2, 2\}$ plotted in the index space.

Completing the process for the remaining indices, we get the following functions (shown in Figure 2.9):

$$N_{1,2}(\xi) = \begin{cases} (1 - \xi)^2 & 0 \leq \xi < 1, \\ 0 & \text{otherwise,} \end{cases} \tag{2.11a}$$

$$N_{2,2}(\xi) = \begin{cases} \xi(1 - \xi) + \frac{1}{2}(2 - \xi)\xi & 0 \leq \xi < 1, \\ \frac{1}{2}(2 - \xi)^2 & 1 \leq \xi < 2, \\ 0 & \text{otherwise,} \end{cases} \tag{2.11b}$$

$$N_{3,2}(\xi) = \begin{cases} \frac{1}{2}\xi^2 & 0 \leq \xi < 1, \\ \frac{1}{2}\xi(2 - \xi) + (2 - \xi)(\xi - 1) & 1 \leq \xi < 2, \\ 0 & \text{otherwise,} \end{cases} \tag{2.11c}$$

$$N_{4,2}(\xi) = \begin{cases} (\xi - 1)^2 & 1 \leq \xi < 2, \\ 0 & \text{otherwise.} \end{cases} \tag{2.11d}$$

To get these results we have had to carefully add the different pieces of these piecewise functions together, which is cumbersome. Fortunately, efficient algorithms exist for basis function evaluation (see, *e.g.*, Piegl and Tiller, 1997). Our purpose in stepping through the equations so pedantically is to remove some of the mystery about the repeated knots. The bottom line is that the equations make sense with repeated knots. The confusion comes from a desire to associate knots with the nodes of classical FEA. We should resist this urge, and understand the B-spline technology for what it is, not merely by analogy with things that are already familiar.

2.1.2.2 Derivatives of B-spline basis functions

The derivatives of B-spline basis functions are efficiently represented in terms of B-spline lower order bases. This should not be surprising in light of the recursive definition of the basis in (2.1) and (2.2). For a given polynomial order p and knot vector Ξ, the derivative of the ith basis function is given by

$$\frac{d}{d\xi} N_{i,p}(\xi) = \frac{p}{\xi_{i+p} - \xi_i} N_{i,p-1}(\xi) - \frac{p}{\xi_{i+p+1} - \xi_{i+1}} N_{i+1,p-1}(\xi). \tag{2.12}$$

We can generalize this to higher derivatives by simply differentiating each side of (2.12) to get

$$\frac{d^k}{d^k\xi} N_{i,p}(\xi) = \frac{p}{\xi_{i+p} - \xi_i} \left(\frac{d^{k-1}}{d^{k-1}\xi} N_{i,p-1}(\xi) \right)$$

$$- \frac{p}{\xi_{i+p+1} - \xi_{i+1}} \left(\frac{d^{k-1}}{d^{k-1}\xi} N_{i+1,p-1}(\xi) \right). \tag{2.13}$$

Expanding (2.13) by means of (2.12) results in an expression purely in terms of lower order functions $N_{i,p-k}, \ldots, N_{i+k,p-k}$. We have

$$\frac{d^k}{d^k\xi} N_{i,p}(\xi) = \frac{p!}{(p-k)!} \sum_{j=0}^{k} \alpha_{k,j} N_{i+j,p-k}(\xi), \tag{2.14}$$

with

$$\alpha_{0,0} = 1,$$

$$\alpha_{k,0} = \frac{\alpha_{k-1,0}}{\xi_{i+p-k+1} - \xi_i},$$

$$\alpha_{k,j} = \frac{\alpha_{k-1,j} - \alpha_{k-1,j-1}}{\xi_{i+p+j-k+1} - \xi_{i+j}} \quad j = 1, \ldots, k-1,$$

$$\alpha_{k,k} = \frac{-\alpha_{k-1,k-1}}{\xi_{i+p+1} - \xi_{i+k}}.$$

The denominator of several of these coefficients can be zero in the presence of repeated knots. Whenever this happens, the coefficient is defined to be zero. Efficient algorithms for these calculations can be found in Piegl and Tiller, 1997.

2.1.3 B-spline geometries

2.1.3.1 B-spline curves

B-spline curves in $\mathbb{R}^d$ are constructed by taking a linear combination of B-spline basis functions, just as in classical FEA. The vector-valued coefficients of the basis functions are referred to as **control points**. These are analogous to nodal coordinates in finite element analysis in that they are the coefficients of the basis functions, but the non-interpolatory nature of the basis does not lead to a concrete interpretation of the control point values. Given n basis

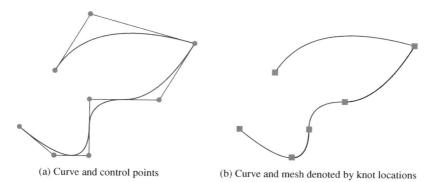

(a) Curve and control points (b) Curve and mesh denoted by knot locations

Figure 2.10 B-spline, piecewise quadratic curve in $\mathbb{R}^2$. (a) Control point locations are denoted by ●. (b) The knots, which define a mesh by partitioning the curve into elements, are denoted by ■. Basis functions and knot vector as in Figure 2.5.

functions, $N_{i,p}$, $i = 1, 2, \ldots, n$, and corresponding control points $\mathbf{B}_i \in \mathbb{R}^d$, $i = 1, 2, \ldots, n$, a piecewise-polynomial **B-spline curve** is given by

$$\mathbf{C}(\xi) = \sum_{i=1}^{n} N_{i,p}(\xi)\mathbf{B}_i. \tag{2.15}$$

Note that the index i in $\mathbf{B}_i$ serves to identify the control point and is not a reference to one of its d components. Piecewise linear interpolation of the control points gives the so-called *control polygon*.

The example shown in Figure 2.10 is built from the quadratic basis functions considered in Figure 2.5. The curve is interpolatory at the first and last control points, a general feature of a curve built from an open knot vector. Note that it is also interpolatory at the sixth control point. As discussed above, this is due to the fact that the multiplicity of the knot $\xi = 4$ is equal to the polynomial order. Note also that the curve is tangent to the control polygon at the first, last, and sixth control points. The curve is $C^{p-1} = C^1$-continuous everywhere except at the location of the repeated knot, $\xi = 4$, where it is $C^{p-2} = C^0$-continuous. Note the difference between the control points, shown in Figure 2.10a, and the images of the knots, shown in Figure 2.10b. It is the knots, mapped into the physical space, that partition the curve into elements.

An affine transformation of a B-spline curve is obtained by applying the transformation directly to the control points. An affine transformation is a mapping $\Phi : \mathbb{R}^3 \to \mathbb{R}^3$ such that for any vector $\mathbf{x} \in \mathbb{R}^3$,

$$\Phi(\mathbf{x}) = \mathbf{A}\mathbf{x} + \mathbf{v} \tag{2.16}$$

for some matrix $\mathbf{A} \in \mathbb{R}^{3\times3}$ and vector $\mathbf{v} \in \mathbb{R}^3$. Affine transformations include translations, rotations, scalings, and uniform stretchings and shearings. The ability to apply an affine transformation to a curve by applying it directly to the control points turns out to be the essential property for satisfying so-called "patch tests," as discussed in Hughes *et al.*, 2005. This property is referred to as **affine covariance**[1].

Many properties of B-spline curves follow directly from the properties of their basis functions. Both Rogers, 2001 and Piegl and Tiller, 1997 discuss many such properties in detail. For example, B-spline curves of degree p have $p - 1$ continuous derivatives in the absence of repeated knots or control points. In general, a curve will have at least as many continuous derivatives across an element boundary as its basis functions have across the corresponding knot value. Another property the curve inherits from its basis is that of locality. Due to the compact support of the B-spline basis functions, moving a single control point can affect the geometry of no more than $p + 1$ elements of the curve.

B-splines obey a strong convex hull property. The non-negativity and partition of unity properties of the basis, combined with the compact support of the functions, lead to the fact that a B-spline curve is completely contained within the convex hull defined by its control points. For a curve of degree p, we define the convex hull as the union of all of the convex hulls formed by $p + 1$ successive control points. Figure 2.11 shows such convex hulls for $p = 1$ through $p = 5$ for a given set of control points. Note, in particular, that the convex hull for a piecewise linear curve is just the control polygon itself. Figure 2.12 shows the corresponding curves that we obtain by pairing these control points with the different bases. As the polynomial order increases, the curves become smoother and the effect of each individual control point is diminished.

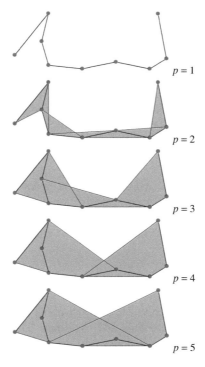

Figure 2.11   Convex hulls for $p = 1$ through $p = 5$.

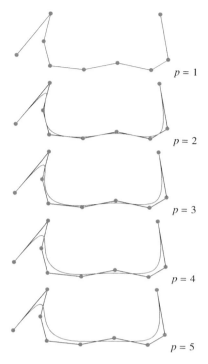

Figure 2.12   B-spline curves for $p = 1$ through $p = 5$ with the control polygon as in Figure 2.11.

B-spline curves also possess a variation diminishing property. No plane has more inter-sections with the curve than it has with the control polygon. This property is particularly striking when compared with the behavior of standard Lagrange polynomials. An exam-ple is illustrated in Figure 2.13a where Lagrange polynomials of orders three, five, and seven interpolate a discontinuity represented by eight data points in $\mathbb{R}^2$. Note that as the order is increased, the amplitude of the oscillations also increases. B-splines behave very differently when the data are viewed as control points. The variation diminishing property leads the B-spline curves in Figure 2.13b to be monotone, a property that proves useful in analysis.

A subtle, yet extremely important, point to recognize about Figure 2.13 is that, for a fixed polynomial order, the Lagrange basis and the B-spline basis we have used in this example span *exactly* the same space. This is because we have used only one element and so we are dealing directly with polynomials, not *piecewise* polynomials. The difference between the oscillatory Figure 2.13a and the monotone Figure 2.13b is in whether we have interpreted the data as nodes in the classical finite element sense, or as control points. It is the pointwise positivity and the non-interpolatory nature of the B-spline basis that makes this latter interpretation possible.

An interesting example of the lack of robustness of Lagrange polynomials is found in Farin, 1999a, which we recreate here. Figure 2.14 shows two attempts at the interpolation of the

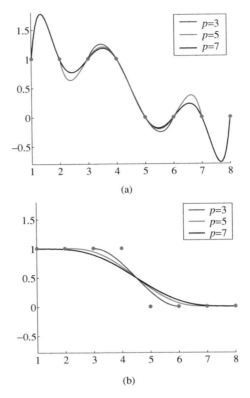

Figure 2.13 (a) Lagrange interpolation oscillates when faced with discontinuous data. (b) NURBS exhibit the variation diminishing property for the same data.

function $\mathbf{f}(\theta) = (2\cos(\theta), \sin(\theta))^T$ in a finite precision environment. First, 21 points are chosen by sampling $\mathbf{f}(\theta)$ at equal intervals of $\theta \in [0, \pi/2]$. To that initial data, random perturbations of a prescribed magnitude are added to each point to simulate noisy data, and then interpolation is performed using a Lagrange basis with $p = 20$. In Figure 2.14a, the input data are accurate to six decimal places, whereas in Figure 2.14b only four digits of accuracy are given. Clearly, the resulting curves differ dramatically. Such fragility makes Lagrange polynomials exceedingly uncommon in geometrical design software.

2.1.3.2 B-spline surfaces

Given a *control net* $\{\mathbf{B}_{i,j}\}$, $i = 1, 2, \ldots, n$, $j = 1, 2, \ldots, m$, polynomial orders p and q, and knot vectors $\Xi = \{\xi_1, \xi_2, \ldots, \xi_{n+p+1}\}$, and $\mathcal{H} = \{\eta_1, \eta_2, \ldots, \eta_{m+q+1}\}$, a tensor product B-spline surface is defined by

$$\mathbf{S}(\xi, \eta) = \sum_{i=1}^{n} \sum_{j=1}^{m} N_{i,p}(\xi) M_{j,q}(\eta) \mathbf{B}_{i,j} \tag{2.17}$$

(a)   (b)

Figure 2.14 Interpolation with Lagrange polynomials. (a) The points to be interpolated are accurate to six digits after the decimal point. (b) The points to be interpolated are accurate to only four digits after the decimal point.

where $N_{i,p}(\xi)$ and $M_{j,q}(\eta)$ are univariate B-spline basis functions of order p and q, corresponding to knot vectors Ξ and $\mathcal{H}$, respectively.

Many of the properties of a B-spline surface are the result of its tensor product nature. The basis is pointwise nonnegative, and forms a partition of unity as $\forall (\xi, \eta) \in [\xi_1, \xi_{n+p+1}] \times [\eta_1, \eta_{m+q+1}]$,

$$\sum_{i=1}^{n} \sum_{j=1}^{m} N_{i,p}(\xi) M_{j,q}(\eta) = \left(\sum_{i=1}^{n} N_{i,p}(\xi) \right) \left(\sum_{j=1}^{m} M_{j,q}(\eta) \right) = 1. \qquad (2.18)$$

The number of continuous partial derivatives in a given parametric direction may be determined from the associated one-dimensional knot vector and polynomial order. The surface again possesses the property of affine covariance and has a strong convex hull property. Interestingly, there is no known variation diminishing property for surfaces, though the convex hull property precludes any two-dimensional analogues of the types of oscillations we saw in Figure 2.13a, thus generalizing the result of Figure 2.13b to multiple dimensions.

The local support of the basis functions also follows directly from the one-dimensional functions that form them. The support of a given bivariate function $\tilde{N}_{i,j;p,q}(\xi, \eta) = N_{i,p}(\xi) M_{j,q}(\eta)$ is exactly $[\xi_i, \xi_{i+p+1}] \times [\eta_j, \eta_{j+q+1}]$. Let us consider a specific example of a biquadratic ($p = q = 2$) surface formed from knot vectors $\Xi = \{0, 0, 0, 0.5, 1, 1, 1\}$ and $\mathcal{H} = \{0, 0, 0, 1, 1, 1\}$, with control points listed in Table 2.1, resulting in the control net and mesh shown in Figure 2.15. For this case, the support of $\tilde{N}_{1,1;2,2}(\xi, \eta)$, is $[\xi_1, \xi_4] \times [\eta_1, \eta_4]$. Similarly, the support of $\tilde{N}_{3,2;2,2}(\xi, \eta)$, for example, is $[\xi_3, \xi_6] \times [\eta_2, \eta_5]$. The support of each of these functions is shown in the index space in Figure 2.16a. By equally spacing each of the knots in the plot, it is easy to see exactly which knot spans each of the functions are supported in, including where they overlap. Such a viewpoint is very useful when developing algorithms (see Appendix A at the end of the book for a discussion of the index space and so-called "NURBS coordinate" in the context of connectivity). Alternatively, we can present the same information in the parameter space, as in Figure 2.16b. Here, we have taken into account the actual knot values. It is clear that we only have two nontrivial elements (elements with positive measure), and therefore only two elements in which calculations need to be performed during analysis. Function $\tilde{N}_{3,2;2,2}(\xi, \eta)$ has support in both of these elements, while $\tilde{N}_{1,1;2,2}(\xi, \eta)$ is only

Table 2.1 Control points for
the biquadratic B-spline surface
depicted in Figure 2.15

i	j	$\mathbf{B}_{i,j}$
1	1	$(0, 0)$
1	2	$(-1, 0)$
1	3	$(-2, 0)$
2	1	$(0, 1)$
2	2	$(-1, 2)$
2	3	$(-2, 2)$
3	1	$(1, 1.5)$
3	2	$(1, 4)$
3	3	$(1, 5)$
4	1	$(3, 1.5)$
4	2	$(3, 4)$
4	3	$(3, 5)$

supported in the leftmost element. Lastly, we can view these elements in the physical space, as in Figure 2.16c, which makes it clear which portions of the actual domain are influenced by each of the basis functions.

In Figure 2.17 we have plotted the actual functions themselves in the physical space. Note that $\tilde{N}_{1,1;2,2}(\xi, \eta)$ takes on positive values on two of the edges, and it is interpolatory in the corner. Alternatively, $\tilde{N}_{3,2;2,2}(\xi, \eta)$ is identically zero on all of the edges. We could not have explicitly told this from looking at the parameter space or the physical space pictures in

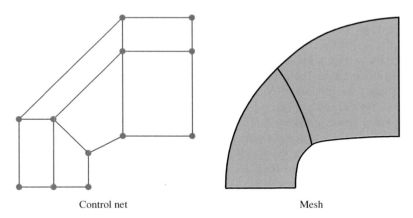

Control net   Mesh

Figure 2.15 The control net and mesh for the biquadratic B-spline surface with $\Xi = \{0, 0, 0, 0.5, 1, 1, 1\}$ and $\mathcal{H} = \{0, 0, 0, 1, 1, 1\}$.

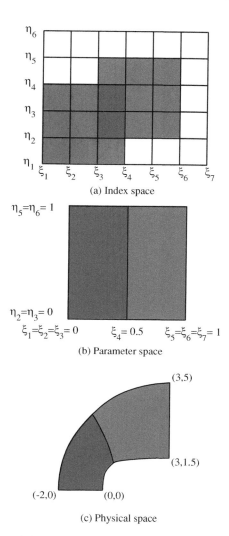

(a) Index space

(b) Parameter space

(c) Physical space

Figure 2.16 The three ways of viewing a B-spline. (a) The index space. The support of $\tilde{N}_{1,1;2,2}(\xi, \eta)$ is shown in red, while the support of $\tilde{N}_{3,2;2,2}(\xi, \eta)$ is in blue. The region in which they overlap is purple. (b) The parameter space. $\tilde{N}_{3,2;2,2}(\xi, \eta)$ is supported in both elements, while $\tilde{N}_{1,1;2,2}(\xi, \eta)$ is only supported in one. (c) The physical space. Again, $\tilde{N}_{3,2;2,2}(\xi, \eta)$ is supported in both elements, while $\tilde{N}_{1,1;2,2}(\xi, \eta)$ is only supported in one.

Figure 2.16b and Figure 2.16c. We could have determined this immediately, however, by looking at the index space in Figure 2.16a.

2.1.3.3 B-spline solids

Tensor product B-spline solids are defined in analogous fashion to B-spline surfaces. Given a *control lattice*[2] $\{\mathbf{B}_{i,j,k}\}, i = 1, 2, \ldots, n, j = 1, 2, \ldots, m, k = 1, 2, \ldots, l$, polynomial orders

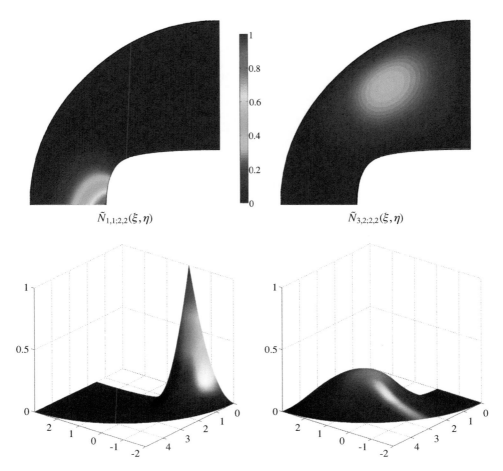

Figure 2.17   Biquadratic functions $\tilde{N}_{1,1;2,2}(\xi, \eta)$ and $\tilde{N}_{3,2;2,2}(\xi, \eta)$ plotted in the physical space, from two different angles.

p, q and r, and knot vectors $\Xi = \{\xi_1, \xi_2, \ldots, \xi_{n+p+1}\}$, $\mathcal{H} = \{\eta_1, \eta_2, \ldots, \eta_{m+q+1}\}$, and $\mathcal{Z} = \{\zeta_1, \zeta_2, \ldots, \zeta_{l+r+1}\}$, a B-spline solid is defined by

$$\mathbf{S}(\xi, \eta, \zeta) = \sum_{i=1}^{n} \sum_{j=1}^{m} \sum_{k=1}^{l} N_{i,p}(\xi) M_{j,q}(\eta) L_{k,r}(\zeta) \mathbf{B}_{i,j,k} \tag{2.19}$$

The properties of a B-spline solid like the one shown in Figure 2.18 are trivariate generalizations of those for B-spline surfaces.

2.1.4 Refinement

One of the most interesting aspects of B-splines is the myriad of ways in which the basis may be enriched while leaving the underlying geometry and its parameterization intact. To

Figure 2.18   A simple B-spline solid.

fully recognize the many possibilities, we must first understand the subtle ways in which the basic mechanisms of B-spline refinement differ from their finite element counterparts. These differences lead to more richness in the overall refinement space. In particular, not only do we have control over the element size and the order of the basis, but we can control the continuity of the basis as well.

2.1.4.1 Knot insertion

The first mechanism by which one can enrich the basis is ***knot insertion***.[3] Knots may be inserted without changing a curve geometrically or parametrically. Given a knot vector $\Xi = \{\xi_1, \xi_2, \ldots, \xi_{n+p+1}\}$, we introduce the notion of an *extended* knot vector $\bar{\Xi} = \{\bar{\xi}_1 = \xi_1, \bar{\xi}_2, \ldots, \bar{\xi}_{n+m+p+1} = \xi_{n+p+1}\}$, such that $\Xi \subset \bar{\Xi}$. As before, the new $n + m$ basis functions are formed by (2.1) and (2.2), now by applying them to the new knot vector $\bar{\Xi}$. The new $n + m$ control points, $\bar{\mathcal{B}} = \{\bar{\mathbf{B}}_1, \bar{\mathbf{B}}_2, \ldots, \bar{\mathbf{B}}_{n+m}\}^{\mathrm{T}}$, are formed from linear combinations of the original control points, $\mathcal{B} = \{\mathbf{B}_1, \mathbf{B}_2, \ldots, \mathbf{B}_n\}^{\mathrm{T}}$, by

$$\bar{\mathcal{B}} = \mathbf{T}^p \mathcal{B} \qquad (2.20)$$

where

$$T_{ij}^0 = \begin{cases} 1 & \bar{\xi}_i \in [\xi_j, \xi_{j+1}) \\ 0 & \text{otherwise} \end{cases} \qquad (2.21)$$

and

$$T_{ij}^{q+1} = \frac{\bar{\xi}_{i+q} - \xi_j}{\xi_{j+q} - \xi_j} T_{ij}^q + \frac{\xi_{j+q+1} - \bar{\xi}_{i+q}}{\xi_{j+q+1} - \xi_{j+1}} T_{ij+1}^q \quad \text{for} \quad q = 0, 1, 2, \ldots, p - 1 \qquad (2.22)$$

Knot values already present in the knot vector may be repeated in this way, thereby increasing their multiplicity, but as described in Section 2.1.2, the continuity of the *basis* will be reduced. However, continuity of the *curve* is preserved by choosing the control points as in (2.20)–(2.22).

An example of knot insertion for a simple, one-element, quadratic B-spline curve is presented in Figure 2.19. The knot vector of the original curve is $\Xi = \{0, 0, 0, 1, 1, 1\}$. The control points,

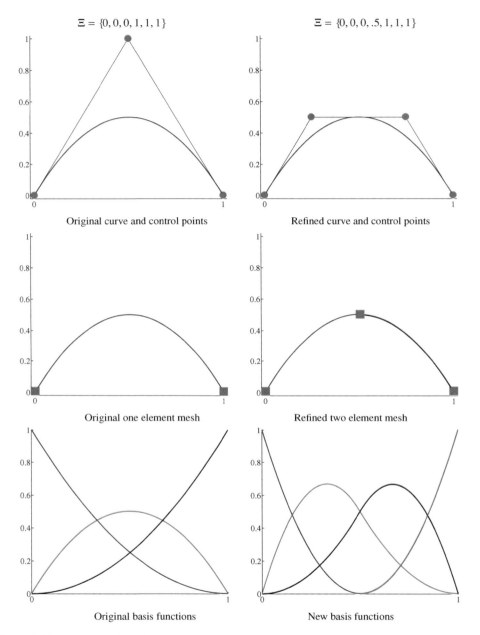

Figure 2.19 Knot insertion. Control points are denoted by ●. The knots, which define a mesh by partitioning the curve into elements, are denoted by ■.

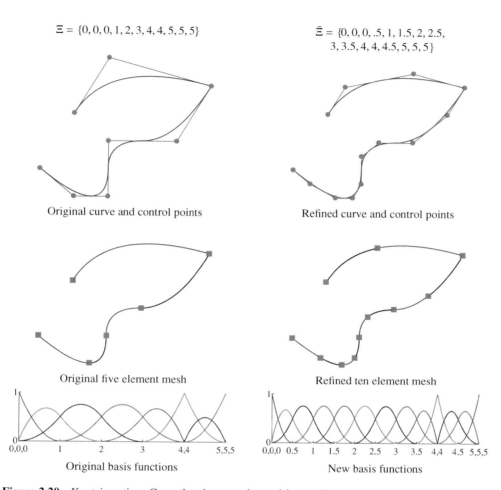

$\Xi = \{0, 0, 0, 1, 2, 3, 4, 4, 5, 5, 5\}$

$\bar{\Xi} = \{0, 0, 0, .5, 1, 1.5, 2, 2.5,$
$3, 3.5, 4, 4, 4.5, 5, 5, 5\}$

Original curve and control points

Refined curve and control points

Original five element mesh

Refined ten element mesh

Original basis functions

New basis functions

Figure 2.20 Knot insertion. Control points are denoted by •. The knots, which define a mesh by partitioning the curve into elements, are denoted by ■. Each element has been evenly split in the parametric domain.

mesh, and basis functions of the unrefined curve are shown on the left. A new knot is inserted at $\bar{\xi} = 0.5$. The new curve, shown on the right, is geometrically and parametrically identical to the original curve, but the control points are changed, the mesh is partitioned, and the basis is richer. There is one more control point, one more element, and one more basis function than in the unrefined case. This process may be repeated to enrich the solution space by adding more basis functions of the same order while leaving the curve unchanged. Figure 2.20 shows the more advanced case of a global refinement of the curve from Figure 2.10.

Insertion of new knot values clearly has similarities with the classical h-refinement strategy in finite element analysis as it splits existing elements into new ones. It differs, however, in the number of new functions that are created, as well as in the continuity of the basis across the newly created element boundaries (C^{p-1} in this case). To perfectly replicate h-refinement, one would need to insert each of the new knot values p times so that the functions will be C^0 across the new boundary. The alternative to inserting new knot values – increasing the multiplicity of existing knot values to decrease the continuity of the basis without creating new elements – does not have an analogue in FEA, as FEA meshes have C^0 element boundaries to begin with. In this way, knot insertion is very closely related, but not identical to h-refinement. We will revisit this idea below.

2.1.4.2 Order elevation

The second mechanism by which one can enrich the basis is **order elevation** (sometimes also called "degree elevation"). As its name implies, the process involves raising the polynomial order of the basis functions used to represent the geometry. Recalling from Section 2.1.1 that the basis has $p - m_i$ continuous derivatives across element boundaries, it is clear that when p is increased, m_i must also be increased if we are to *preserve* the discontinuities in the various derivatives already existing in the original curve. During order elevation, the multiplicity of each knot value is increased by one, but no new knot values are added. As with knot insertion, neither the geometry nor the parameterization are changed.

The process for order elevation begins by replicating existing knots until their multiplicity is equal to the polynomial order, thus effectively subdividing the curve into many Bézier curves by knot insertion (see Rogers, 2001 or Farin, 1999b for a discussion of Bézier curves; we may think of them as one-element B-spline curves). The next step is to elevate the order of the polynomial on each of these individual segments. Lastly, excess knots are removed to combine the segments into one, order-elevated, B-spline curve. Several efficient algorithms exist which combine the steps so as to minimize the computational cost of the process. For a thorough treatment, see Piegl and Tiller, 1997.

An example of order elevation for a one-element curve is depicted in Figure 2.21. The original control points, mesh, and quadratic basis functions, shown on the left, are the same as considered in Figure 2.19. This time the *multiplicity* of the knots is increased by one but, as stated above, no new knot *values* are added. For this simple case, the number of control points and the number of basis functions each increase by one. The locations of the control points change, but the elevated curve is geometrically and parametrically identical to the original curve. There are now four cubic basis functions. Figure 2.22 shows this process on the more complex example considered in Figure 2.20. The multiplicities of the knots have been increased but no new elements created. Note that the locations of control points for these order-elevated curves are different than those in the h-refinement examples (cf. Figures 2.19 and 2.20).

Order elevation clearly has much in common with the classical p-refinement strategy in finite element analysis as it increases the polynomial order of the basis. The major difference is that p-refinement always begins with a basis that is C^0 everywhere, while order elevation is compatible with any combination of continuities that exist in the unrefined B-spline mesh. This flexibility leads us to a new higher-order technique that is unique to isogeometric analysis.

2.1.4.3 k-refinement: higher order *and* higher continuity

As we have seen, the two primitive refinement operations for B-splines are knot insertion and order elevation. Knot insertion is similar to h-refinement, but for it to be a perfect analogue each new knot value would have to be inserted with multiplicity $m_i = p$ to ensure a C^0 basis everywhere. Similarly, if we begin with a mesh in which all of the functions are already C^0

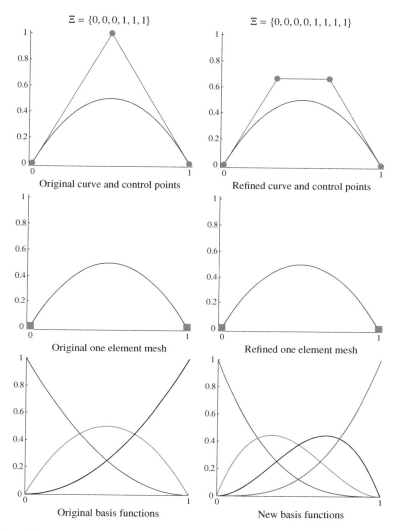

Figure 2.21 Order elevation. Control points are denoted by •. The knots, which define a mesh by partitioning the curve into elements, are denoted by ■.

across element boundaries, order elevation coincides exactly with the traditional notion of p-refinement. Knot insertion and order elevation, however, provide us with more to work with than do the two standard notions of refinement.

As mentioned above, we can insert new knot values with multiplicities equal to one to define new elements across whose boundaries functions will be C^{p-1}. We can also repeat existing knot values to lower the continuity of the basis across existing element boundaries. This makes knot insertion a more flexible process than simple h-refinement. Similarly, we have a more flexible higher-order refinement as well. It stems from the fact that the processes of order elevation and knot insertion do not commute. If a unique knot value, $\bar{\xi}$, is inserted between

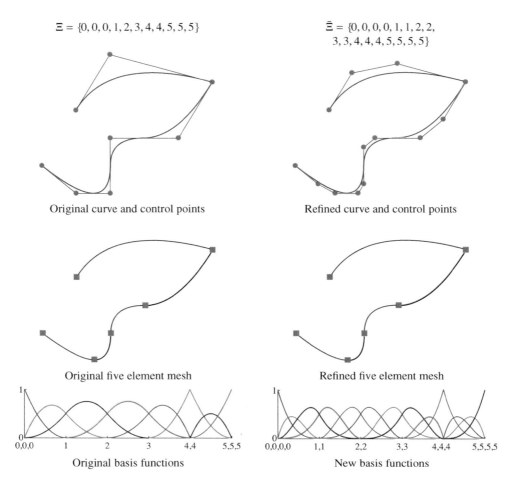

Figure 2.22   Order elevation. Control points are denoted by •. The knots, which define a mesh by partitioning the curve into elements, are denoted by ■. Note the increased multiplicity of internal knots. This is done to preserve discontinuities in the appropriate derivatives of the curve.

two distinct knot values in a curve of order p, the number of continuous derivatives of the basis functions at $\bar{\xi}$ is $p - 1$. If we subsequently elevate the order to q, the multiplicity of every distinct knot value (including the knot just inserted) is increased so that discontinuities in the pth derivative of the basis are preserved. That is, the basis still has $p - 1$ continuous derivatives at $\bar{\xi}$, although the polynomial order is now q. If, instead, we elevated the order of the original, coarsest curve to q and only then inserted the unique knot value $\bar{\xi}$, the basis would have $q - 1$ continuous derivatives at $\bar{\xi}$. We refer to this latter procedure as **k-refinement**. We know of no analogous practice in standard finite element analysis.

   It is important that we point out that this notion of k-refinement is *not* the same as the "k-convergence" described in Kagan *et al.*, 1998 in which the position of the knots is altered. It bears more in common with the "k-version finite element method" of Surana *et al.*, 2002 in that k refers to continuity, but the motivations are different. The increased continuity in Surana *et al.*, 2002 is required so that a least-squares finite element approach is possible. Such an approach requires that the solution space have the same number of continuous derivatives as found in the highest order derivative of the differential operator. Our motivations for using basis functions of higher continuity are efficiency and robustness of the solution space in a classical Galerkin finite element formulation of the problem (see Chapter 3).

The concept of k-refinement is potentially a superior approach to high-precision analysis than p-refinement. In traditional p-refinement there is a very inhomogeneous structure to arrays due to the different basis functions associated with surface, edge, vertex and interior nodes. In addition, there is a proliferation in the number of nodes because C^0-continuity is maintained in the refinement process. In k-refinement, there is a homogeneous structure within patches and growth in the number of control variables is limited.

Consider a classical p-refinement process such as is seen in Figures 2.23b and 2.24a. Assume the initial domain consists of one element and $p + 1$ basis functions (assuming an open knot

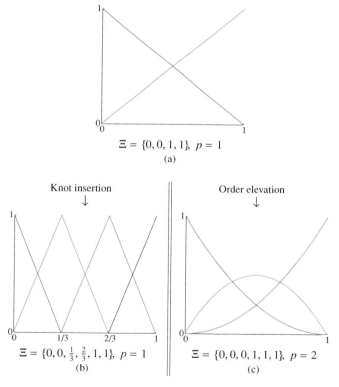

$$\Xi = \{0, 0, 1, 1\}, \ p = 1$$
(a)

Knot insertion   Order elevation
↓   ↓

$$\Xi = \{0, 0, \tfrac{1}{3}, \tfrac{2}{3}, 1, 1\}, \ p = 1$$
(b)

$$\Xi = \{0, 0, 0, 1, 1, 1\}, \ p = 2$$
(c)

Figure 2.23 When refining a coarse, low-order mesh to create a fine, higher-order mesh, one may choose between a p- or k-refinement strategy. Here we see the initial step for each case. (a) Base case of one linear element. (b) Classic p-refinement approach: knot insertion is performed first to create many low-order elements. Subsequent order elevation will preserve the C^0-continuity across element boundaries. (c) New k-refinement approach: order elevation is performed on the coarsest discretization. Subsequent knot insertion will result in a basis which is C^{p-1} across the newly created element boundaries. See the results of p- and k-refinement for several different polynomial orders in Figure 2.24.

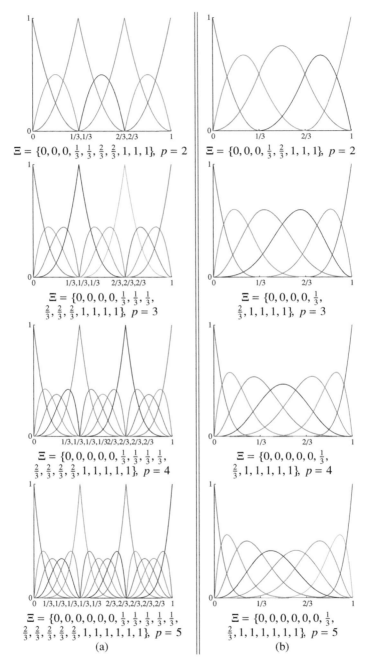

Figure 2.24 Three element, higher-order meshes for p- and k-refinement. (a) The p-refinement approach results in many functions that are C^0 across element boundaries. (b) In comparison, k-refinement results in a much smaller number of functions, each of which is C^{p-1} across element boundaries.

vector), which we then refine by inserting new knot values until we have $n - p$ elements and n basis functions, all C^{p-1}. We then perform order elevation, maintaining continuity at the $p - 1$ level. This requires replicating each distinct knot value, adding a basis function in each element and so increasing the total number of basis functions by $n - p$ to $2n - p$. After a total of r order elevations of this type, we have $(r + 1)n - rp$ basis functions, where p is still the order of the original basis. This is seen to be a large number of functions when one considers that in most cases of practical interest the number of elements will be quite a bit larger than the order of the basis. For comparison, consider beginning with the same one-element domain and proceeding by k-refinement, as in Figures 2.23c and 2.24b. That is, order elevate r times adding only *one* basis function at each refinement, then insert knots until we have $n - p$ elements as before. The final number of basis functions is $n + r$, each having $r + p - 1$ continuity. This amounts to an enormous savings as $n + r$ is considerably smaller than $(r + 1)n - rp$. Additionally, keep in mind that in d dimensions these numbers are raised to the d power. Graphical comparisons are shown in Figure 2.25. Note that the mesh, defined by the knot *locations*, is fixed and is the same for p- and k-refinements.

Observe that k-refinement, as we have defined it, is not really "refinement" in the traditional sense in that it does not lead to a sequence of nested spaces. Consider again the examples of Figure 2.24. As we p-refine in Figure 2.24a from quadratics to quintics, each set of basis functions is capable of representing every function that could be represented by any of the bases of lower order. The space was being "enriched" as something is gained at each step, but nothing is lost. Alternatively, the k-refinement process in Figure 2.24b does not possess this property. This is obvious if we only consider the continuity. A general function of order p has discontinuities in the pth derivative, but every function of order $p + 1$ has p continuous derivatives. While the higher-order bases in this sequence have better approximation properties, they cannot represent the same set of functions as the lower-order bases in the sequence. This should not be seen as a shortcoming of the approach, but it is a difference between k-refinement and the more traditional h- and p-refinements.

It is also important to note that "pure" k-refinement, where all functions maintain maximal C^{p-1} continuity across element boundaries, is only possible if the coarsest mesh is comprised of a single element. If the initial mesh places constraints on the continuity across certain element boundaries, these constraints will exist on all meshes. In general, though some such constraints will exist, the number of elements desired for analysis will be much higher than the number needed for modeling the geometry. Refinements may be performed such that the functions have $p - 1$ continuous derivatives across these new element boundaries and the benefits of k-refinement will still be significant.

2.1.4.4 The *hpk*-refinement space

As we have shown, knot insertion and order elevation are the primitive operations by which classical h- and p-refinements, as well as the new k-refinement, can be implemented. Recognizing their flexibility as compared with classical refinement procedures makes feasible the notion of an *hpk*-refinement space. Recalling that B-spline curves may have no more than $p - 1$ continuous derivatives across an element boundary, the set of possible refinements may be characterized as in Figure 2.26. Pure k-refinement keeps h fixed but increases the continuity along with the polynomial order, as in Figure 2.27a. Pure p-refinement increases the polynomial order while the basis remains C^0, as in Figure 2.27b. Increasing the multiplicity of

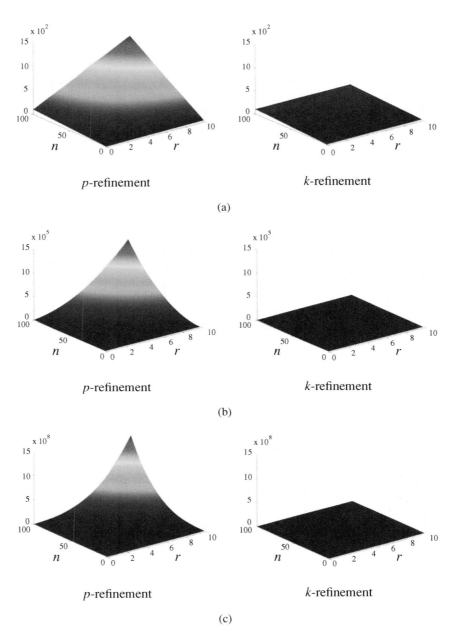

Figure 2.25 Comparison of control variable growth. (a) The one-dimensional case with n initial control points. (b) The two-dimensional case with n^2 initial control points. (c) The three-dimensional case with n^3 initial control points.

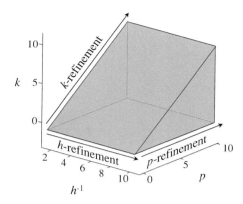

Figure 2.26 The *hpk*-space. The set of all allowable refinements is contained in the region shown in green. Note that this region extends in the direction of the arrows.

existing knot values decreases the continuity without introducing new elements, as in Figure 2.27c. Inserting new knot values with a multiplicity of *p* results in classical *h*-refinement, whereby new elements are introduced that have C^0 boundaries, shown in Figure 2.27d. Inserting new knot values with a multiplicity of 1 decreases *h* without decreasing the minimum continuity already found in the mesh, as in Figure 2.27e. Considering all of the aforementioned techniques results in a multitude of refinement options beyond simple *h*-, *p*- and *k*-refinement; see Figure 2.27f.

2.2 Non-Uniform Rational B-Splines

The step from the non-rational B-splines that we have been discussing thus far to Non-Uniform Rational B-Splines (NURBS) is a significant one because we gain the ability to exactly represent a wide array of objects that cannot be exactly represented by polynomials, many of which are ubiquitous in engineering design. To best appreciate how to work with NURBS entities we must understand them from both a geometric perspective and an algebraic one. The former viewpoint gives us insight and intuition that will prove invaluable in designing meshes, proving theorems, and a host of other activities related to isogeometric analysis. The latter viewpoint is particularly useful in designing algorithms and creating software, and will be the setting in which we most frequently work. Both are essential for cultivating a broad understanding of NURBS technology.

2.2.1 The geometric point of view

A NURBS entity in $\mathbb{R}^d$ is obtained by the projective transformation of a B-spline entity in $\mathbb{R}^{d+1}$. In particular, conic sections, such as circles and ellipses, can be *exactly* constructed by projective transformations of piecewise quadratic curves – one of the defining features of isogeometric analysis. A full discussion of projective geometry is beyond the scope of this book, but a good introduction in the context of NURBS can be found in Farin, 1999b. For our purposes, it suffices to consider the example illustrated in Figure 2.28 in which a circle in $\mathbb{R}^2$ is constructed from a piecewise quadratic B-spline curve in $\mathbb{R}^3$. The transformation is applied by projecting every point in the curve onto the $z = 1$ plane by a ray through the origin.

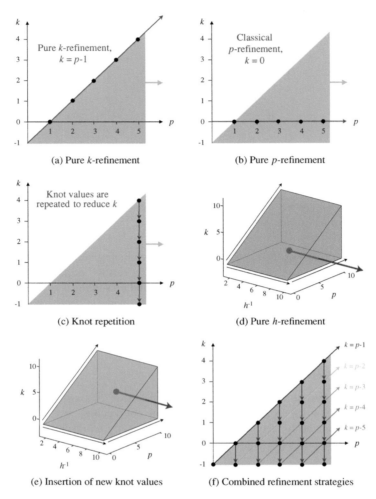

(a) Pure k-refinement   (b) Pure p-refinement

(c) Knot repetition   (d) Pure h-refinement

(e) Insertion of new knot values   (f) Combined refinement strategies

Figure 2.27 The hpk-space. (a) In pure k-refinement, the locations of the element boundaries (and thus element size, h) are fixed. As the polynomial order, p, is increased, the continuity of the functions across element boundaries, k, is increased such that $k = p - 1$ at all levels of refinement. (b) In pure p-refinement, the locations of the element boundaries (and thus element size, h) are fixed. As the polynomial order, p, is increased, the continuity of the functions across element boundaries is fixed at $k = 0$ for all levels of refinement. (c) Repetition of existing knot values decreases the continuity across the corresponding element boundary without creating new elements or changing the polynomial order. The basis has $p - m_i$ continuous derivatives across knot ξ_i, where m_i is the multiplicity of that knot value. (d) If we insert new knot values with multiplicity of p, new elements are created and the basis remains C^0 across all element boundaries. In this way classical h-refinement is exactly replicated. (e) Insertion of new knot values with a multiplicity of 1 results in a splitting of elements, and thus a decrease in h (shown in the figure as an increase in h^{-1}). The basis has $p - 1$ continuous derivatives across these new element boundaries, and so the (possibly lower) minimum continuity already existing in the mesh is unchanged, as is the polynomial order. (f) Combining knot insertion and order elevation in various permutations allows us to traverse the entire allowable refinement space.

We get the control points for the NURBS curve by performing exactly the same projective transformation to the control points of the B-spline curve. In this context the B-spline, $\mathbf{C}^w(\xi)$, is called the "projective curve" with its associated "projective control points," $\mathbf{B}_i^w$, while the terms "curve" and "control points" are reserved for the NURBS objects $\mathbf{C}(\xi)$ and $\mathbf{B}_i$, respectively.

With a given projective B-spline curve and its associated projective control points in hand, the control points for the NURBS curve are obtained by the following relations:

$$(\mathbf{B}_i)_j = (\mathbf{B}_i^w)_j / w_i, \quad j = 1, \ldots, d \tag{2.23}$$

$$w_i = (\mathbf{B}_i^w)_{d+1} \tag{2.24}$$

where $(\mathbf{B}_i)_j$ is the jth component of the vector $\mathbf{B}_i$ and w_i is referred to as the ith **weight**. In Figure 2.28a, the weights are the z-components of the projective control points. These values, in general the $d + 1$ components of projective control points in $\mathbb{R}^{d+1}$, are positive in most applications of engineering interest. We will consider them to be positive throughout this book. Dividing the projective control points by the weights is equivalent to applying the projective transformation to them. We would like to apply the same transformation to every point in the curve (*e.g.*, for a projective curve in $\mathbb{R}^3$, we would like to divide every point in the curve by its

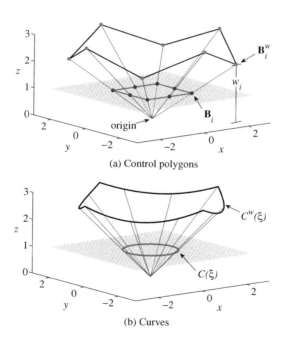

(a) Control polygons

(b) Curves

Figure 2.28 A circle in $\mathbb{R}^2$ constructed by the projective transformation of a piecewise quadratic B-spline in $\mathbb{R}^3$. (a) Projective transformation of "projective control point" $\mathbf{B}_i^W$ yields control point $\mathbf{B}_i$. Weight w_i is the z-component of B_i^w. (b) Projective transformation of the B-spline curve $C^w(\xi)$ yields the NURBS curve $C(\xi)$.

height). We accomplish this by defining the **weighting function**,

$$W(\xi) = \sum_{i=1}^{n} N_{i,p}(\xi)w_i, \tag{2.25}$$

where $N_{i,p}(\xi)$ is the standard B-spline basis function. In $\mathbb{R}^3$, $W(\xi) = z(\xi)$ is the height of the curve as a function of the parameter ξ. We can now define the NURBS curve as

$$(\mathbf{C}(\xi))_j = \frac{(\mathbf{C}^w(\xi))_j}{W(\xi)}, \quad j = 1, \ldots, d. \tag{2.26}$$

As $\mathbf{C}^w(\xi)$ and $W(\xi)$ are both piecewise polynomial functions, the curve $\mathbf{C}(\xi)$ is a piecewise **rational function** – within each element it is a polynomial divided by another polynomial. In this NURBS setting, the two polynomials have the same order, and so we frequently refer to the "order of the NURBS curve," meaning that of the B-spline from which it was generated. Phrases such as "a quadratic NURBS curve" are common and should be interpreted in this sense.

In Figure 2.28b, the B-spline curve clearly has four points of only C^0-continuity. To achieve this with quadratic functions, the knot values at these locations have multiplicities of two. Now note that the circle itself has no obvious points of reduced continuity. This is not an uncommon scenario. Frequently, the maximum level of continuity is restricted by the shape of the projective curve rather than the curve itself. For the example at hand, there is no reasonable way to construct a circle without some knots at which the continuity is C^0, even though there is nothing obvious about the circle itself to indicate it.

It is interesting to observe that if one were to multiply all of the projective control points by a constant (the simplest affine transformation of the projective curve), the resulting NURBS curve would be unchanged. This is because each point of the projective curve would move along its ray through the origin, but not onto a different ray. To achieve an affine transformation of a NURBS object, we apply the affine transformation directly to its control points (as we would hope), while leaving the weights fixed. Though each weight is *associated* with a specific control point, it is important that we do not think of is as a *component* of the control point. This is an easy mistake to make as most NURBS data structures will store the weight as the fourth component of the array that stores its associated control point.

Though this projective geometric interpretation of the situation may seem daunting, rest assured that we rarely explicitly use it in practice. We will build familiar objects, such as cylinders or spheres in $\mathbb{R}^3$, from simple templates or by using CAD packages rather than explicitly determining four-dimensional objects that yield the desired result under transformation. The main reason that it is important to understand the underlying nature of NURBS is to recognize that everything that we have discussed thus far for B-splines still holds true. This is due to the fact that NURBS are built directly from B-splines. In particular, one can refine the NURBS objects by applying the desired combination of *hpk*-strategies to the projective B-spline objects themselves.

2.2.2 The algebraic point of view

Though the geometric viewpoint provides us with some intuition about the NURBS objects we are working with and how they are constructed, it is the algebraic perspective that allows us to manipulate them most directly. Part of the power of B-splines is the ability to intuitively change their shape by adjusting the control points. We would like to manipulate NURBS in exactly the same intuitive fashion. To do this, we need to construct a basis for the NURBS space from knot vectors, and to build curves, surfaces, and solids from linear combinations of basis functions and control points. In this way, everything that we have learned about B-splines will also be true of NURBS.

As we have seen above, the weighting function of (2.25) is a scalar, piecewise polynomial function for the $d + 1$ component of the projective curve. From the geometric point of view, we have used it to project a B-spline curve from $\mathbb{R}^{d+1}$ into $\mathbb{R}^d$. From the algebraic point of view, it is more productive to use it to construct a basis for the NURBS space directly so that we may build geometries and meshes in $\mathbb{R}^d$ while remaining blissfully ignorant of the projective geometry lurking behind the scenes. This NURBS basis is given by

$$R_i^p(\xi) = \frac{N_{i,p}(\xi)w_i}{W(\xi)} = \frac{N_{i,p}(\xi)w_i}{\sum_{\hat{i}=1}^{n} N_{\hat{i},p}(\xi)w_{\hat{i}}}, \tag{2.27}$$

which is clearly a piecewise rational function. Using (2.27) in conjunction with the control points of (2.23) leads to an equation for a NURBS curve,

$$\mathbf{C}(\xi) = \sum_{i=1}^{n} R_i^p(\xi)\mathbf{B}_i, \tag{2.28}$$

that is form identical to that for B-splines. In practice, we will always use (2.28) and not (2.26), although they are equivalent. Rational surfaces and solids are defined analogously in terms of the rational basis functions

$$R_{i,j}^{p,q}(\xi, \eta) = \frac{N_{i,p}(\xi)M_{j,q}(\eta)w_{i,j}}{\sum_{\hat{i}=1}^{n} \sum_{\hat{j}=1}^{m} N_{\hat{i},p}(\xi)M_{\hat{j},q}(\eta)w_{\hat{i},\hat{j}}}, \tag{2.29}$$

$$R_{i,j,k}^{p,q,r}(\xi, \eta, \zeta) = \frac{N_{i,p}(\xi)M_{j,q}(\eta)L_{k,r}(\zeta)w_{i,j,k}}{\sum_{\hat{i}=1}^{n} \sum_{\hat{j}=1}^{m} \sum_{\hat{k}=1}^{l} N_{\hat{i},p}(\xi)M_{\hat{j},q}(\eta)L_{\hat{k},r}(\zeta)w_{\hat{i},\hat{j},\hat{k}}}. \tag{2.30}$$

These rational basis functions bear much in common with their polynomial progenitors. In particular, the continuity of the functions, as well as their support, follows directly from the knot vectors exactly as before. The basis still constitutes a partition of unity, and it is pointwise nonnegative. These properties taken together again result in a strong convex hull property for the NURBS functions.

Note that the weights play an important role in defining the basis, but they are divorced from any explicit geometric interpretation in this setting, and we are free to choose control points independently from their associated weights. Also note that if the weights are all equal, then $R_i^p(\xi) = N_{i,p}(\xi)$ and the curve is again a polynomial. Thus, B-splines are a special case of NURBS.

2.2.2.1 Derivatives of NURBS basis functions

As the NURBS basis functions are constructed from the B-spline basis functions, the derivatives of rational functions will clearly depend on the derivatives of their non-rational counterparts as well. Simply applying the quotient rule to (2.27) yields

$$\frac{d}{d\xi} R_i^p(\xi) = w_i \frac{W(\xi)N'_{i,p}(\xi) - W'(\xi)N_{i,p}(\xi)}{(W(\xi))^2}, \tag{2.31}$$

where $N'_{i,p}(\xi) \equiv \frac{d}{d\xi} N_{i,p}(\xi)$ and

$$W'(\xi) = \sum_{\hat{i}=1}^{n} N'_{\hat{i},p}(\xi) w_{\hat{i}}. \tag{2.32}$$

In practice, this is how we typically compute these derivatives. We have an efficient algorithm for the derivatives of the non-rational basis (see Piegl and Tiller, 1997, chapter 3, pp. 91–100), and we use it to compute those of the rational functions using the quotient rule.

An expression is also available for higher-order derivatives of NURBS basis functions. Following Piegl and Tiller, 1997, let us simplify notation by defining

$$A_i^{(k)}(\xi) = w_i \frac{d^k}{d\xi^k} N_{i,p}(\xi), \quad \text{(no sum on } i) \tag{2.33}$$

where we *do not sum on the repeated index*, and let

$$W^{(k)}(\xi) = \frac{d^k}{d\xi^k} W(\xi). \tag{2.34}$$

Higher-order derivatives of these rational functions may be expressed in terms of lower-order derivatives as

$$\frac{d^k}{d\xi^k} R_i^p(\xi) = \frac{A_i^{(k)}(\xi) - \sum_{j=1}^{k} \binom{k}{j} W^{(j)}(\xi) \frac{d^{(k-j)}}{d\xi^{(k-j)}} R_i^p(\xi)}{W(\xi)}, \tag{2.35}$$

where

$$\binom{k}{j} = \frac{k!}{j!(k-j)!}. \tag{2.36}$$

2.3 Multiple patches

In almost all practical circumstances, it will be necessary to describe domains with multiple NURBS patches. For example, if different material or physical models are to be used in different parts of the domain, it might simplify things to describe these subdomains by different patches. Also, if different subdomains are to be assembled in parallel on a multiple processor machine, it is convenient from the point of view of data structures to not have a single patch split between different processors. Most common is the case where the domain simply differs topologically from a cube. The tensor product structure of the parameter space of a patch makes it poorly suited for representing complex, multiply connected domains. Such geometries can frequently be handled quite simply by using multiple patches (see, *e.g.*, Figure 2.29).

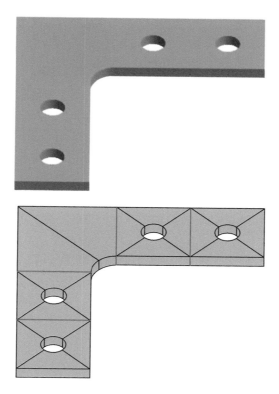

Figure 2.29 The bracket on the top is exactly and concisely represented by five simple NURBS patches (patch boundaries are shown in red, element boundaries in black). The patches match geometrically and parametrically on the internal faces where they meet.

Remark

In the isogeometric concept, the geometry is exactly preserved as the mesh is refined. This means that the precise fillet and hole radii in Figure 2.29 will be maintained during refinement. The reader knowledgeable of elasticity theory will realize that the exact stress concentrations induced by the fillet and hole radii will be attained upon convergence. This may be contrasted with the case of finite element mesh refinement. The difficulty of meshing around sharp fillets and small holes motivates removing these features, as shown in Figure 2.30, but this produces entirely incorrect stress concentrations. That due to the holes is eliminated, and the right-angle replacing the fillet leads to infinite stresses. Feature removal is common practice in creating FEA models. Results must be interpreted with extreme caution, because the solution can change dramatically. An attractive property of isogeometric analysis is that small features, such as fillets and holes, can be retained in the model.

Even in cases where a cube can be mapped into the desired object, doing so might introduce such extreme mesh distortion and widely varying Jacobians within elements that analysis will be adversely affected. Figure 2.31b (from Hughes *et al.*, 2005) shows the amount of mesh distortion needed to represent the shell with stiffener of Figure 2.31a with a single NURBS

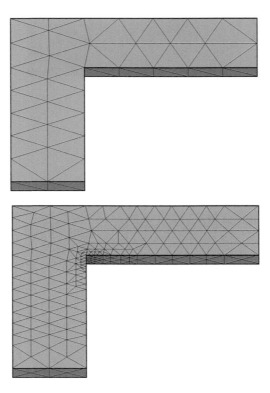

Figure 2.30 The removal of holes and fillets from the model facilitates the use of an automatic mesh generator. The values of the stress, however, will be entirely incorrect. Moreover, automatic mesh refinement may attempt to resolve the spurious stress singularity, resulting in a proliferation of degrees-of-freedom near the reentrant corner.

patch. A mesh using multiple patches, shown in Figure 2.31c, exhibits far less distortion and yields a much more "natural" mesh.

2.4 Generating a NURBS mesh: a tutorial

To complete this chapter on NURBS geometry, it seems appropriate to step through one simple, albeit nontrivial, example of actually generating a NURBS geometry from scratch. Though mesh generators and CAD systems obviate some of the details of such an exercise, getting one's hands dirty at least once does provide some insight into what goes on under the hood of such software. Understanding the details of such a process is the first step toward being able to make sense of any problems that may arise, as well as toward expanding the technology.

Let us attempt to build a NURBS model of the pipe with a 90° elbow bend shown in Figure 2.32. The pipe has an inner radius of 1, outer radius of 2, and the circular bend of the elbow has a radius of 3 along the center-line of the pipe. This object is easily described by a single NURBS patch. We will proceed by laying out a set of general steps for geometric design one at a time, applying them to the modeling of this particular object as we go.

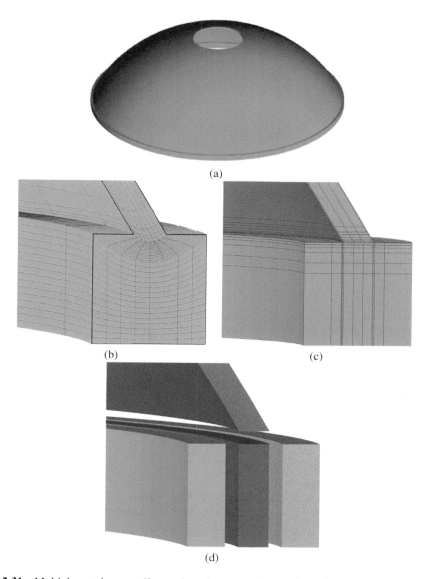

(a)

(b)   (c)

(d)

Figure 2.31 Multiple patches usually produce better quality meshes (from Hughes *et al.*, 2005). (a) The stiffened shell can be modeled using a single NURBS patch. (b) A detail of the stiffener reveals that such a mapping produces severe mesh distortion that is unavoidable when using a single patch. (c) Allowing the shell and the stiffener to be modeled by different patches creates a much more natural mesh. The patch boundaries are shown in red. Analysis on this mesh will be described in Chapter 4. (d) Each of these unique patches has its own parameter space.

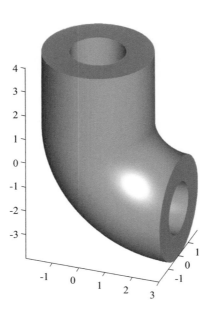

Figure 2.32 A pipe with an elbow bend. This object lends itself to a simple description using NURBS, but it is completely outside of the space of standard finite element geometries.

2.4.1 Preliminary considerations

There are many ways to go about building a geometrical model from scratch. Each geometer undoubtedly approaches the process in a different fashion. In our experience, it is fruitful to begin by identifying some basic features before even starting to assign polynomial orders, knot vectors, or control points. The major features to look for are:

1. Corners and other points to be interpolated
2. Edges and other lines of reduced continuity
3. Geometric primitives and lower-dimensional NURBS objects
4. Extrusions, surfaces of revolution, symmetries, or other tensor-product-like features

Starting at the beginning of the list, let us address each type of feature one at a time, and look for them in Figure 2.32. Corners are a natural place to begin as the parameter space is a cube (there is no loss of generality in assuming it to be a cube, as dividing an entire knot vector by a constant does not change the resulting geometry in any way at all, and so we may always normalize the knot vectors such that the parameter space is the unit cube). Additionally, the use of open knot vectors means that the basis will interpolate the corners, and so identifying them can give us a few control points immediately. Unfortunately, the pipe in Figure 2.32 does not possess any corners, and this first step does not help for this example.

The next thing to look for is any place where the continuity is obviously decreased. The most obvious thing would be a crease in the geometry, that is, a sharp edge other than the image of one of the edges of the parametric cube. For the pipe example, no such creases exist. The astute observer will note that there is a discontinuity in the curvature where the cylindrical portion of the pipe meets the bend. The experienced NURBS practitioner will realize that this will actually call for C^0-continuity of the basis. Moreover, there will be additional continuity restrictions in the circumferential direction as well (recall Figure 2.28b). For our purposes, it suffices to say that we do not see any creases. We will momentarily postpone the discussion of the decreased continuity until after we have discussed circles in more detail.

The third step is to identify simple objects that we already know how to construct that may be part of the overall geometry of interest. We can look for "geometrical primitives" such as polynomials or conic sections that we might have templates for. Recall that the use of open knot vectors means that each face of the NURBS patch will actually be a NURBS surface, and each edge of those surfaces is a NURBS curve. Thus we can be on the lookout for one-dimensional objects that we may already know how to model. In this case, we clearly see the presence of circular edges in the pipe, and so we will need to know how to construct a circle.

The last step is to look for places in which a NURBS curve or surface has been swept along a path defined by another NURBS curve. Such extrusions are very common in engineering design, and identifying them makes the job of modeling much easier by effectively reducing a three-dimensional problem into two problems of lower dimension. In the case at hand, we can see that the annular cross section of the pipe is swept along a path to define the solid geometry. That path is composed of a straight section (in the upper, cylindrical portion of the pipe) and a circular section (in the elbow). By first calculating the control points for the annulus, then those for the path along which it is to be swept, and finally sweeping the control points for the annulus along the control polygon of the path, we will avoid a lot of redundant work.

Foreshadowing the next chapter, it is best to keep in mind that we are generating a mesh to be used in analysis, not just creating art. For this reason it is prudent as a final preliminary step to consider how to avoid excessive distortion of the elements in the geometry we are about to create. Also, we may want to avoid creating a geometry that will have singularities in the mapping, or in its inverse. These are only guidelines, not strict rules. For example, the mesh in Figure 2.33 has a singularity in the inverse of the geometrical mapping along its axis, and yet has been used in analysis without difficulty. One face of each of the elements adjacent to the axis has been degenerated by placing multiple control points at the same location, and thus many parameter values map to the same point in physical space. Such a mapping is clearly not invertible. Still, isogeometric fluid and fluid–structure interaction analyses of arterial blood flow have been performed quite successfully on meshes topologically identical to this one (see Bazilevs et al., 2006b). This is due to the fact that the quadrature points utilized never fall on the singularity itself.

2.4.1.1 A template for a circle

A circle is one of the most common objects in engineering design. There are many ways to construct circles using NURBS. Some of the more exotic techniques involve negative weights or control points at infinity, neither of which are desirable for our purposes. The approach shown here is one of the simplest. For a more thorough treatment, see Piegl and Tiller, 1997.

Arcs of less that 180° can be constructed from a single quadratic NURBS element, as shown in Figure 2.34. The use of the open knot vector $\Xi = \{0, 0, 0, 1, 1, 1\}$ means that the first and last of the three elements in the basis will be interpolatory. Thus, the first and last control points, $\mathbf{B}_1$ and $\mathbf{B}_3$, respectively, will lie at the endpoints of the desired arc. We select the associated weights, w_1 and w_3, to be equal to one. The remaining control point lies at the

Figure 2.33 The geometrical mapping of this solid cylinder has a degeneracy along its axis. This results in a singularity in the inverse mapping. In practice, however, such meshes have been used successfully.

intersection of the tangent lines passing through the other two points (now we see why this technique only works for arcs of less than 180°). Its weight, w_2, is the cosine of half of the angle subtended by the arc. That is, if $\angle \mathbf{B}_1 \mathbf{C} \mathbf{B}_3 = \theta$, where $\mathbf{C}$ is the center of the circle, then $w_2 = \cos(\theta/2)$.

Arcs greater than 180° may be constructed from multiple smaller arcs. These do not have to be separate patches entirely, but the basis must be no more than C^0-continuous where the arcs

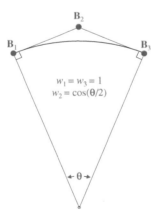

Figure 2.34 To build a circular arc from quadratic NURBS, place the first and third control points at the endpoints of the arc. The second control point is at the intersection of the tangent lines passing through these control points. The first and third weights are equal to one, while the second weight is equal to the cosine of half of the angle subtended by the arc. The associated knot vector is $\Xi = \{0, 0, 0, 1, 1, 1\}$.

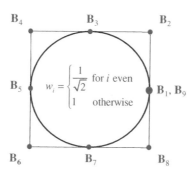

Figure 2.35 The NURBS mesh for a circle comprised of four 90° arcs. All of the control points lie on a square in which the circle is inscribed. Five of the control points (one repeated value) are on the circle itself and have a weight of 1, while the remaining control points are at the corners of the square and have a weight of $\frac{1}{\sqrt{2}}$.

meet. The reason for this is best understood by referring back to Figure 2.28b. This is a case where continuity is being restricted by the projective B-spline curve, not the actual geometry we are trying to create. The continuity in the geometry is preserved by the appropriate selection of the control points.

In Figure 2.35 we have created a complete circle from four 90° arcs. We could represent each of the arcs using separate patches such as the one we have just created. This, however, is inefficient as there would be redundant control points where the arcs meet. Instead of having two control points at each of these locations, one from each patch, we can use a single patch with multiple elements and C^0-continuity at each element boundary. In this case, this is accomplished using $\Xi = \{0, 0, 0, 1, 1, 2, 2, 3, 3, 4, 4, 4\}$. The resulting geometry is identical to that in the multiple patch case, and the redundant control points are no longer present. Note that we have closed the circle by placing the first and the last control point at the same location[4].

2.4.2 Selection of polynomial orders

Returning to the pipe geometry, the first thing to determine is what polynomial orders will be needed. In general, we will want to use the lowest polynomial order possible in each of the parametric directions. Analysis may frequently demand higher orders than geometric design (*e.g.*, higher-order functions may be needed to avoid locking in structural analysis), but it is best to work with the lowest order possible during design. This will provide the widest array of options when it finally does become time for analysis.

For the pipe in Figure 2.32, we will definitely need at least quadratic NURBS in the circumferential direction in order to replicate the circular features in the cross section. We will make this the ξ-direction in the parametric space and set $p = 2$. Note that we have no special features that require modeling through the thickness of the pipe, the η-direction. Linear functions will suffice, and so $q = 1$. Lastly, we will need quadratic NURBS in the axial direction (the ζ-direction in the parameter space) to model the circular geometry of the elbow, making $r = 2$. These are the lowest orders that can be used, and so there is no ambiguity in their selection.

2.4.3 Selection of knot vectors

Determining the knot vector is not as difficult as it may at first appear. The decision will be made by determining how many elements are necessary and what level of continuity is required across each element boundary. Often, we can also gain insight from the templates that we are using. For most purposes, integer knot values are perfectly sufficient. If a knot vector in [0, 1] is preferable for some reason, we may proceed by assigning integer values and simply divide by the greatest value once we are finished, recalling that such an operation has no bearing on the resulting geometry (nor does adding a constant).

For the pipe, the only geometry in the circumferential direction is circular. If we intend to follow the template of the above section on circles and use four arcs of 90°, then we can use the exact same knot vector used previously, thus $\Xi = \{0, 0, 0, 1, 1, 2, 2, 3, 3, 4, 4, 4\}$.

In the η-direction, we have no features to model. Here we are only defining the thickness. A single linear element will do, and so $\mathcal{H} = \{0, 0, 1, 1\}$.

The ζ-direction requires a bit more care. As is represented schematically in Figure 2.36, we may think of the whole pipe as being comprised of a cylindrical section adjoining an elbow. Each of these requires only one element in the axial direction, and we have already determined that the order of the NURBS basis will be $r = 2$. We could build the two objects separately using $\mathcal{Z}_1 = \{0, 0, 0, 1, 1, 1\}$ for the cylinder and $\mathcal{Z}_2 = \{1, 1, 1, 2, 2, 2\}$ for the elbow. We do not, however, want to use two distinct patches (each with identical control points on the surface where they meet). As we did with the circle above, we can avoid such redundancy and use a single patch with two elements in this direction, with C^0-continuity between them. The appropriate knot vector in this case is $\mathcal{Z}_2 = \{0, 0, 0, 1, 1, 2, 2, 2\}$.

Figure 2.36   The pipe has two basic sections: a cylinder and an elbow. We must choose a knot vector that respects both of these distinct pieces, while allowing us to join them into one geometrical object.

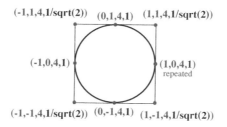

Figure 2.37   Control points for the inner edge at the top of the pipe. Note that the weights (in blue) are frequently stored as the fourth component of the control point that they are associated with, but we do *not* consider them to be part of the control point itself.

2.4.4 Selection of control points

Only now that all of the other pieces of the puzzle are in place are we ready to assign the actual control points. The easiest place to start is normally the corners of an object as they will be interpolated, but for the pipe example, we do not have any true corners. What we do have is the template for a circle. More over, we know that the solid geometry is an extrusion formed by a NURBS surface being swept along a NURBS curve. Thus, let us begin by using the template to construct the surface to be extruded.

Knowing that the inner radius of the pipe is 1, we can directly apply the template of Figure 2.34 to obtain control points for the inner edge at the top of the pipe (at height $z = 4$), as in Figure 2.37. Similarly, we can apply the template to the outer edge, whose radius we know to be 2. As the through-thickness discretization consists of a single linear element, this is all of the information needed to represent the annular cross section, shown with in Figure 2.38.

To form the cylindrical portion of the pipe, we need only to sweep the annulus in a straight line downward from the top, as in Figure 2.39. We accomplish this by sweeping its control points in exactly the same way. We know the control points for the top surface, at $z = 4$. The control points for the bottom of the cylindrical section will be identical, except with $z = 1$. If we were using linear elements in the ζ-direction, this would be the whole story, but we are

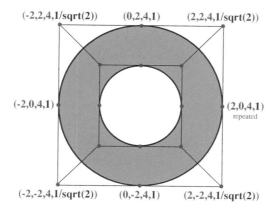

Figure 2.38   The annular surface at the top of the pipe. The control points for the outer edge are shown. Note that the weights are the same as in Figure 2.37.

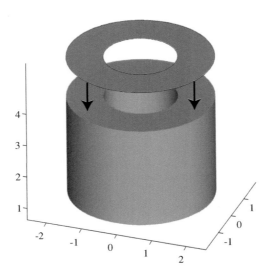

Figure 2.39   We form the cylindrical portion of the pipe by sweeping the cross section straight downward.

using quadratics (recall that quadratics were not necessary for this portion of the domain, but
they will be necessary when we get to the elbow). This means that we have an additional level
of control points between the top and bottom of the cylinder (each level corresponds to one of
the three basis functions in the ζ-direction of this element). Placing them directly between the
other two levels, at $z = 2.5$, leads to a linear parameterization of this part of the domain, and
so that is what we will do. The resulting control lattice is shown in Figure 2.40.

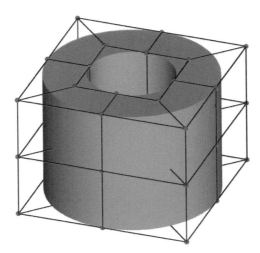

Figure 2.40   The control lattice for the cylindrical portion of the pipe. Note that there are three levels
of control points, one corresponding to each of the three quadratic NURBS functions in the ζ-direction
of this element.

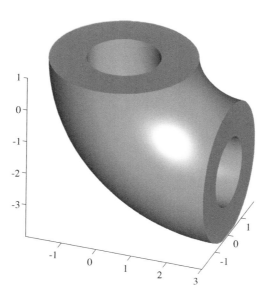

Figure 2.41 The elbow of the pipe is obtained by extruding the annulus of Figure 2.38 along the circular arc of Figure 2.42.

The elbow portion of the pipe (Figure 2.41) is formed by sweeping the annulus of Figure 2.38 along a 90° circular arc. As we see in Figure 2.42, the control polygon for such a curve makes an "L", forming a right angle. The weight of the point at the corner, as we would expect, is $1/\sqrt{2}$. The construction of the control lattice for the elbow exhibits an outer product structure wherein each point of the control net of the cross section follows the path of the control polygon for a circle whose radius varies depending on where the control point is relative to the axis of the pipe. Moreover, the first plane of control points is multiplied by the weight of the first control point in the circular arc, the second level of control points is multiplied by the weight of the second control point in the arc, and likewise with the third. The result is shown in Figure 2.43.

Joining the cylinder and the elbow together, we get the control lattice for the entire pipe, shown in Figure 2.44. The resulting eight element mesh, exactly encapsulating the pipe

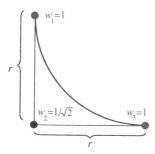

Figure 2.42 A 90° circular arc with radius r. The positions of the control points depend on the radius, but the weights do not. This is just a special case of the general template seen in Figure 2.34.

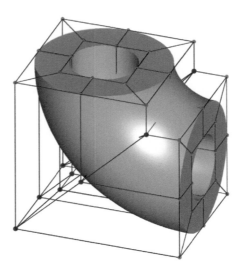

Figure 2.43   The control lattice for the elbow is obtained by "extruding" the control net of the annulus along the control polygon of the circular curve. The weights of the control points in blue are those of the annulus, multiplied by the weight of the second control point of the circular curve, $\frac{1}{\sqrt{2}}$.

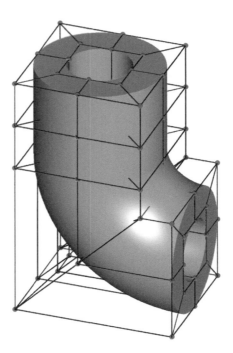

Figure 2.44   The control lattice for the pipe.

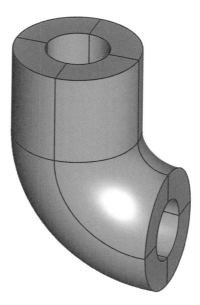

Figure 2.45   The mesh for the pipe consists of eight elements.

geometry, is shown in Figure 2.45. The control points, polynomial orders, and knot vectors used in constructing this NURBS model of the pipe are all tabulated in Appendix 2.A at the end of the chapter.

2.5 Notation

Now that we have introduced NURBS and seen that B-splines are simply a special case of NURBS, and we have looked at curves, surface, and solids in several spatial dimensions, it behooves us to consolidate our notation and terminology a bit. Henceforth, we will always refer to "NURBS" even when we may mean "B-splines," as every B-spline is also a NURBS. Additionally, we will simply write $N(\xi)$ to refer to any basis function. It is to be understood that this could be a univariate, bivariate, or trivariate, non-rational or rational basis function, possibly comprised of a tensor-product of univariate functions of differing orders. We may even suppress the explicit dependence on the parameter for the sake of brevity when it is not needed. Lastly, to avoid distinguishing between curves, surfaces, and solids, we will refer generically to a point in the domain at parameter value ξ as $\mathbf{x}(\xi)$. When the details of a NURBS object under consideration are not stated or made obvious by the context, the reader may always conclude that the discussion generalizes to all possible cases.

Some additional notation is also warranted. Let us denote the domain in the physical space (*i.e.*, the geometry) by Ω. Similarly, let us denote the domain in the parameter space by $\hat{\Omega}$. Thus, $\mathbf{x} : \hat{\Omega} \to \Omega$ *is* the geometrical mapping, taking points in the parameter space and returning the corresponding points in the physical space. Unless otherwise specified, we assume this mapping to be invertible, and so $\mathbf{x}^{-1} : \Omega \to \hat{\Omega}$ takes points in the physical domain and identifies their corresponding parameter values.

Another point in need of elaboration is the definition of elements. We have already stated that we consider elements to be the images of knot spans under the NURBS mapping. We will denote these knot spans in the parameter space by $\hat{\Omega}^e$, and their image in the physical space as Ω^e, where e runs from $1, \ldots, n_{el}$, with n_{el} being the total number of elements in the mesh. Confusion is possible when we consider the case of repeated knot values. In our current implementations, we count every knot span, regardless of its measure, as an element. This is the most convenient approach for the sake of bookkeeping and algorithm development as the support of basis functions is always well defined in terms of the elements when we count this way (see Appendix A at the end of the book for an example). One could say that such an approach counts the number of elements in the index space. Alternatively, it is equally valid to consider only elements of positive measure, particularly if we intend to only use open knot vectors. This would be equivalent to counting elements in the parameter space (or physical space as the mapping is assumed invertible) instead of in the index space. It is usually more intuitive and convenient to take this parameter space view. The reason is that the number of elements can frequently be determined by simply looking at the mesh. This is the approach we will take for the remainder of the text. In the end, either approach is equally valid as long as it is consistent with the data structure being used. None of the discussion in this book will be dependent upon one interpretation over another.

Lastly, let us refine the usage of indices. In future chapters, the index i, as well as j, k, and l, will be reserved for components of vectors in physical space, and as such will take on values from $1, \ldots, d$. The index A will be used to identify the basis functions. It is taken to run from $1, \ldots, n_{np}$, where n_{np} is the total number of basis functions in the mesh[5]. The same is true of control points; again A is used and again it runs from $1, \ldots, n_{np}$, as may be expected since the number of control points is equal to the number of basis functions. When additional indices are needed in the same role, the letters B, C, and D will be used. On each element, we can simplify things by recognizing that only a limited number of basis functions, denoted by n_{en} will have support on the element. For the NURBS basis, n_{en} is fixed and does not depend on the specific element under consideration. Thus we will frequently identify basis functions, control points, and control variables (see Chapter 3) on a given element Ω^e, not by their *global* index A, but by a *local* index a, which runs from $1, \ldots, n_{en}$. When additional indices are needed in this same role, the letters b, c, and d will be used. This notation is in keeping with standard finite element software data structures. For a thorough discussion of the global and local indices, as well as data structures for relating them on an element by element basis, see Chapter 3.

Appendix 2.A: Data for the bent pipe

The solid geometry for the pipe with an elbow bend featured in Section 2.4 uses a trivariate NURBS basis. The three parametric directions, ξ, η, and ζ, correspond to the circumferential, radial, and axial directions, respectively. The corresponding polynomial orders and knot vectors are given in Table 2.A.1 and the control points are given in Table 2.A.2.

Table 2.A.1 Polynomial orders and knot vectors for the bent pipe

Direction	Order	Knot vector
ξ	$p = 2$	$\Xi = \{0, 0, 0, 1, 1, 2, 2, 3, 3, 4, 4, 4\}$
η	$q = 1$	$\Xi = \{0, 0, 1, 1\}$
ζ	$r = 2$	$\Xi = \{0, 0, 0, 1, 1, 2, 2, 2\}$

Table 2.A.1 Control points for the bent pipe

i	k	$B_{i,1,k}$	$B_{i,2,k}$	$w_{i,1,k}$	$w_{i,2,k}$
1	1	$(1, 0, 4)$	$(2, 0, 4)$	1	1
1	2	$(1, 0, 2.5)$	$(2, 0, 2.5)$	1	1
1	3	$(1, 0, 1)$	$(2, 0, 1)$	1	1
1	4	$(1, 0, -1)$	$(2, 0, 0)$	$1/\sqrt{2}$	$1/\sqrt{2}$
1	5	$(3, 0, -1)$	$(3, 0, 0)$	1	1
2	1	$(1, 1, 4)$	$(2, 2, 4)$	$1/\sqrt{2}$	$1/\sqrt{2}$
2	2	$(1, 1, 2.5)$	$(2, 2, 2.5)$	$1/\sqrt{2}$	$1/\sqrt{2}$
2	3	$(1, 1, 1)$	$(2, 2, 1)$	$1/\sqrt{2}$	$1/\sqrt{2}$
2	4	$(1, 1, -1)$	$(2, 2, 0)$	$1/2$	$1/2$
2	5	$(3, 1, -1)$	$(3, 2, 0)$	$1/\sqrt{2}$	$1/\sqrt{2}$
3	1	$(0, 1, 4)$	$(0, 2, 4)$	1	1
3	2	$(0, 1, 2.5)$	$(0, 2, 2.5)$	1	1
3	3	$(0, 1, 1)$	$(0, 2, 1)$	1	1
3	4	$(0, 1, -2)$	$(0, 2, -2)$	$1/\sqrt{2}$	$1/\sqrt{2}$
3	5	$(3, 1, -2)$	$(3, 2, -2)$	1	1
4	1	$(-1, 1, 4)$	$(-2, 2, 4)$	$1/\sqrt{2}$	$1/\sqrt{2}$
4	2	$(-1, 1, 2.5)$	$(-2, 2, 2.5)$	$1/\sqrt{2}$	$1/\sqrt{2}$
4	3	$(-1, 1, 1)$	$(-2, 2, 1)$	$1/\sqrt{2}$	$1/\sqrt{2}$
4	4	$(-1, 1, -3)$	$(-2, 2, -4)$	$1/2$	$1/2$
4	5	$(3, 1, -3)$	$(3, 2, -4)$	$1/\sqrt{2}$	$1/\sqrt{2}$
5	1	$(-1, 0, 4)$	$(-2, 0, 4)$	1	1
5	2	$(-1, 0, 2.5)$	$(-2, 0, 2.5)$	1	1
5	3	$(-1, 0, 1)$	$(-2, 0, 1)$	1	1
5	4	$(-1, 0, -3)$	$(-2, 0, -4)$	$1/\sqrt{2}$	$1/\sqrt{2}$
5	5	$(3, 0, -3)$	$(3, 0, -4)$	1	1
6	1	$(-1, -1, 4)$	$(-2, -2, 4)$	$1/\sqrt{2}$	$1/\sqrt{2}$
6	2	$(-1, -1, 2.5)$	$(-2, -2, 2.5)$	$1/\sqrt{2}$	$1/\sqrt{2}$
6	3	$(-1, -1, 1)$	$(-2, -2, 1)$	$1/\sqrt{2}$	$1/\sqrt{2}$
6	4	$(-1, -1, -3)$	$(-2, -2, -4)$	$1/2$	$1/2$
6	5	$(3, -1, -3)$	$(3, -2, -4)$	$1/\sqrt{2}$	$1/\sqrt{2}$
7	1	$(0, -1, 4)$	$(0, -2, 4)$	1	1
7	2	$(0, -1, 2.5)$	$(0, -2, 2.5)$	1	1
7	3	$(0, -1, 1)$	$(0, -2, 1)$	1	1
7	4	$(0, -1, -2)$	$(0, -2, -2)$	$1/\sqrt{2}$	$1/\sqrt{2}$
7	5	$(3, -1, -2)$	$(3, -2, -2)$	1	1
8	1	$(1, -1, 4)$	$(2, -2, 4)$	$1/\sqrt{2}$	$1/\sqrt{2}$
8	2	$(1, -1, 2.5)$	$(2, -2, 2.5)$	$1/\sqrt{2}$	$1/\sqrt{2}$
8	3	$(1, -1, 1)$	$(2, -2, 1)$	$1/\sqrt{2}$	$1/\sqrt{2}$
8	4	$(1, -1, -1)$	$(2, -2, 0)$	$1/2$	$1/2$
8	5	$(3, -1, -1)$	$(3, -2, 0)$	$1/\sqrt{2}$	$1/\sqrt{2}$
9	1	$(1, 0, 4)$	$(2, 0, 4)$	1	1
9	2	$(1, 0, 2.5)$	$(2, 0, 2.5)$	1	1
9	3	$(1, 0, 1)$	$(2, 0, 1)$	1	1
9	4	$(1, 0, -1)$	$(2, 0, 0)$	$1/\sqrt{2}$	$1/\sqrt{2}$
9	5	$(3, 0, -1)$	$(3, 0, 0)$	1	1

Notes

1. This is frequently referred to as "affine invariance" in the geometry literature
2. The terms control polygon, control net, and control lattice are precise in one, two, and three dimensions, respectively. It is often our practice, however, to simplify the matter and use the term control net regardless of the number of dimensions in which we are working. This will be the case throughout the book, except when there is a benefit in distinguishing between the cases. All of these are equivalent to the term "control mesh" introduced in the previous chapter.
3. In the CAD literature "knot insertion" refers to inserting a single knot into a knot vector, whereas "knot refinement" refers to inserting multiple knots simultaneously. Here, we make no distinction and use "knot insertion" to refer to both cases.
4. This repetition of control points could be removed by abandoning open knot vectors in this instance. Though this would be nice in theory, all of our current software is designed under the assumption of open knot vectors.
5. The subscript np denotes "nodal points," a term wholly inappropriate for isogeometric analysis in which we do not have "nodes." Its present usage is meant to keep the notation consistent with that of the finite element text of Hughes, 2000.

3

NURBS as a Basis for Analysis: Linear Problems

In the previous chapter we introduced NURBS in their natural setting. They have been a mainstay of geometric design for many years due to their flexibility and precision. In this chapter we bring them into the setting of analysis – an arena to which their unique properties are also ideally suited. As a basis for analysis, NURBS generalize and improve upon the traditional piecewise polynomial basis functions, providing unprecedented accuracy and robustness across a wide array of applications. The power of this combination of geometric and analytic capabilities is at the very heart of isogeometric analysis. We will consider linear problems in this chapter. Nonlinear problems will be discussed in Chapter 7.

3.1 The isoparametric concept

The root idea behind isogeometric analysis is that the basis used to exactly model the geometry will also serve as the basis for the solution space of the numerical method (Figure 3.1). This notion of using the same basis for geometry and analysis is called the ***isoparametric concept***, and it is quite common in classical finite element analysis. The fundamental difference between this new concept of isogeometric analysis and the old concept of isoparametric finite element analysis is that, in classical FEA, the basis chosen to approximate the *unknown* solution fields is then used to approximate *known* geometry. Isogeometric analysis turns this idea around and selects a basis capable of exactly representing the known geometry and uses it as a basis for the fields we wish to approximate. In a sense, we are reversing the isoparametric arrow such that it points from the geometry toward the solution space, rather than vice versa; see Figure 3.2. This logical shift allows us to utilize all of the information that we possess. Fortuitously, we will see that the NURBS basis also possesses many properties that are quite desirable when approximating solution fields *independently* of any geometrical considerations.

The reliance of traditional FEA on polynomials is, at least in part, because of their simplicity. They are easy to program, easy to understand, easy to prove theorems with, and have well known approximation properties. As long as they are used as the basis of the solution space, convergence rates and other similar mathematical apparatus are reasonably straightforward to obtain. This is not to say that proving theorems about other bases is impossible. On the

Isogeometric Analysis: Toward Integration of CAD and FEA by J. A. Cottrell, T. J. R. Hughes, Y. Bazilevs
© 2009, John Wiley & Sons, Ltd

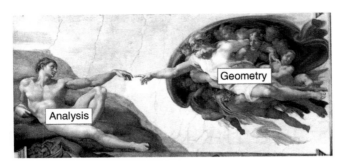

Figure 3.1 The isoparametric concept links analysis with geometry. Traditional FEA has been slow to acknowledge the power and importance of geometry – a sin isogeometric analysis avoids. *The Creation of Adam*, Michelangelo, circa 1511, Fresco, 480 × 230 cm, Sistine Chapel, Vatican City (http://en.wikipedia.org/wiki/Michelangelo).

contrary, it is the isoparametric concept itself that allows us to work confidently with more exotic bases. Though precise results for non-polynomial bases do exist – for example, several theorems regarding convergence for NURBS based isogeometric analysis have already been proved in Bazilevs *et al.*, 2006a and are discussed in Appendix 3.B at the end of this chapter – the most basic convergence requirements in many numerical methods are achieved by *any* reasonably smooth isoparametric basis that is also a partition of unity.

As seen, for example, in Hughes, 2000, sufficient conditions for a basic convergence proof for a wide class of problems are satisfied by a basis that is

- C^1 on the element interiors,
- C^0 on the element boundaries,
- complete.

The requirements of C^1-continuity on the element interiors and C^0-continuity on the element boundaries are not at all restrictive. Most bases that we might consider are C^∞ on the element interiors and (with the exception of Discontinuous Galerkin methods) have at least C^0-continuity on the element boundaries. The third condition, ***completeness***, requires that, on any given element Ω^e, the basis be capable of representing all linear functions. That is, given a basis $\{N_a\}_{a=1}^{n_{en}}$ for the solution space, completeness demands that there exist coefficients d_a

Classical FEA: Geometry $\Longleftarrow$ Fields
                                                                imposed
                                                                  on
Isogeometric Analysis: Geometry $\Longrightarrow$ Fields

Figure 3.2 Reversing the isoparametric arrow. Classical finite element analysis imposes its chosen solution space onto the description of the geometry. Isogeometric analysis begins with a basis capable of representing the exact geometry and imposes it on the solution fields.

such that, for arbitrary constants C_0, C_1, C_2, and C_3,

$$u^h|_{\Omega^e} \equiv \sum_{a=1}^{n_{en}} N_a d_a = C_0 + C_1 x + C_2 y + C_3 z. \tag{3.1}$$

This last property is satisfied by any isoparametric basis that is also a partition of unity. To see this, simply note that for each point $\mathbf{x} \in \Omega^e$ there exists a parameter ξ such that

$$\mathbf{x}(\xi) \equiv \begin{Bmatrix} x(\xi) \\ y(\xi) \\ z(\xi) \end{Bmatrix} = \sum_{a=1}^{e_{en}} N_a(\xi) \begin{Bmatrix} x_a^e \\ y_a^e \\ z_a^e \end{Bmatrix}, \tag{3.2}$$

where x_a^e, y_a^e, and z_a^e are simply the components of the ath vector-valued coefficient defining the geometry in element Ω^e (these could be nodes, control points, etc., depending on the specific basis being used). As the basis is a partition of unity, at that same point ξ we have

$$\sum_{a=1}^{e} N_a(\xi) \equiv 1. \tag{3.3}$$

Inserting (3.2) and (3.3) into (3.1) and solving for d_a yields

$$d_a = C_0 + C_1 x_a^e + C_2 y_a^e + C_3 z_a^e. \tag{3.4}$$

Thus, the isoparametric concept and the partition of unity are enough to ensure completeness. Moreover, they are vital to ensuring that isogeometric analysis will result in convergent methods for many different choices of element technology, NURBS included.

3.1.1 Defining functions on the domain

Interpreting the $N_a(\xi)$'s as NURBS functions and thus recognizing $\{x_a^e, y_a^e, z_a^e\}^{\mathrm{T}}$ as the components of control point $\mathbf{B}_a$, we see that the geometrical mapping $\mathbf{x} : \hat{\Omega} \to \Omega$ is defined exactly as in (3.2). We can build other functions over the entire parametric domain in similar fashion. For example, let $\hat{u}^h : \hat{\Omega} \to \mathbb{R}$ be defined by

$$\hat{u}^h(\xi) \equiv \sum_{A=1}^{n_{np}} N_A(\xi) d_A. \tag{3.5}$$

The coefficients d_A are called **control variables**. As with control points, the non-interpolatory nature of the basis prevents strictly interpreting the control variables as we can do with nodal values in FEA. We can define the function over the domain in the physical space by considering a composition with the inverse of the geometrical mapping such that $u^h : \Omega \to \mathbb{R}$ is given by

$$u^h = \hat{u}^h \circ \mathbf{x}^{-1}. \tag{3.6}$$

In practice, we will take advantage of the fact that the geometrical mapping is invertible, and we will not distinguish between u^h and $\hat{u}^h$, writing u^h to refer to the function regardless of which coordinates we are working in.

The properties of the function u^h follow from the basis, just as was the case with other NURBS objects previously considered. The maximum level of continuity across an element boundary, for example, is determined by the continuity of the basis across the corresponding knot span. If the level of resolution is insufficient, the basis may be refined. Recall from Chapter 2 that such a refinement leaves both the geometry and its parameterization unchanged. This means that the geometrical mapping itself is unchanged. Thus refinement may proceed as needed for analysis without regard for the geometry, which is exact from the coarsest mesh onward.

3.2 Boundary value problems (BVPs)

As an example of solving a differential equation posed over the domain defined by a NURBS geometry, let us consider Laplace's equation. The goal is to find $u : \bar{\Omega} \rightarrow \mathbb{R}$ such that

$$\Delta u + f = 0 \quad \text{in } \Omega, \tag{3.7a}$$

$$u = g \quad \text{on } \Gamma_D, \tag{3.7b}$$

$$\nabla u \cdot \mathbf{n} = h \quad \text{on } \Gamma_N, \tag{3.7c}$$

$$\beta u + \nabla u \cdot \mathbf{n} = r \quad \text{on } \Gamma_R, \tag{3.7d}$$

where $\overline{\Gamma_D \bigcup \Gamma_N \bigcup \Gamma_R} = \Gamma \equiv \partial\Omega$, $\Gamma_D \bigcap \Gamma_N \bigcap \Gamma_R = \emptyset$, and $\mathbf{n}$ is the unit outward normal vector on $\partial\Omega$. The functions $f : \Omega \rightarrow \mathbb{R}$, $g : \Gamma_D \rightarrow \mathbb{R}$, $h : \Gamma_N \rightarrow \mathbb{R}$, and $r : \Gamma_R \rightarrow \mathbb{R}$ are all given, as is the constant β. Equation (3.7) constitutes the **strong form** of the boundary value problem (BVP). The boundary conditions given in (3.7b), (3.7c), and (3.7d) represent the three major types of boundary conditions one is likely to encounter. These are Dirichlet conditions, Neumann conditions, and Robin conditions, respectively. They will be discussed in detail in Section 3.4.

For a sufficiently smooth domain, and under certain restrictions on g, h, and r, a unique solution u satisfying (3.7) is known to exist, but an analytical expression will usually be impossible to obtain. However, we may seek an approximate solution of the form of (3.5). We generically refer to techniques for doing so as **numerical methods**. Different numerical methods are simply different techniques for finding d_A such that $u^h \approx u$. Several different numerical methods that could be implemented in an isogeometric analysis framework are presented below.

3.3 Numerical methods

There are several classes of numerical methods that lend themselves to isogeometric analysis. The primary one, Galerkin finite element analysis, is the approach that has been utilized for most of the examples contained in subsequent chapters. The other techniques to be described – collocation, least-squares finite element analysis, and meshless methods – all can be implemented using NURBS. In fact, a NURBS-based approach may have significant advantages over some of the more traditional implementations of these numerical methods.

3.3.1 Galerkin

In many circles, the term "Galerkin's method" is for all intents and purposes synonymous with finite element analysis. While FEA has grown considerably beyond the classical Galerkin method itself, its roots still lie in the approach. Here we present the Bubnov–Galerkin[1] method that underlies most of modern finite element analysis.

3.3.1.1 A weak form of the problem

The technique begins by defining a weak, or variational, counterpart of (3.7). To do so, we need to characterize two classes of functions. The first is to be composed of candidate, or trial solutions. From the outset, these functions will be required to satisfy the Dirichlet boundary condition of (3.7b), as will be discussed in Section 3.4.

To define the trial and weighting spaces formally, let us first define the space of square integrable functions on Ω. This space, called $L^2(\Omega)$, is defined as the collection of all functions $u : \Omega \rightarrow \mathbb{R}$ such that

$$\int_\Omega u^2 \, d\Omega < +\infty. \tag{3.8}$$

Let us consider a multi-index $\boldsymbol{\alpha} \in \mathbb{N}^d$ where d is the number of spatial dimensions in the space. For $\boldsymbol{\alpha} = \{\alpha_1, \ldots, \alpha_d\}$, we define $|\boldsymbol{\alpha}| = \sum_{i=1}^d \alpha_i$. We now have a concise way to represent derivative operators. Let $D^{\boldsymbol{\alpha}} = D_1^{\alpha_1} D_2^{\alpha_2} \ldots D_d^{\alpha_d}$, where $D_i^j = \frac{\partial^j}{\partial x_i^j}$. So that certain expressions to be employed in the formulation make sense, we shall require that the derivatives of the trial solutions be square-integrable. Specifically, if $u : \Omega \rightarrow \mathbb{R}$ is a trial solution, then we must insist that

$$\int_\Omega \nabla u \cdot \nabla u \, d\Omega < +\infty. \tag{3.9}$$

Such a function is said to be in the Sobolev space $H^1(\Omega)$, which is characterized by

$$H^1(\Omega) = \{u \,|\, D^{\boldsymbol{\alpha}} u \in L^2(\Omega), |\boldsymbol{\alpha}| \leq 1\}. \tag{3.10}$$

We may now define the collection of **trial solutions**, denoted by $\mathcal{S}$, as all of the functions which have square-integrable derivatives and that also satisfy

$$u|_{\Gamma_D} = g. \tag{3.11}$$

This is written as

$$\mathcal{S} = \{u \,|\, u \in H^1(\Omega), u|_{\Gamma_D} = g\}. \tag{3.12}$$

The second collection of functions in which we are interested is called the **weighting functions**. This collection is very similar to the trial functions, except that we have the

homogeneous counterpart of the Dirichlet boundary condition. That is, the weighting functions
are denoted by a set $\mathcal{V}$ defined by

$$\mathcal{V} = \{w \mid w \in H^1(\Omega), w|_{\Gamma_D} = 0\}. \tag{3.13}$$

We may now obtain a variational statement of the BVP by multiplying (3.7a) by an arbitrary
test function $w \in \mathcal{V}$ and integrating by parts, incorporating (3.7c) and (3.7d) as needed. The
resulting weak form of the problem is now: Given f, g, h, and r, find $u \in \mathcal{S}$ such that for all
$w \in \mathcal{V}$

$$\int_\Omega \nabla w \cdot \nabla u \, d\Omega + \beta \int_{\Gamma_R} wu \, d\Gamma = \int_\Omega wf \, d\Omega + \int_{\Gamma_N} wh \, d\Gamma + \int_{\Gamma_R} wr \, d\Gamma. \tag{3.14}$$

Note that all of the unknown information, namely u, is contained on the left-hand side of the
equation, while all of the given data, h and r, are contained on the right-hand side.

We now see why $H^1(\Omega)$ was the appropriate space in which to work. Despite the fact that
the strong form of the equation (3.7) required u to have well defined second derivatives, the
weak form from which the numerical method is built (3.14) only requires that first derivatives
be square-integrable.

This weak form may be rewritten as

$$a(w, u) = L(w) \tag{3.15}$$

where

$$a(w, u) = \int_\Omega \nabla w \cdot \nabla u \, d\Omega + \beta \int_{\Gamma_R} wu \, d\Gamma, \tag{3.16}$$

and

$$L(w) = \int_\Omega wf \, d\Omega + \int_{\Gamma_N} wh \, d\Gamma + \int_{\Gamma_R} wr \, d\Gamma. \tag{3.17}$$

A few properties of $a(\cdot, \cdot)$ and $L(\cdot)$ are worth noting. The first is the symmetry of $a(\cdot, \cdot)$. It
follows directly from its definition that $a(w, u) = a(u, w)$. Also, $a(\cdot, \cdot)$ is bilinear and $L(\cdot)$ is
linear. That is, for all constants C_1 and C_2,

$$a(C_1 u + C_2 v, w) = C_1 a(u, w) + C_2 a(v, w), \tag{3.18}$$

$$L(C_1 u + C_2 v) = C_1 L(u) + C_2 L(v). \tag{3.19}$$

This concise notation, or variants thereof, is quite common in the finite element literature.
For problems other than the Laplace equation, the details vary, but the basic form remains. It
captures the essential mathematical features of the variational method (as well as suggesting
features of a finite element implementation) that are more general than the details of the
equation itself.

The solution to (3.14), or equivalently (3.15), is called a **weak solution**. Under appropriate
regularity assumptions it can be shown that the weak solution and the strong solution of (3.7)
are equivalent; see Hughes, 2000.

3.3.1.2 Galerkin's method

Galerkin's method consists of constructing finite-dimensional approximations of $\mathcal{S}$ and $\mathcal{V}$, denoted $\mathcal{S}^h$ and $\mathcal{V}^h$, respectively. Strictly speaking, these will be subsets such that

$$\mathcal{S}^h \subset \mathcal{S}, \tag{3.20}$$

$$\mathcal{V}^h \subset \mathcal{V}. \tag{3.21}$$

Furthermore, these will be associated with subsets of the space spanned by the isoparametric basis.

We can further characterize $\mathcal{S}^h$ by recognizing that if we have a *given* function $g^h \in \mathcal{S}^h$ such that $g^h|_{\Gamma_D} = g$, then for every $u^h \in \mathcal{S}^h$ there exists a unique $v^h \in \mathcal{V}^h$ such that

$$u^h = v^h + g^h. \tag{3.22}$$

This clearly will not be possible for an arbitrary function g, but at present let us assume that such a g^h exists. Section 3.4 will discuss the general case at length.

We can now write a variational equation of the form of (3.15). The Galerkin form of the problem is: Given g^h, h, and r, find $u^h = v^h + g^h$, where $v^h \in \mathcal{V}^h$, such that for all w^h in $\mathcal{V}^h$

$$a(w^h, u^h) = L(w^h). \tag{3.23}$$

Recalling (3.22) and the bilinearity of $a(\cdot, \cdot)$, we can rewrite (3.23) as

$$a(w^h, v^h) = L(w^h) - a(w^h, g^h). \tag{3.24}$$

In this latter form, the unknown information is on the left-hand side, while everything on the right-hand side is given, as before. (3.23) and (3.24) are sometimes referred to as the Bubnov-Galerkin method.

Remark

A related method, the Petrov–Galerkin method, assumes a weighting space that is different than $\mathcal{V}^h$, that is, $v^h \in \mathcal{V}^h$ but $w^h \in \tilde{\mathcal{V}}^h \neq \mathcal{V}^h$. The use of Petrov's name seems to emanate from Mikhlin, 1964. Boris Galerkin (see Figure 3.3) published his seminal paper in 1915 (see http//en.wikipedia.org/wiki/Boris_Galerkin).

Figure 3.3   Boris Galerkin.

3.3.1.3 Matrix equations

The finite-dimensional nature of the function spaces used in Galerkin's method leads to a coupled system of linear algebraic equations. Let the solution space consist of all linear combinations of a given set of NURBS functions $N_A : \hat{\Omega} \to \mathbb{R}$, where $A = 1, \ldots, n_{np}$. Recall that the support of the functions is highly localized and that very few functions are non-zero on the boundary of the domain. Without loss of generality, we may assume a numbering for these functions such that there exists an integer $n_{eq} < n_{np}$ such that

$$N_A|_{\Gamma_D} = 0 \quad \forall A = 1, \ldots, n_{eq}. \tag{3.25}$$

Thus, for all $w^h \in \mathcal{V}^h$, there exist constants $c_A, A = 1, \ldots, n_{eq}$ such that

$$w^h = \sum_{A=1}^{n_{eq}} N_A c_A. \tag{3.26}$$

Furthermore, the function g^h (frequently called a "lifting") is given similarly by coefficients $g_A, A = 1, \ldots, n_{np}$. In practice, we will always choose g^h such that $g_1 = \ldots = g_{n_{eq}} = 0$ as they have no effect on its value on Γ_D, and so

$$g^h = \sum_{A=n_{eq}+1}^{n_{np}} N_A g_A. \tag{3.27}$$

Finally, recalling again (3.22), for any $u^h \in \mathcal{S}^h$ there exist $d_A, A = 1, \ldots, n_{eq}$ such that

$$u^h = \sum_{A=1}^{n_{eq}} N_A d_A + \sum_{B=n_{eq}+1}^{n_{np}} N_B g_B = \sum_{A=1}^{n_{eq}} N_A d_A + g^h. \tag{3.28}$$

We may insert (3.26) and (3.28) into (3.24), and take advantage of linearity to obtain the expression

$$\sum_{A=1}^{n_{eq}} c_A \left(\sum_{B=1}^{n_{eq}} a(N_A, N_B) d_B - L(N_A) + a(N_A, g^h) \right) = 0. \tag{3.29}$$

As the c_A's are arbitrary (recall that (3.29) is to hold for all $w^h \in \mathcal{V}^h$), it follows that the term in parentheses must vanish. Thus, for $A = 1, \ldots, n_{eq}$,

$$\sum_{B=1}^{n_{eq}} a(N_A, N_B) d_B = L(N_A) - a(N_A, g^h). \tag{3.30}$$

Proceeding to define

$$K_{AB} = a(N_A, N_B), \tag{3.31}$$

$$F_A = L(N_A) - a(N_A, g^h), \tag{3.32}$$

and

$$\mathbf{K} = [K_{AB}], \tag{3.33}$$

$$\mathbf{F} = \{F_A\}, \tag{3.34}$$

$$\mathbf{d} = \{d_A\}, \tag{3.35}$$

for $A, B = 1 \ldots, n_{eq}$, we can rewrite (3.30) as the matrix problem

$$\mathbf{Kd} = \mathbf{F}. \tag{3.36}$$

Due to the finite element method's historical origins in structural analysis, the following terminology is frequently applied independently of the actual problem being solved

$$\mathbf{K} = \text{stiffness matrix},$$

$$\mathbf{F} = \text{force vector},$$

$$\mathbf{d} = \text{displacement vector}.$$

Solving (3.36) for the d_A's for $A = 1, \ldots, n_{eq}$ as

$$\mathbf{d} = \mathbf{K}^{-1}\mathbf{F}, \tag{3.37}$$

and then inserting them back into (3.28) lets us finally write the solution u^h as

$$u^h = \sum_{A=1}^{n_{eq}} N_A d_A + \sum_{B=n_{eq}+1}^{n_{np}} N_B g_B. \tag{3.38}$$

3.3.1.4 Assembling the system

It is important to note that $\mathbf{K}$ is a sparse matrix. This is a result of the fact that the support of each basis function is highly localized. Thus, for many combinations of A and B in the $n_{eq} \times n_{eq}$ global stiffness matrix, $K_{AB} = a(N_A, N_B) = 0$. We can take advantage of this fact in order to reduce the amount of work necessary in building and solving the algebraic system.

If we look back at any of the pictures of basis functions from Chapter 2, we note that the maximum number of functions with support on any given element is always fixed by the order of the polynomial. That is, for each element in the patch, the maximum number of functions that are not identically equal to zero throughout the patch is the same regardless of which element is under consideration. Let us denote this number of *local shape functions* by n_{en}. Thus, if we were to build an $n_{en} \times n_{en}$ *element stiffness matrix*, $\mathbf{k}^e$, by posing the problem over a single element, this matrix would always be dense. The term *local stiffness matrix* is also common, and we use it interchangeably.

The above approach is exactly the one we take in practice. The process of building the global stiffness matrix and force vector is called *assembly*. Instead of looping through all of the global shape functions, taking global integrals to build $\mathbf{K}$ one entry at a time, we will loop through the elements, building element stiffness matrices as we go. Every entry of each of these dense element stiffness matrices will then be added to the appropriate spot in the global stiffness matrix. In this way, we need not expend effort integrating functions over regions in which we know *a priori* that they are zero[2].

This process is made simple by a **connectivity array** that links every local shape function number to a global shape function number. Let us call the connectivity array IEN. For each element number e from $1, \ldots, n_{el}$, and local function number a from $1, \ldots, n_{en}$, there is a global function number A from $1, \ldots, n_{eq}$ such that $\text{IEN}(a, e) = A$. That is, local function N_a of element Ω^e and global function N_A are exactly the same. This allows us to build the global stiffness matrix from a sequence of local ones. Similarly, the global force vector **F** is assembled from the local force vectors $\mathbf{F}^e$. Along the way, we are only performing integration on functions that are non-zero. See Appendix A at the end of the book for a more detailed discussion of connectivity arrays. The assembly process is described in detail in Hughes, 2000.

The actual integration is performed by Gaussian quadrature. As seen in Figure 3.4, integrals are pulled back, first onto the parametric domain and then onto a bi-unit **parent element**, and integration is performed using a classical change of variables formulation. Having already denoted coordinates in the physical space by **x** and coordinates in the parameter space by $\boldsymbol{\xi}$, let us denote coordinates in the parent element by $\tilde{\boldsymbol{\xi}}$. Similarly, we have already denoted the element in the physical space by Ω^e and in the parameter space by $\hat{\Omega}^e$, and so let us denote the parent element by $\tilde{\Omega}^e$. The mapping $\mathbf{x} : \hat{\Omega} \to \Omega$ from the parameter space to the physical space is defined in (3.2). We now introduce an affine mapping $\phi : \tilde{\Omega}^e \to \hat{\Omega}^e$ from the parent element to the element in the parameter space. The pullback from the physical space to the parent element, needed to perform quadrature, is achieved using the composition of the inverses of the two mappings, as in Figure 3.4. At each quadrature point in the parent element, we must evaluate the basis functions, their gradients, and the Jacobian determinant of the pullback. This is done via a **shape function routine**[3]. The details of these computations, as well as pseudo-code for a shape function routine for NURBS, are contained in Appendix 3.A at the end of this chapter. See Hughes, 2000, for additional details of numerical integration.

Even though the NURBS functions are not necessarily polynomials, Gaussian quadrature seems to be very effective for integrating them. We can use the same Gauss rule for a pth order NURBS function as one would use for a polynomial of the same order. Though this approach to integration is only approximate, it is important to note that integrating the classical polynomial functions by quadrature on elements with curved sides is only an approximation as well.

Once Galerkin's method has been applied and an approximation, u^h, has been obtained, it is fair to inquire as to just how good of an approximation it is. Results for classical FEA and isogeometric analysis are discussed in Appendix 3.B at the end of this chapter. It turns out that, for elliptic problems such as the one considered in this section, the solution is optimal in a very natural sense; see chapter 4 of Hughes, 2000.

3.3.2 Collocation

A much simpler numerical method for approximating solutions to differential equations such as (3.7) is called **collocation**. Its simplicity follows from the fact that the differential equation is only enforced at a discrete set of points, thus making evaluation and assembly much faster. The trade-off for this simplicity is that it is much more difficult to perform error analysis of

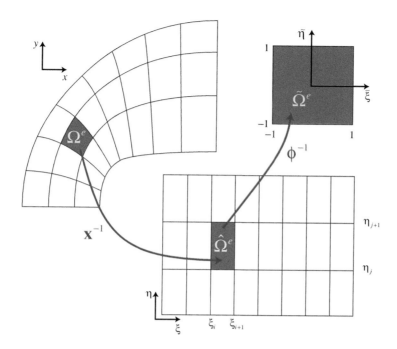

Figure 3.4 Integration is performed by Gaussian quadrature on one element at a time. The physical element is pulled back first to the parametric domain through the geometrical mapping, then through a second mapping, this one affine, to the parent element. Standard change of variables rules apply.

the method, particularly on general domains. However, its efficiency makes it an attractive alternative to FEA.

A full treatment of collocation is beyond the scope of this book. Practitioners will find this brief introduction woefully lacking as many collocation techniques are much more sophisticated than what we present. Our goal is to give a very simple introduction to how the approach might be used in an isogeometric analysis setting, not to give a full representation of the current state of the art.

The goal of collocation is to generate a solution that obeys the differential equation at a set of discrete points called the ***collocation points***. This is still an isoparametric method in that we look for a solution in the space of functions spanned by the basis from which we have built the geometry. That is, we will again look for a solution having the form of (3.28). For simplicity, let us assume that $f = 0$, $\Gamma_D \equiv \Gamma$ and thus $\Gamma_N = \Gamma_R = \varnothing$.

In the simplest of all collocation methods, we will introduce a set of collocation points $\zeta_A \in \Omega$ at which to enforce the differential equation. The collocation problem is simply: Find u^h such that

$$\Delta u^h(\zeta_A) = 0, \tag{3.39}$$

for $A = 1, \ldots, n_{eq}$. Note that, as with the Galerkin finite element case, we have built the Dirichlet boundary condition directly into the solution space, and thus have no equation explicitly pertaining to it as we did in the strong form of the equation (3.7). If we substitute (3.28) into (3.39) and take advantage of the linearity of the Laplace operator, we may rewrite the system as

$$\sum_{B=1}^{n_{eq}} \Delta N_B(\zeta_A) d_B = -g^h(\zeta_A).$$ (3.40)

We may again recast the problem in the language of matrix algebra. To do so, we define

$$K_{AB} = \Delta N_B(\zeta_A)$$ (3.41)

$$F_A = -g^h(\zeta_A)$$ (3.42)

and

$$\mathbf{K} = [K_{AB}],$$ (3.43)

$$\mathbf{F} = \{F_A\},$$ (3.44)

$$\mathbf{d} = \{d_A\},$$ (3.45)

for $A, B = 1 \ldots, n_{eq}$. Now (3.40) may be rewritten as

$$\mathbf{Kd} = \mathbf{F}.$$ (3.46)

If $\mathbf{K}$ is nonsingular, then (3.46) will have a unique solution

$$\mathbf{d} = \mathbf{K}^{-1}\mathbf{F},$$ (3.47)

and u^h will again be exactly as in (3.38). Whether or not we can invert the matrix depends on the locations of the collocation points. The selection of these points is clearly not arbitrary. It is important that the total number of collocation points be equal to the number of degrees-of-freedom in the solution space, but if all n_{eq} of them were placed in a single element in which only a small fraction of the total number of basis functions were non-zero, many of the columns of $\mathbf{K}$ would be filled with zeros.

While selecting the appropriate locations of the collocation points has been the topic of much research, there is a fairly simple solution that seems to work for the NURBS functions. It was first suggested for curve and surface interpolation in Lim, 1999, and appeared again in the context of solving BVPs in Kwok *et al.*, 2001. The idea is to associate each of the n_{eq} collocation points with one of the n_{eq} basis functions. Each ζ_A is placed at exactly the location of the maximum of function N_A, see Figure 3.5. Not only does this guarantee a nonsingular matrix $\mathbf{K}$, but its conditioning, as well as the smoothness of the results, compare quite favorably with other methods. This simple approach is not perfect, however, as the stability of the resulting system can still be a problem; see Kwok *et al.*, 2001. Selection of optimal collocation points is still an open problem.

The only issue remaining is that of selecting the order of NURBS functions to be used. Linear functions will clearly be insufficient as the second derivatives are required in the assembly of the stiffness matrix. The second derivatives of the linear functions are all zero on element interiors and are Dirac layers on the element boundaries. In such cases, the stiffness matrix

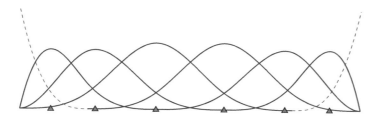

Figure 3.5 For NURBS functions, one approach to the selection of the collocation points is to place them at the maxima of the basis functions. Each of the collocation points, shown here as red triangles, are located at parameter values at which a function achieves its maximum. The first and last function, which have support on the boundary, are not included because they correspond to the lifting, g^h. That is, we do not solve for their coefficients; they are given as g_A's. Another promising possibility is to locate the collocation points at the *Greville abscissae* (see Farin, 1999a) as proposed by Johnson, 2005a, 2005b.

would be unusable. Quadratic functions could be used, as long as the collocation points never corresponded to knot values as the second derivatives would be undefined at the knots. Cubics and higher are the best option. Not only could the collocation points be chosen at the knots if necessary, but the continuous second derivative might make the the quality of the result less sensitive to the location of the collocation points.

3.3.3 Least-squares

An approach that is similar in style to Galerkin FEA is the technique of least-squares finite element analysis[4]. As opposed to multiplying the strong form of the equation (3.7) by a weighting function and integrating, as we did in the Galerkin case, here we first apply the differential operator to the weighting function itself before multiplying and integrating.

In defining the spaces of trial solutions and weighting functions, we must consider the effect of the differential operator acting on the weighting function. The space $H^1(\Omega)$ was sufficient for Galerkin finite elements because integrating by parts allowed us to shift one of the derivatives off of the trial solution and onto the weighting function. In the case of least-squares, the weighting function already has the same number of derivatives applied to it as does the trial solution, and so the integration by parts will not be performed. Thus, we must require that the spaces have, not just square integrable first derivatives, but square integrable second derivatives. The Sobolev space $H^2(\Omega)$ is just such a space. It is defined by

$$H^2(\Omega) = \{u \,|\, D^\alpha u \in L^2(\Omega), |\alpha| \leq 2\}. \tag{3.48}$$

We define the trial solution space and weighting function spaces as

$$\mathcal{S} = \{u \mid u \in H^2(\Omega), u|_{\Gamma_D} = g\} \tag{3.49}$$

and

$$\mathcal{V} = \{w \mid w \in H^2(\Omega), w|_{\Gamma_D} = 0\}, \tag{3.50}$$

respectively.

Assuming again that $\Gamma_D = \Gamma$, the resulting weak problem for a least-squares approach is: Find $u \in S$ such that for all $w \in V$

$$\int_\Omega \Delta w \Delta u \, d\Omega = \int_\Omega \Delta w f \, d\Omega. \tag{3.51}$$

We write this using bilinear form $a(\cdot, \cdot)$ as

$$a(w, u) = L(w), \tag{3.52}$$

where

$$a(w, u) = \int_\Omega \Delta w \Delta u \, d\Omega, \tag{3.53}$$

and

$$L(w) = \int_\Omega \Delta w f \, d\Omega. \tag{3.54}$$

As with Galerkin FEA, we will seek a numerical solution by working with finite dimensional subspaces $S^h \subset S$ and $V^h \subset V$ that are spanned by the NURBS basis. We assume that there exists a lifting $g^h \in S^h$ such that $g^h|_{\Gamma_D} = g$, and thus for every $u^h \in S^h$ we have a unique decomposition $u^h = v^h + g^h$, where $v^h \in V^h$. Thus, the least-squares finite element problem is now: Find $u^h = v^h + g^h$, where $v^h \in V^h$, such that for all $w^h \in V^h$

$$a(w^h, v^h) = L(w^h) - a(w^h, g^h). \tag{3.55}$$

From here we proceed exactly as in the Galerkin case to assemble and solve a matrix system for the coefficients of the solution.

Recall that the solution and weighting spaces here are different than in the Galerkin case. The need for square-integrable second derivatives demands the use of C^1-continuous basis functions. This is clearly not a problem for the NURBS basis, as long as the functions are quadratic or higher order. However, at the interface between patches, constructed from open knot vectors, there will only be C^0 continuity. Here the C^1 continuity can be enforced weakly by employing the so-called "continuous/discontinuous Galerkin (CDG) method" of Engel et al., 2002. See also Hughes and Garikipati, 2004; Wells et al., 2006; Wells and Dung, 2007; and Dung and Wells, 2008.

Proponents of the least-squares approach assert several desirable features of the method, many of which strike at topics that are beyond the scope of this book. At the very least, it is worth mentioning that the resulting stiffness matrix is symmetric and positive definite in all cases, whereas for the Galerkin approach it depends on the differential operator being used. (Note that in the example of the Laplace equation, both techniques result in a symmetric positive definite matrix.) For a good overview of least-squares methods, see Bochev and Gunzburger, 1998.

3.3.4 Meshless methods

So called "element-free" or "mesh-free" methods have generated a substantial amount of research interest in recent years (see, *e.g.*, Nayroles *et al.*, 1992; Belytschko *et al.*, 1994; Liu *et al.*, 1995; Duarte and Oden, 1996; Melenk and Babuska, 1996), though they are as-of-yet rarely seen in industrial practice. There are differences in the approaches that are taken and the names that are used. Usually, the underlying numerical method is Galerkin finite element analysis. What distinguishes the method is the relationship between the geometry, the basis of the solution space, and the quadrature technique being used.

The basic concept of a meshless method is schematically depicted in Figure 3.6. Here the geometry of the domain is typically represented by a boundary description. The basis for the solution space has nothing to do with the geometric description, and hence this is *not* an isoparametric approach. Here a set of radial basis functions is being generated by a set of nodes scattered in an unstructured fashion throughout the domain. Lastly, a "background cell structure" is used for the purposes of integration. Quadrature rules are applied on the intersection of the cells with the domain.

In some ways this is the opposite approach to avoiding the problems of mesh generation as that taken by isogeometric analysis. Rather than uniting the geometry with the solution space as much as possible, meshless methods give them complete autonomy so that one might be considered without the other. While attractive in theory, such an approach is not without its problems. Sakurai, 2006, describes the difficulty of integration, the treatment of Dirichlet boundary conditions, and the low computational efficiency as major drawbacks of the method,

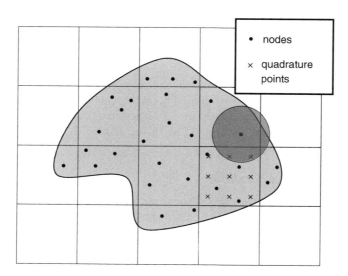

Figure 3.6   Meshless methods separate the tasks of geometry representation, solution representation, and quadrature. A boundary representation is used to describe the geometry. An unrelated basis is introduced to represent the solution. In this case, we have used an unstructured distribution of nodes, each with an associated radial basis function. The support of one such function is seen in red. In order to integrate the functions, a background cell structure is often introduced, with quadrature points placed within each cell.

but goes on to say that "the greatest disadvantage of [meshless methods] caused by abandoning elements is that the ability to define the geometry of the analysis object has been lost." Of course, this is exactly the area in which isogeometric analysis shines. Still, meshless methods are enjoying much popularity in the academic community and are worth being aware of.

There are at least two spline-based techniques in the literature that fall under the umbrella of meshless methods. The first is that of Hollig, 2003, in which the background cells are simply a B-spline mesh in the parameter space. A basis is generated from this grid, but it is not used to define a geometrical mapping. Instead, an unrelated description of the geometry is imposed directly in this parameter space, similar to that in Figure 3.6 (though B-spline functions are used, instead of the radial basis functions shown). Once the boundary is identified, the B-spline basis is augmented to construct so-called "weighted, extended B-splines" or web-splines, which are used as the basis of the solution space. Integration is performed using the background cells, as described above.

A different approach, far more in keeping with the concept of isogeometric analysis, is that of Natekar *et al.*, 2004. They perform shape optimization (a true union of design and analysis) using a NURBS description of the geometry and solution space. Though quadrature is performed on the knot spans, as in Section 3.3.1, rather than on some background cell structure, the authors refer to this as a meshless method. The reason for this meshless designation seems to be related to the nature of the constructive solid geometry approach they employ. In this technique, the domain is defined by the unions and intersections of multiple, frequently overlapping, NURBS patches. Thus, there are regions of the domain with multiple parametric descriptions that must be reconciled. This complicates quadrature and prevents the technique from being truly isoparametric in the traditional sense, though the same descriptions are employed for both the geometry and the solution space.

We may also mention the recent work of Gonzalez *et al.*, 2008.

3.4 Boundary conditions

Turning our attention back to NURBS based isogeometric analysis in a Galerkin FEA setting, let us take a closer look at the implementation of boundary conditions. The three major types of boundary conditions are each represented in (3.7). They are Dirichlet conditions, Neumann conditions, and Robin conditions, given in (3.7b), (3.7c), and (3.7d), respectively. They each present their own set of challenges.

3.4.1 Dirichlet boundary conditions

Dirichlet conditions as in (3.7b) are also known as "essential boundary conditions." This reflects the fact that a standard variational statement of the problem, as in (3.15), requires that these conditions be built directly into the solution space. That is, there is no place in the Galerkin formulation of the problem in which to impose these conditions, and for that reason we build them directly into the space.

In Section 3.3.1.2 we assumed that there existed a function $g^h \in \mathcal{S}^h$ such that $g^h|_{\Gamma_D} = g$, and we referred to this function as a lifting. In practice, this will frequently be the case, but there will also often be instances in which a lifting is only an approximation of g. In either case, we refer to this process of imposing the Dirichlet conditions by building them directly into the space by means of a lifting as ***strong imposition*** of the boundary condition. The alternative of ***weak imposition*** of the Dirichlet condition will also be discussed.

3.4.1.1 Strongly imposed Dirichlet conditions

The case of $g = 0$ is referred to as having "homogeneous Dirichlet conditions," and it is the most frequently encountered case in practice. Assuming g^h is of the form of (3.27), homogeneous boundary conditions are easily built into the solution space by letting $g_A = 0$ for $A = n_{eq} + 1, \ldots, n_{np}$ – recall that we always set $g_1, \ldots, g_{n_{eq}} = 0$ regardless of the value of g as they have no effect on the boundary. Similarly, if g is a constant, the partition of unity property of the basis ensures that we need only set the $g_{n_{eq}+1}, \ldots, g_{n_{np}}$ to that constant value. Other functions that are in the NURBS space (e.g., linear functions) may be set by selecting the control variables appropriately.

In many instances we will have no reason to believe that g actually exists in the NURBS space, and so the lifting will only be an approximation $g^h|_{\Gamma_D} \approx g$. In such a case, there are several ways in which to proceed. Classical finite elements typically interpolate g at the nodes. In isogeometric analysis, we may interpolate g with the control points, but as the basis itself is non-interpolatory this may result in a slightly smeared g^h. It is still a viable option, frequently yielding better results than FEA (recall Figure 2.13). If this approximation is unacceptable, a better lifting may be found by running a curve or surface fitting algorithm to get, for example, a least-squares fit of the prescribed Dirichlet data. Either way, once a lifting g^h is found, it is built into the space and used as though it were exact.

3.4.1.2 Weakly imposed Dirichlet conditions

An alternative to the traditional approach of using a lifting is to impose the boundary conditions weakly by adding terms to the variational equation to enforce them as Euler–Lagrange conditions. This can be done regardless of whether or not an appropriate lifting exists. In Bazilevs and Hughes, 2007, it was shown that weakly imposing boundary conditions in problems associated with boundary layer phenomena can help eliminate some of the spurious oscillations encountered with traditional strongly imposed conditions.

Let the trial solution and weighting spaces both be $H^1(\Omega)$. Note that we do not enforce $w|_{\Omega_D} = 0$ in this case. The weak form of (3.7a) that naturally arises from integration by parts is simply

$$-\int_\Omega \nabla w \cdot \nabla u \, d\Omega + \int_\Gamma w \nabla u \cdot \mathbf{n} \, d\Gamma + \int_\Omega wf \, d\Omega = 0. \tag{3.56}$$

We will augment this equation by the addition of two extra terms, which together serve to penalize errors in the enforcement of the boundary condition and to ensure that the method is still optimally convergent. The resulting formulation is

$$-\int_\Omega \nabla w \cdot \nabla u \, d\Omega + \int_\Gamma w \nabla u \cdot \mathbf{n} \, d\Gamma + \int_\Omega wf \, d\Omega$$
$$+ \int_{\Gamma_D} \gamma \, (\nabla w \cdot \mathbf{n}) \, (u - g) \, d\Gamma + \int_{\Gamma_D} \frac{C}{h_e} w \, (u - g) \, d\Gamma = 0, \tag{3.57}$$

where h_e is an element length scale, C is a constant, and $\gamma = \pm 1$. A discussion of the proper selection of h_e, C, and the sign of γ is beyond the scope of the current discussion; for details see Bazilevs and Hughes, 2007. Note that this formulation remains consistent, as the exact solution u is also a solution to (3.57).

The result of this approach is that the boundary condition is never enforced exactly, although the approximation will improve as the mesh is refined. This ability to satisfy the Dirichlet boundary condition approximately can be a significant advantage if it allows for greater accuracy on the interior of the domain. This approach can also be employed to weakly enforce solution compatibility between non-conforming portions of the mesh, as we will see in several different settings throughout this book.

3.4.2 Neumann boundary conditions

Neumann boundary conditions of the form of (3.7c) are frequently referred to as "natural boundary conditions." This is because of the way they automatically arise in the variational statement of a problem. Let us assume for the moment that $\Gamma_R = \varnothing$ and that the Dirichlet conditions are being strongly imposed. Multiplying by a test function and integrating leads us to

$$
\begin{aligned}
0 &= \int_\Omega w \left(\Delta u + f \right) d\Omega \\
&= -\int_\Omega \nabla w \cdot \nabla u \, d\Omega + \int_\Gamma w \nabla u \cdot \mathbf{n} \, d\Gamma + \int_\Omega w f \, d\Omega \\
&= -\int_\Omega \nabla w \cdot \nabla u \, d\Omega + \int_{\Gamma_N} w \nabla u \cdot \mathbf{n} \, d\Gamma + \int_\Omega w f \, d\Omega
\end{aligned}
\tag{3.58}
$$

where in the third line we have used the fact that the weighting space is defined such that $w|_{\Gamma_D} = 0$.

The integration by parts has completely naturally introduced a boundary integral over Γ_N that refers explicitly to the condition that we would like to impose. Using (3.7c) we simply replace $\nabla u \cdot \mathbf{n}$ with the value we are imposing, h, resulting in

$$
-\int_\Omega \nabla w \cdot \nabla u \, d\Omega + \int_\Gamma w h \, d\Gamma + \int_\Omega w f \, d\Omega = 0.
\tag{3.59}
$$

The effect is a weak imposition of the Neumann condition. That is, we only expect it to be approximately satisfied. We do, however, expect the accuracy with which this condition is satisfied to improve under refinement along with the accuracy of the solution on the interior of the domain. Unlike the choice of weak imposition of Dirichlet conditions, the terms needed to impose the Neumann condition originated from the variational formulation of the problem itself. In the Dirichlet case, the terms had to be introduced somewhat artificially, though their effect is similar.

3.4.3 Robin boundary conditions

Robin boundary conditions of the form of (3.7d) are very similar to Neumann conditions. They are treated naturally, working from the second line of (3.58) as in the Neumann case. For brevity, let us assume for the moment that $\Gamma_N = \varnothing$. We rearrange (3.7d) to obtain

$$
\nabla u \cdot \mathbf{n} = r - \beta u,
\tag{3.60}
$$

which we insert into (3.58) along with $w|_{\Gamma_D} = 0$ to arrive at

$$-\int_\Omega \nabla w \cdot \nabla u \, d\Omega + \int_\Omega wf \, d\Omega + \int_{\Gamma_R} wr \, d\Gamma - \beta \int_{\Gamma_R} wu \, d\Gamma = 0. \qquad (3.61)$$

Though we have replaced one expression involving the unknown u for another, we have enforced the relationship between them weakly.

3.5 Multiple patches revisited

3.5.1 Local refinement

In Section 3 of Chapter 2 we discussed the need for modeling domains using multiple patches. We always assume compatible discretizations for the geometry, meaning that on the coarsest mesh, mappings and parameterizations on the adjoining patch faces are identical. Each control point on a face is in one-to-one correspondence with a control point from the adjoining face, likewise for the control variables of the solution. In many instances, this relationship will be preserved as we refine. To make the assembly of the stiffness matrices and force vectors as simple as possible, the connectivity array will identify the equivalent local control variables on each face with a single control variable in the global array. By identifying them as a single entity for analysis purposes, we simplify the logic and decrease the total amount of work needed. The result is that the two patches are joined as though they were one.

Identifying two control points at the same location in physical space as being a single entity is fairly trivial. Slightly subtler is what this implies for the basis functions. The situation is shown in Figure 3.7. When two bases generated using open knot vectors are brought together, the effect is indistinguishable from the case of one knot vector with a knot repeated p times, as long as the coefficients of the two joining functions are the same. This is, of course, exactly what we have assured by identifying the two control variables as one.

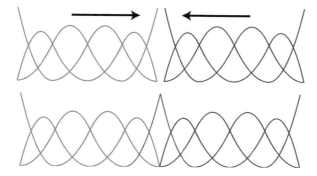

Figure 3.7 If two knot vectors formed from open knot vectors are brought together, they can be made to act as one if the coefficients (*i.e.*, control variables) of the two functions on their interface are always equal to each other. The result is indistinguishable from the case of a single knot vector with a C^0 boundary at the interface.

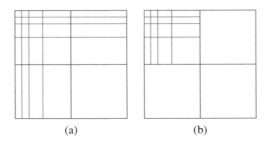

(a)   (b)

Figure 3.8 (a) Global refinement employing the continuous Galerkin method. (b) Local refinement employing the discontinuous Galerkin method or constraint equations at the patch level. With constraint equations, at least C^0-continuity can be attained across patches, and higher-order continuity can be achieved in certain cases if desired.

Another reason for using multiple patches is that it makes local refinement possible. The situation is represented in Figure 3.8. Even with multiple patches, if we want the control points of the two patches on their interface to be in one-to-one correspondence, we need to have matching knot vectors. This means that refinements of one patch must necessarily propagate from that patch to the next. If instead we are to allow knots to be inserted on one side and not the other (*i.e.*, local refinement), we may proceed as follows.

Consider the two B-spline[5] patches that meet on an interface, as shown in Figure 3.9. On the coarsest mesh, we assume that the control points and knot vectors in the plane of the face are identical on both patches, thus ensuring that the patches match geometrically and parametrically on that shared face. Using superscripts 1 and 2 to identify the patch numbers, a subscript f to denote control points on the face where the patches meet, and a subscript n to denote control points *not* on that face, we may write the control points for Patches 1 and 2 as

$$\mathbf{B}^1 = \begin{pmatrix} \mathbf{B}_n^1 \\ \mathbf{B}_f^1 \end{pmatrix} \quad \text{and} \quad \mathbf{B}^2 = \begin{pmatrix} \mathbf{B}_n^2 \\ \mathbf{B}_f^2 \end{pmatrix}, \tag{3.62}$$

respectively, where

$$\mathbf{B}_f^2 = \mathbf{B}_f^1. \tag{3.63}$$

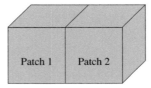

Figure 3.9 The two patches share a common interface. On the coarsest mesh, their control points on that interface are in one-to-one correspondence, trivially enforcing C^0 continuity.

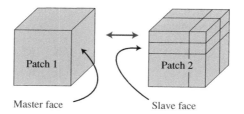

Figure 3.10 Patch 2 is refined by knot insertion and the one-to-one correspondence of the interface control points is lost. Constraint equations may be employed to ensure that continuity is maintained.

If we now refine the basis of Patch 2 by knot insertion (Figure 3.10), then we have the following new set of control points for Patch 2:

$$\tilde{\mathbf{B}}^2 = \tilde{\mathbf{T}}\mathbf{B}^2 = \begin{pmatrix} \tilde{\mathbf{T}}_n & 0 \\ 0 & \tilde{\mathbf{T}}_f \end{pmatrix} \begin{pmatrix} \mathbf{B}_n^2 \\ \mathbf{B}_f^2 \end{pmatrix}, \tag{3.64}$$

where $\tilde{\mathbf{T}}$ is the multi-dimensional generalization of the extension operator defined in (2.22). As before, it is sparse and its values are entirely defined by the knot vectors and the polynomial order. The block diagonal structure follows from the fact that we are using open knot vectors. When open knot vectors are used, each face of a NURBS solid is influenced only by the control points on that face. Put simply, each face of the NURBS solid is a NURBS surface.

Combining (3.63) and (3.64), we see that C^0-continuity of the geometry is maintained by the relationship

$$\tilde{\mathbf{B}}_f^2 = \tilde{\mathbf{T}}_f \mathbf{B}_f^1. \tag{3.65}$$

Building on the approach of Kagan et al., 2003[6], it follows that for the solution space to enforce the same continuity constraints, we need the control variables to obey precisely the same relationship. Let

$$\mathbf{u}^1 = \begin{pmatrix} \mathbf{u}_n^1 \\ \mathbf{u}_f^1 \end{pmatrix} \quad \text{and} \quad \mathbf{u}^2 = \begin{pmatrix} \mathbf{u}_n^2 \\ \mathbf{u}_f^2 \end{pmatrix} \tag{3.66}$$

be the control variables on Patch 1 and the refined Patch 2, respectively. Then C^0-continuity of the solution across the interface between the patches may be maintained by enforcing the constraint

$$\mathbf{u}_f^2 = \tilde{\mathbf{T}}_f \mathbf{u}_f^1. \tag{3.67}$$

From an implementational point of view, the two patches may be assembled locally to create the two local problems

$$\mathbf{K}^1 \mathbf{u}^1 = \mathbf{b}^1 \tag{3.68}$$

and

$$\mathbf{K}^2 \mathbf{u}^2 = \mathbf{b}^2 \tag{3.69}$$

for the control points on either patch. Consistent with the partitioning of the control variables in (3.66), we partition the stiffness matrices as

$$\mathbf{K}^1 = \begin{pmatrix} \mathbf{K}^1_{nn} & \mathbf{K}^1_{nf} \\ \mathbf{K}^1_{fn} & \mathbf{K}^1_{ff} \end{pmatrix} \quad \text{and} \quad \mathbf{K}^2 = \begin{pmatrix} \mathbf{K}^2_{nn} & \mathbf{K}^2_{nf} \\ \mathbf{K}^2_{fn} & \mathbf{K}^2_{ff} \end{pmatrix}. \tag{3.70}$$

Before solving, we must assemble problems (3.68) and (3.69) into one global problem accounting for the behavior of both patches, as well as their interaction. We should have three coupled blocks of equations: one corresponding to weighting functions with support in Patch 1 that vanish on the face shared by the two patches, one corresponding to weighting functions with support on either or both patches that do *not* vanish on the shared face, and one corresponding to weighting functions with support on Patch 2 that vanish on the shared face. We begin by expanding (3.68) using the partitioning of (3.70) to get

$$\mathbf{K}^1_{nn}\mathbf{u}^1_n + \mathbf{K}^1_{nf}\mathbf{u}^1_f = \mathbf{b}^1_n \tag{3.71}$$

and

$$\mathbf{K}^1_{fn}\mathbf{u}^1_n + \mathbf{K}^1_{ff}\mathbf{u}^1_f = \mathbf{b}^1_f. \tag{3.72}$$

Inserting (3.67) into (3.69) and expanding yields

$$\mathbf{K}^2_{nn}\mathbf{u}^2_n + \mathbf{K}^2_{nf}\tilde{\mathbf{T}}_f\mathbf{u}^1_f = \mathbf{b}^2_n \tag{3.73}$$

and

$$\mathbf{K}^2_{fn}\mathbf{u}^2_n + \mathbf{K}^2_{ff}\tilde{\mathbf{T}}_f\mathbf{u}^1_f = \mathbf{b}^2_f. \tag{3.74}$$

Note that (3.71) is the block of equations corresponding to weighting functions in Patch 1 that vanish on the shared face. Similarly, (3.73) is the block of equations corresponding to weighting functions in Patch 2 that vanish on the shared face. Now (3.72) and (3.74) both correspond to weighting functions with support on the shared face and as such we would like to add them together to get a final expression for that block. Unfortunately, they contain different numbers of equations. This is because we assembled the two patches independently. We correctly generated the equations in (3.72) by testing against functions in the "master" weighting space associated with Patch 1, but we generated the equations in (3.74) by testing against all of the functions in the larger "slave" weighting space on Patch 2 without regard for the constraint. The basis functions of the slave *solution* space on Patch 2 corresponding to the shared face are restricted to act only in the linear combinations defined by $\tilde{\mathbf{T}}_f$ that result in functions existing in the master solution space. So too must the functions in the slave *weighting* space act only in such linear combinations as replicate functions in the master weighting space. This constraint may be enforced by now premultiplying (3.74) by $\tilde{\mathbf{T}}^T_f$, thus constraining the weighting functions and reducing the number of equations to match that of (3.72):

$$\tilde{\mathbf{T}}^T_f\mathbf{K}^2_{fn}\mathbf{u}^2_n + \tilde{\mathbf{T}}^T_f\mathbf{K}^2_{ff}\tilde{\mathbf{T}}_f\mathbf{u}^1_f = \tilde{\mathbf{T}}^T_f\mathbf{b}^2_f. \tag{3.75}$$

We may now express the global system comprised of (3.71), (3.73), and $((3.72)+(3.75))$ as

$$\mathbf{Ku} = \mathbf{b}, \tag{3.76}$$

where

$$\mathbf{K} = \begin{pmatrix} \mathbf{K}_{nn}^1 & \mathbf{K}_{nf}^1 & 0 \\ \mathbf{K}_{fn}^1 & (\mathbf{K}_{ff}^1 + \tilde{\mathbf{T}}_f^{\mathrm{T}} \mathbf{K}_{ff}^2 \tilde{\mathbf{T}}_f) & \tilde{\mathbf{T}}_f^{\mathrm{T}} \mathbf{K}_{fn}^2 \\ 0 & \mathbf{K}_{nf}^2 \tilde{\mathbf{T}}_f & \mathbf{K}_{nn}^2 \end{pmatrix}, \tag{3.77}$$

$$\mathbf{u} = \begin{pmatrix} \mathbf{u}_n^1 \\ \mathbf{u}_f^1 \\ \mathbf{u}_n^2 \end{pmatrix}, \tag{3.78}$$

and

$$\mathbf{b} = \begin{pmatrix} \mathbf{b}_n^1 \\ \mathbf{b}_f^1 + \tilde{\mathbf{T}}_f^{\mathrm{T}} \mathbf{b}_f^2 \\ \mathbf{b}_n^2 \end{pmatrix}. \tag{3.79}$$

We may recover $\mathbf{u}_f^2$ via (3.67) after solving (3.76).

This approach ensures C^0-continuity in the solution across the patch boundary when one patch is a knot-refined version of the other patch on their common interface. Higher continuity has also been implemented by applying similar constraint equations in the normal direction. As long as the geometries are compatible, the patch boundary may be seen as the result of inserting a knot into some "meta-patch" $p + 1$ times. It should be noted that these are strong, exact constraints, not approximations. An approach that would allow for weak enforcement of continuity, as well as allowing for local order elevation is to use discontinuous Galerkin techniques at the patch level. That is, weakly enforce continuity of appropriate fluxes across patch boundaries while strongly enforcing them across element boundaries within the patch. See, for example, Cockburn, 2004 for an overview of the discontinuous Galerkin method.

3.5.2 Arbitrary topologies

Due to the tensor product structure of B-splines and NURBS it might be assumed that isogeometric analysis utilizing B-splines and NURBS is restricted to block-structured discretizations in which patches play the role of the blocks. This is *not* the case. The full topological generality of finite elements can be achieved by assuming the patches consist of single elements as the following examples illustrate.

C^0 Lagrange elements are widely used in finite element analysis. They are constructed from tensor products of Lagrange interpolatory polynomials (see Hughes, 2000). A B-spline basis of Bernstein polynomials can be constructed to produce C^0 elements with exactly the same span. Such elements are called Bézier elements, and they are identical to B-spline patches comprised of a single element. These Bézier elements possess all the usual properties of B-splines, namely, the convex hull and variation diminishing properties in terms of the control points. See Section 2.1.3 and Figure 2.13b in Chapter 2. The elements depicted in Figure 2.13b are in fact Bézier elements. The construction of bivariate C^0 Bézier elements is illustrated in Figure 3.11 and typical basis functions are presented in Figure 3.12. It should be clear that by using the concept of a patch consisting of one Bézier element, we can construct unstructured meshes of isogeometric Bézier elements with completely arbitrary topology, just as in finite element analysis. We need to define the data processing arrays in the usual way to achieve C^0

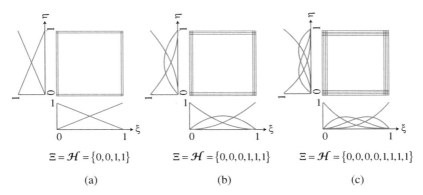

$$\Xi = \mathcal{H} = \{0,0,1,1\} \qquad \Xi = \mathcal{H} = \{0,0,0,1,1,1\} \qquad \Xi = \mathcal{H} = \{0,0,0,0,1,1,1,1\}$$

(a)   (b)   (c)

Figure 3.11 Construction of bivariate C^0 Bézier elements. The basis functions are the tensor products of the one-dimensional basis functions illustrated. The spans of these elements are the same as for C^0 Lagrange elements. The control points of the Bézier elements at the corners are interpolated but all others are not, in contrast with Lagrange elements. One can transform Lagrange elements to Bézier elements, and vice versa, through a simple linear transformation of the control points. (a) Four control point bilinear Bézier element. This element is identical to the four-node bilinear Lagrange element. (b) Nine control point biquadratic Bézier element. (c) Sixteen control point bicubic Bézier element. Notation $\Xi = \{\xi_i\}$, $i = 1, 2, \ldots, 2(p + 1)$, and $\mathcal{H} = \{\eta_j\}$, $j = 1, 2, \ldots, 2(p + 1)$, are the knot vectors, where p is the polynomial order.

continuity (see Section 2.3 of Chapter 2 and Section 3.5.1 of this chapter). The span of Bézier finite element spaces is identical to the span of C^0 Lagrange elements spaces, although their basis functions are different. See Figure 3.13. It is important to observe that the ability to assemble meshes of this type emanates from the nested loop structure that we advocate for isogeometric analysis, which consists of an outer patch loop and an element loop within each patch. A further outer loop can be used to define element groups (see the description of the DLEARN program in Hughes, 2000) or substructures. See Figure 3.14.

Our intent in discussing the analogy between Lagrange and Bézier elements is to emphasize anything that can be done with the former is still possible in an isogeometric setting, though the flexibility of the NURBS technology allows for many more possibilities (two of the most important being exact representations of conic sections and the use of C^{p-1}-continuous basis functions).

By using the element degeneration concept (see Hughes, 2000, chapter 3), a variety of other element shapes, such as triangles, tetrahedra, wedges, and pyramids, can also be constructed. Patches may even be viewed as unstructured by using T-spline discretizations (see Sederberg *et al.*, 2003, 2004). We may also mention that smooth, spline-based triangles and tetrahedra can also be directly constructed (Lai and Schumaker, 2007).

3.6 Comparing isogeometric analysis with classical finite element analysis

We have seen that isogeometric analysis, in this case NURBS based Galerkin finite elements, is quite similar in its structure to classical FEA. In short, the only difference is the basis being used. Of course, this "small" change has huge implications, many of which will be made clear

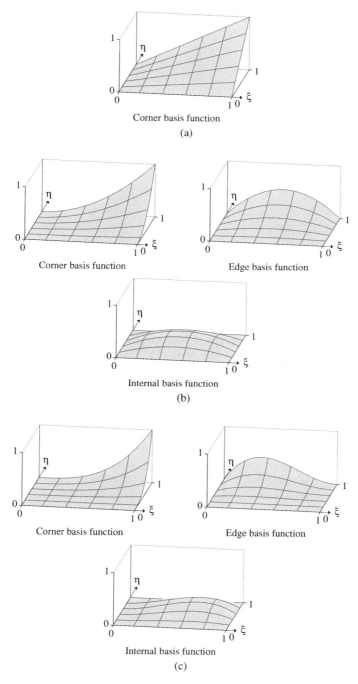

Figure 3.12 (a) Bilinear Bézier element (identical to the standard bilinear element). There are four corner basis functions. (b) Biquadratic Bézier element. There are four corner basis functions, four edge basis functions, and one internal basis function. (c) Bicubic Bézier element. There are four corner elements, eight edge basis functions, and four internal basis functions.

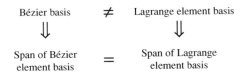

Figure 3.13   Relationships of Bézier and Lagrange basis functions and spans of basis functions.

in future chapters when we look at specific applications. In this section, we look at the major similarities and differences between isogeometric analysis and FEA. It will be important to bear these comparisons in mind as we examine the numerous examples throughout the remainder of the book.

3.6.1   Code architecture

Let us first consider the architecture of a classical FEA code. The flowchart for a typical example of such a piece of software is given in Figure 3.15. The program begins with the data defining the boundary value problem, the mesh, and all of the geometrical data being read from files. Once these data have been read, the connectivity information can be generated (though sometimes this will be read in from an external file as well) and the memory is allocated for all of the major global arrays, which are subsequently initialized to zero. Once these preprocessing steps are completed, assembly of the system begins, following the process described in Section 3.3.1.4. There is a loop through all of the elements in the mesh. Within each element, the element stiffness matrix and element force vector are initialized, and then the code enters a loop through the quadrature points. At each quadrature point, a routine is called that will evaluate all of the basis functions and any necessary derivatives. It is helpful, for the moment, to think of this routine as a black box. If we know the number of local basis functions, it is not important what those functions are or how they are evaluated. It is only important that we have a routine from which we can obtain those values when they are needed. With these values in hand, we proceed to build the local stiffness matrix and force vector. After we have been through each quadrature point and fully assembled the local arrays, we

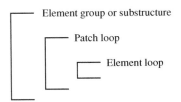

Figure 3.14   Program architecture of the assembly algorithm in isogeometric analysis. The patch loop does not have a direct analog in finite element analysis, although it might be considered analogous to a macro-element loop. If each patch consists of a single element, we have the assembly algorithm that is standard in finite element analysis (see Hughes, 2000).

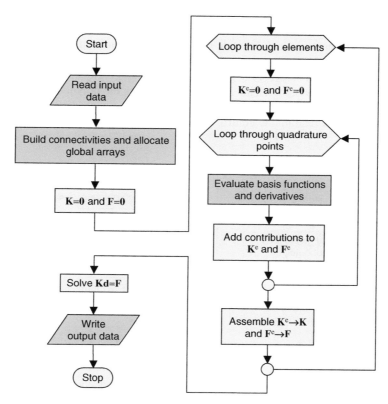

Figure 3.15 Flowchart of a classical finite element code. Such a code can be converted to a single-patch isogeometric analysis code by replacing the routines shown in green.

use the connectivity information to add their contributions to the global stiffness matrix and force vector, and then move on to the next element. After all of the elements are assembled, the global arrays are complete. We then solve the system, write the result to a file, postprocess, and we are finished. See Hughes, 2000 for further details.

To convert an existing finite element code to a single-patch isogeometric analysis code, the only portions of the code that require modification are the ones shown in green in Figure 3.15. Clearly, the input will change as the file format will depend on the specific element technology being used. The precise forms of the connectivity arrays and the global matrices also depend on the basis. The structured nature of the NURBS mesh means that the IEN array (see Section 3.3.1.4) can be calculated automatically from the knot vectors and polynomial orders. Next, the "black box" that evaluated the basis functions must be updated to evaluate the NURBS functions. This is why we emphasized the modular nature of this routine previously: the type of information about the basis that it provides to the routine that calls it is exactly the same, but that information should now correspond to the NURBS basis. Lastly, the output must be written, and the format of that output will be specific to the NURBS basis.

3.6.1.1 A multiple patch code

A "multi-patch" isogeometric analysis code can be made to conform with the flowchart in
Figure 3.15. In practice, however, it makes more sense to consider the slight modification
shown in Figure 3.16. In this case, we begin by inputting enough global information to build
the global connectivities, as before. This information includes the polynomial orders and the
knot vectors for each of the patches, but it does not require the control points. We can save
time and memory by not reading the control points until they are needed. If local refinement

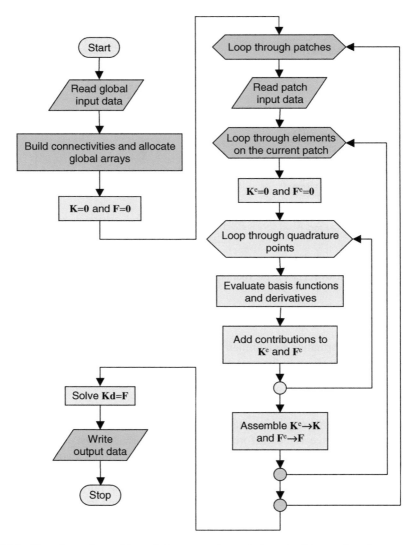

Figure 3.16 Flowchart of a multi-patch isogeometric analysis code. The routines in green represent
differences from the single-patch code.

has taken place, we must build the **T** matrices to be used in the enforcement of continuity, as in Section 3.3. Again, the knot vectors and polynomial orders are all that is required. Global arrays are allocated and initialized as before.

At this point the code enters a loop through the patches. The reason for making this loop explicit is that the control points defining the geometry are relevant to only one patch at a time. We can input this information within the loop, reading only the information relevant to the patch we are currently working with. We now loop through the elements on the current patch. Everything then proceeds exactly as before until after the global system is solved. Lastly, the output is written to files, typically in a format that makes it easy to identify control variables with the patch that they correspond to, and so this routine will be specific to the multiple-patch setting.

The only other potential source of complexity is if local refinement it to be applied, as described in the previous section. This can either be implemented during assembly, or within the solver. In either case, modifications of the appropriate routine will be required.

3.6.2 Similarities and differences

Throughout our discussion of NURBS, we have made comparisons with classical FEA functions and geometries. Some of the most notable differences are summarized in Table 3.1. The first and most important difference is that isogeometric analysis employs the exact geometry at all levels of discretization, whereas FEA uses piecewise polynomial approximations, even for such common objects as conic sections. This geometric exactness not only affects the accuracy of computed solutions, but even the analysis process as a whole as refinement requires no external description of the geometry, unlike in FEA. The second striking difference between the methods is that neither the control points nor the control variables of isogeometric analysis are interpolated, unlike nodal points and nodal variables. This means that we cannot strictly interpret these entities by themselves, but only in conjunction with the basis functions. Solutions in both cases, however, are linear combinations of coefficients and basis functions, and so nothing about the mathematical structure of the Galerkin method or its implementation differs between the methods. Other major differences relate to the properties possessed by the bases; see Table 3.1.

Isogeometric analysis and classical FEA have many similarities as well, a few of which are summarized in Table 3.2. They are both isoparametric implementations of Galerkin's method, and as such they have a very similar code architecture. Both methods use compactly supported basis functions, and the bandwidth of matrices corresponding to a given polynomial order are the same for the two methods (recall Figure 2.4). Both bases obey the partition of unity property and affine transformations are achieved by applying them directly to the vector valued coefficients that define the geometry.

Appendix 3.A: Shape function routine

The *shape function routine* is one of the fundamental components of any finite element code. Given an element number, e, and quadrature points on the parent element, $(\tilde{\xi}_1, \tilde{\xi}_2, \tilde{\xi}_3) \equiv (\tilde{\xi}, \tilde{\eta}, \tilde{\zeta}) \in [-1, 1]^3$, the shape function routine must evaluate each of the local basis functions (*i.e.*, each function with support in the element) at the given quadrature point, as well as any required derivatives. Additionally, the Jacobian determinant of the mapping must be calculated

Table 3.1 Differences between NURBS based isogeometric analysis and finite element analysis

Isogeometric analysis	Finite element analysis
Exact geometry	Approximate geometry
Control points	Nodal points
Control variables	Nodal variables
Basis does *not* interpolate control points and variables	Basis interpolates nodal points and variables
NURBS basis	Polynomial basis
High, easily controlled continuity	C^0-continuity, always fixed
hpk-refinement space	*hp*-refinement space
Pointwise positive basis	Basis not necessarily positive
Convex hull property	No convex hull property
Variation diminishing in the presence of discontinuous data	Oscillatory in the presence of discontinuous data

in order to perform integration. In this appendix, we give an example of a shape function routine for a NURBS based isogeometric analysis code. It relies heavily on NURBS coordinates and the IEN array. For a thorough discussion of these concepts, see Appendix A at the end of the book.

The shape function routine we present assumes the existence of another routine entitled **Bspline_basis_and_deriv** that will calculate all of the relevant univariate B-spline basis functions and their parametric derivatives. For example, let element $\hat{\Omega}^e = [\xi_i, \xi_{i+1}] \times [\eta_j, \eta_{j+1}] \times [\zeta_k, \zeta_{k+1}]$. For the ξ-direction, **Bspline_basis_and_deriv** will return a vector of $p + 1$ function values corresponding to the $p + 1$ functions that are nonzero on $[\xi_i, \xi_{i+1}]$. Specifically, the first entry will be N_i, the second N_{i+1} and so forth; likewise for the vector of derivatives. With these univariate, non-rational function values in hand, the trivariate, rational NURBS functions, R are calculated using (2.30). The derivatives with respect to the parametric coordinates, $\boldsymbol{\xi}$ are calculated by (2.31). To obtain derivatives with respect to the physical coordinates, $(x_1, x_2, x_3) \equiv (x, y, z)$, one must apply the chain rule in the form

$$\frac{\partial R}{\partial x_i} = \frac{\partial R}{\partial \xi_j} \frac{\partial \xi_j}{\partial x_i}. \tag{3.A.1}$$

Table 3.2 Common features shared by isogeometric analysis and finite element analysis

Isogeometric analysis and finite element analysis
Isoparametric concept
Galerkin's method
Code architecture
Compactly supported basis
Bandwidth of matrices
Partition of unity
Affine covariance
Patch tests are satisfied

Thus, the gradient of the mapping, $\partial \mathbf{x}/\partial \boldsymbol{\xi}$ must be calculated, along with its inverse. The mapping will be inverted using the external **inverse_Cramer** routine that uses Cramer's rule to compute the inverse of a matrix.

Lastly, as the parts of the code that call this shape function routine will be performing the numerical integration *in the parent element*, the Jacobian determinant of the mapping from the parent element to the physical space, J must be calculated. It is given by

$$J = \left| \frac{d\mathbf{x}}{d\tilde{\boldsymbol{\xi}}} \right| = \left| \frac{d\mathbf{x}}{d\boldsymbol{\xi}} \frac{d\boldsymbol{\xi}}{d\tilde{\boldsymbol{\xi}}} \right|. \tag{3.A.2}$$

The actual determinant is calculated in the external routine **determinant**.

Our shape function routine is presented in Algorithms 1–3. For the sake of clarity, we have divided the routine into three parts, though in practice they would usually all be within the same function. Part I, in Algorithm 1, initializes all of the variables to zero. Then the NURBS coordinates are determined using the INC and IEN arrays. The parametric coordinates are then calculated from the knot vectors and the parent element coordinates of the quadrature point. For example, with $\hat{\Omega}^e = [\xi_i, \xi_{i+1}] \times [\eta_j, \eta_{j+1}] \times [\zeta_k, \zeta_{k+1}]$, and thus NURBS coordinates (i, j, k), we can calculate $(\xi, \eta, \zeta) \in \hat{\Omega}^e$ from $(\tilde{\xi}, \tilde{\eta}, \tilde{\zeta}) \in \tilde{\Omega}^e$ as

$$\xi = \xi_i + (\tilde{\xi} + 1)\frac{(\xi_{i+1} - \xi_i)}{2} = \frac{((\xi_{i+1} - \xi_i)\tilde{\xi} + (\xi_{i+1} + \xi_i))}{2}, \tag{3.A.3}$$

$$\eta = \eta_j + (\tilde{\eta} + 1)\frac{(\eta_{j+1} - \eta_j)}{2} = \frac{((\eta_{j+1} - \eta_j)\tilde{\eta} + (\eta_{j+1} + \eta_j))}{2}, \tag{3.A.4}$$

$$\zeta = \zeta_k + (\tilde{\zeta} + 1)\frac{(\zeta_{k+1} - \zeta_k)}{2} = \frac{((\zeta_{k+1} - \zeta_k)\tilde{\zeta} + (\zeta_{k+1} + \zeta_k))}{2}. \tag{3.A.5}$$

Part II, in Algorithm 2, calculates the values of the basis functions and their derivatives with respect to the parametric coordinates. These calculations follow directly from the definitions in Chapter 2. Part III, in Algorithm 3, determines the derivatives with respect to the physical coordinates. It does so by first calculating the gradient of the mapping, and then using it in conjunction with the parametric derivatives of Algorithm 2 and the chain rule as in (3.A.1). The Jacobian determinant is also calculated as in (3.A.2).

Algorithm 1: Shape function routine: Part I

Data: The quadrature points $\tilde{\xi}$, element number e, polynomial orders (p, q, and r), control
net **B**, knot vectors Ξ, $\mathcal{H}$, and $\mathcal{Z}$, and connectivity arrays INC and IEN must be
included as inputs. The number of local shape functions n_{en} must also be given. Note
that the weights are stored as the fourth component of each control point.

Result: The vector of local shape functions R and an array of their derivatives dR_dx, and the
Jacobian determinant J will be returned.

```
// Initializations:

R[nen]=0;                    // Array of trivariate NURBS basis functions
dR_dx[nen][3]=0;             // Trivariate NURBS function derivatives
                             //     w.r.t. physical coordinates
J=0;                         // Jacobian determinant

// Local variable initializations:

ni, nj, nk = 0;                   // NURBS coordinates
xi, eta, zeta = 0;                // Parametric coordinates
N[p+1], M[q+1], L[r+1] = 0;       // Arrays of univariate B-spline
                                  //     basis functions
dN_dxi[p+1] = 0;                  // Univariate B-spline
dM_deta[q+1] = 0;                 //     function derivatives w.r.t.
dL_dzeta[r+1] = 0;                //     appropriate parametric coordinates
dR_dxi[nen][3] = 0;               // Trivariate NURBS function derivatives
                                  //     w.r.t. parametric coordinates
dx_dxi[3][3] = 0;                 // Derivative of physical coordinates
                                  //     w.r.t. parametric coordinates
dxi_dx[3][3] = 0;                 // Inverse of dx_dxi
dxi_dtildexi[3][3] = 0;           // Derivative of parametric coordinates
                                  //     w.r.t. parent element coordinates
J_mat[3][3] = 0;                  // Jacobian matrix
i, j, k, aa, bb, cc = 0;          // Loop counters
loc_num = 0;                      // Local basis function counter
sum_xi, sum_eta, sum_zeta, sum_tot = 0;    // Dummy sums for calculating
                                           //     rational derivatives

// NURBS coordinates; convention consistent with Algorithm 7
ni = INN[IEN[e][1]][1];
nj = INN[IEN[e][1]][2];
nk = INN[IEN[e][1]][3];

// Calculate parametric coordinates from parent element coordinates
//     Knot vectors KV_Xi, KV_Eta, and KV_Zeta and
//     parent element coordinates xi_tilde, eta_tilde, zeta_tilde
//     are given as input
xi = ((KV_Xi[ni+1]-KV_Xi[ni])*xi_tilde...
    ... + (KV_Xi[ni+1]+KV_Xi[ni])) / 2;
eta = ((KV_Eta[nj+1]-KV_Eta[nj])*eta_tilde...
    ... + (KV_Eta[nj+1]+KV_Eta[nj])) / 2;
zeta = ((KV_Zeta[nk+1]-KV_Zeta[nk])*zeta_tilde...
    ... + (KV_Zeta[nk+1]+KV_Zeta[nk])) / 2;
```

Algorithm 2: Shape function routine: Part II

```
// Calculate univariate B-spline functions using (2.1) and (2.2)
//    and their derivatives using (2.12)
call Bspline_basis_and_deriv(ni,p,KV_Xi; N,dN_dxi);        // xi-dir.
call Bspline_basis_and_deriv(nj,q,KV_Eta; M,dM_deta);      // eta-dir.
call Bspline_basis_and_deriv(nk,r,KV_Zeta; L,dL_dzeta);    // zeta-dir.

// Build numerators and denominators
for k = 0 to r do
    for j = 0 to q do
        for i = 0 to p do

            loc_num = loc_num+1;              // Local basis function number

            R[loc_num] = N[p+1-i]*M[q+1-j]*L[r+1-k]...
                ... * B[ni-i][nj-j][nk-k][4];      // Function numerator
            sum_tot = sum_tot + R[loc_num];        // Function denominator

            dR_dxi[loc_num][1] = dN_dxi[p+1]*M[q+1-j]*L[r+1-k]...
                ... * B[ni-i][nj-j][nk-k][4];          // Derivative num.
            sum_xi = sum_xi + dR_dxi[loc_num][1];      // Derivative denom.
            dR_dxi[loc_num][2] = N[p+1]*dM_deta[q+1-j]*L[r+1-k]...
                ... * B[ni-i][nj-j][nk-k][4];          // Derivative num.
            sum_eta = sum_eta + dR_dxi[loc_num][2];    // Derivative denom.
            dR_dxi[loc_num][3] = N_dx[p+1]*M[q+1-j]*dL_dzeta[r+1-k]...
                ... * B[ni-i][nj-j][nk-k][4];          // Derivative num.
            sum_zeta = sum_zeta + dR_dxi[loc_num][1]; // Derivative denom.
        end
    end
end

// Divide by denominators to complete definitions of functions
//    and derivatives w.r.t. parametric coordinates
for loc_num = 1 to nen do
    R[loc_num] = R[loc_num]/sum_tot;

    dR_dxi[loc_num][1] = (dR_dxi[loc_num][1]*sum_tot...
        ... - R[loc_num]*sum_xi) / sum_tot²
    dR_dxi[loc_num][2] = (dR_dxi[loc_num][2]*sum_tot...
        ... - R[loc_num]*sum_eta) / sum_tot²
    dR_dxi[loc_num][3] = (dR_dxi[loc_num][3]*sum_tot...
        ... - R[loc_num]*sum_zeta) / sum_tot²
end
```

Algorithm 3: Shape function routine: Part III

```
// Gradient of mapping from parameter space to physical space
loc_num=0;
for k = 0 to r do
    for j = 0 to q do
        for i = 0 to p do
            loc_num=loc_num+1;
            for aa = 1 to 3 do
                for bb = 1 to 3 do
                    dx_dxi[aa][bb] = dx_dxi[aa][bb]...
                        ... + B[ni-i][nj-j][nk-k][aa]*dR_dxi[loc_num][bb];
                end
            end
        end
    end
end

// Compute inverse of gradient
call inverse_Cramer(dx_dxi; dxi_dx);

// Compute derivatives of basis functions
//     with respect to physical coordinates
for loc_num = 1 to nen do
    for aa = 1 to 3 do
        for bb = 1 to 3 do
            dR_dx[loc_num][aa] = dR_dx[loc_num][aa]...
                ... + dR_dxi[loc_num][bb]*dxi_dx[bb][aa];
        end
    end
end

// Gradient of mapping from parent element to parameter space
dxi_dtildexi[1][1] = (KV_xi[ni+1]-KV_xi[ni])/2;
dxi_dtildexi[2][2] = (KV_eta[nj+1]-KV_eta[nj])/2;
dxi_dtildexi[3][3] = (KV_zeta[nk+1]-KV_zeta[nk])/2;

for aa = 1 to 3 do
    for bb = 1 to 3 do
        for cc = 1 to 3 do
            J_mat[aa][bb] = J_mat[aa][bb]...
                ... + dx_dxi[aa][cc]*dxi_dtildexi[cc][bb];
        end
    end
end

// Compute Jacobian determinant
call determinant(J_mat; J);
```

Appendix 3.B: Error estimates

FEA

Well established *a priori* approximation results exist for classical finite elements applied to elliptic problems (see, for example, the classic text by Ciarlet, 1978). Recall from above that a Sobolev space of order r is defined by

$$H^r(\Omega) = \{\mathbf{u} | D^\alpha \mathbf{u} \in L^2(\Omega), |\alpha| \leq r\}. \tag{3.B.1}$$

The norm associated with $H^r(\Omega)$ is given by

$$\|\mathbf{u}\|_r^2 = \sum_{|\alpha| \leq r} \int_\Omega \left(D^\alpha \mathbf{u}\right) \cdot \left(D^\alpha \mathbf{u}\right) \, d\mathbf{x}. \tag{3.B.2}$$

In classical FEA, the fundamental error estimate for the elliptic boundary value problem, expressed as a bound on the difference between the exact solution, $\mathbf{u}$, and the FEA solution, $\mathbf{u}^h$, takes the form

$$\|\mathbf{u} - \mathbf{u}^h\|_m \leq Ch^\beta \|\mathbf{u}\|_r, \tag{3.B.3}$$

where $\| \cdot \|_m$ and $\| \cdot \|_r$ are the norms corresponding to Sobolev spaces $H^m(\Omega)$ and $H^r(\Omega)$, respectively, h is a characteristic length scale related to the size of the elements in the mesh, $\beta = \min(p + 1 - m, r - m)$ where p is the polynomial order of the basis, and C is a constant that does not depend on $\mathbf{u}$ or h.

The term of interest in (3.B.3) is h^β. The mesh parameter, h, can be defined in several ways, with the specific definition affecting C. A fairly general definition is the diameter of the smallest circle (in two dimensions) or sphere (in three dimensions) that is large enough to circumscribe any element in the mesh. The **order of convergence**, β, expresses how the error changes under refinement of the mesh. In particular, if we use h-refinement to bisect each of the elements in the mesh (*i.e.*, h is replaced with $h/2$), we would expect the error to decrease by a factor of $(1/2)^\beta$.

NURBS

The extremely technical details of the process of obtaining a result analogous to (3.B.3) for NURBS can be found in Bazilevs *et al.*, 2006a. Here we present the basic ideas, but encourage the interested reader to consult the original publication.

For classical FEA polynomials, the result in (3.B.3) is obtained by first establishing the interpolation properties of the basis. Let Π_m be the projection operator from $H^m(\Omega)$ into the space spanned by the FEA basis. Then the optimal interpolate is the function

$$\eta^h = \Pi_m \mathbf{u} \tag{3.B.4}$$

such that

$$\|\mathbf{u} - \eta^h\|_m \leq \|\mathbf{u} - \mathbf{v}^h\|_m \quad \forall \mathbf{v}^h \in \mathcal{S}^h, \tag{3.B.5}$$

where $\mathcal{S}^h$ is the finite element space. To establish just how good this optimal approximation is (*i.e.*, to determine how can $\|\mathbf{u} - \boldsymbol{\eta}^h\|_m$ be bounded), we obtain a bound on each element, and then sum over all of the elements to get a global result. With this interpolation result in hand, the second step in the process is to relate the result of the Galerkin finite element method, $\mathbf{u}^h$, to the optimal interpolate, $\boldsymbol{\eta}^h$. In particular, it can be shown that the order of convergence of the finite element solution is the same as for the optimal interpolate. Taken together, these two results yield the the bound (3.B.3), which states that (up to a constant) Galerkin's method gives us the optimal result.

When we seek an analogous result for NURBS, we face several difficulties. The first is that the approximation properties of this rational basis are harder to determine than are those of a standard polynomial basis. In particular, note that the weights are determined by the geometry and so are out of our control when we attempt to approximate a field over that geometry and cannot be adjusted to improve the result. The second difficulty originates from the large support of the spline functions. Standard interpolation estimates seek to find a best fit within each element and then aggregate these results to obtain an approximation over the entire domain. This is non-trivial with the spline functions because the support of each function spans several elements, and so we cannot determine optimal values for the control variables by looking at each element individually. The issue is further complicated by the possibility of differing levels of continuity (and thus differing sizes of the the supports of the functions) throughout the domain.

To overcome the fact that the basis is rational rather than polynomial, we first note that the parameter space $\hat{\Omega}$ can be considered to be the unit cube $[0, 1]^d$. No generality is lost in this assumption as dividing a knot vector by a constant or adding a constant does not change the resulting physical domain in any way. Let us recall the definition of the rational basis from Chapter 2:

$$R_i(\xi) = \frac{N_i(\xi)w_i}{W(\xi)}, \tag{3.B.6}$$

with

$$W(\xi) = \sum_{i=1}^{n} N_i(\xi)w_i. \tag{3.B.7}$$

The important thing to note is that the weighting function[7], $W(\xi)$, does not change as we h-refine the mesh (it does not change under p-refinement either, though this is not the case we are interested in at present). While both the weights and the basis functions change, they do so in such a way as to leave $W(\xi)$ unaltered. Similarly, the geometrical mapping from the parameter space into the physical space, $\mathbf{F} : \hat{\Omega} \rightarrow \Omega$, does not change as we insert new knot values. See Figure 3.B.1. It remains exactly the same at all levels of refinement. To take advantage of this fact, we consider the function we wish to approximate, $\mathbf{u} : \Omega \rightarrow \mathbb{R}^\ell$. As the geometrical mapping is one-to-one, we can pull this back to the parametric domain to define $\hat{\mathbf{u}} = \mathbf{u} \circ \mathbf{F}^{-1} : \hat{\Omega} \rightarrow \mathbb{R}^\ell$. Lastly, we can lift the image of the function using the weighting function to define $\tilde{\mathbf{u}} = \{W\hat{\mathbf{u}}, W\} : \hat{\Omega} \rightarrow \mathbb{R}^{\ell+1}$. Recalling that we obtain the rational basis in $\mathbb{R}^d$ by a projective transformation (equivalent to dividing by W) of a B-spline basis in $\mathbb{R}^{d+1}$, we see that the ability of the rational NURBS basis to approximate $\mathbf{u}$ on Ω is intimately related

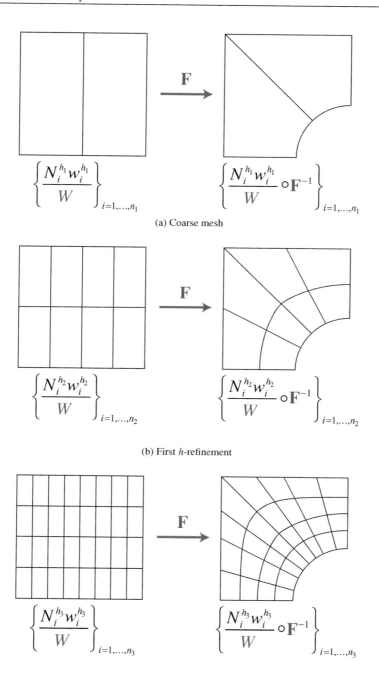

(a) Coarse mesh

(b) First h-refinement

(c) Second h-refinement

Figure 3.B.1 As we h-refine the mesh, the basis functions N_i and weights w_i change, but the geometrical mapping $\mathbf{F}$ and the weighting function W are *completely fixed* at the coarsest level of discretization. They do not change under refinement.

to the ability of the underlying B-spline basis to approximate $\tilde{\mathbf{u}}$ on $\hat{\Omega}$. Thus we have reduced the problem of understanding a rational basis on a general domain to that of understanding a polynomial basis on the unit cube.

The second hurdle is more technical. The fact that each function has support over many elements and that the continuity across the various element boundaries can vary from one boundary to the next greatly complicates matters compared with the classical case. Bazilevs et al., 2006a address this difficulty by proving approximation results in so-called "bent" Sobolev spaces in which the continuity varies throughout the domain. They prove a sequence of lemmas leading up to an approximation result that includes not only the norm in these bent Sobolev spaces of the function $\mathbf{u}$ being approximated, but also the gradient of the mapping, $\nabla\mathbf{F}$. This last term presents no problem because, as already discussed, it does not change as the mesh is refined, and thus does not affect the rate of convergence. The resulting approximation result is: Let k and l be integer indices such that $0 \le k \le l \le p + 1$, and let $u \in H^l(\Omega)$; then

$$\sum_{e=1}^{n_{el}} |u - \Pi_k u|^2_{H^k(\Omega^e)} \le C \sum_{e=1}^{n_{el}} h_e^{2(l-k)} \sum_{i=0}^{l} \|\nabla\mathbf{F}\|^{2(i-l)}_{L^\infty(\mathbf{F}^{-1}(\Omega^e))} |u|^2_{H^i(\Omega^e)}. \tag{3.B.8}$$

The constant C depends on p and the shape (but not size) of the domain Ω, as well as the shape regularity of the mesh. The factors involving the gradient of the mapping render the estimate dimensionally consistent.

Finally, with the approximation result of (3.B.8) in hand, establishing the manner in which the isogeometric analysis solution, $\mathbf{u}^h$, relates to the optimal interpolate, $\boldsymbol{\eta}^h$, proceeds exactly as in the classical case. Combining these results yields the desired result: *The isogeometric analysis solution obtained using NURBS of order p has the same order of convergence as we would expect in a classical FEA setting using classical basis functions with a polynomial order of p.* This is an exceptionally strong result as it is independent of the order of continuity that the mesh possesses. That is, bisecting all of the elements in an FEA mesh (thus cutting the mesh parameter from h to $h/2$) requires the introduction of many more degrees-of-freedom than does bisection of the same number of NURBS elements while maintaining $p - 1$ continuity (see Section 2.1.4 of the previous Chapter). This means that NURBS can converge at the same rate as FEA polynomials, while remaining much more efficient.

Notes

1. The "Bubnov" in Bubnov–Galerkin signifies the fact that the weighting and trial functions come from the same underlying space. This is in contrast to a Petrov–Galerkin method in which the weighting and solution spaces may have little in common.
2. This prior knowledge lets us use a sparse storage format for the global matrices. See Hughes, 2000.
3. The term "shape function" and "basis function" are often used interchangeably.
4. The term "least-squares" refers to the nature of the formulation. There is no reason to believe that the resulting solution will be a least-squares fit of the unknown exact solution.

5. We will discuss the B-spline case here, but it is crucial to note that if we were to use NURBS rather than B-splines, all of the relationships in this section must hold for the *projective* control points and *projective* control variables.
6. In Kagan *et al.*, 2003, a similar approach was taken for B-spline *surfaces*. Here we extend that to NURBS *solids*.
7. Do not confuse this use of the term "weighting function" with the unrelated use of the same terminology in Galerkin's method.

4

Linear Elasticity

As a first example of an application of NURBS-based isogeometric analysis, let us consider the problem of linear elasticity. This is a very classical subject with a rich history, and it is one that is particularly well suited to examination by isogeometric analysis – not only for the geometrical accuracy that it provides, but also for the high quality of the stress fields resulting from the use of C^1-continuous bases. We will restrict ourselves at present to the case of linear elastostatics and equilibrium solutions.

Before examining the equations, let us introduce some simplifying notation. First, the reader is reminded that indices i, j, k, and l take on values $1, \ldots, d$, where d is the number of spatial dimensions. Unlike the previous chapter's example of the Laplace equation, in linear elasticity the solution field will be vector-valued, with u_i referring to the ith component of vector $\mathbf{u}$. Moreover, differentiation will be denoted by a comma (e.g., $u_{i,j} = u_{i,x_j} = \partial u_i / \partial x_j$). Lastly, we will employ a summation convention applying to i, j, k, and l, in which repeated indices imply summation (e.g., in $\mathbb{R}^3$, $u_{i,jj} = u_{i,11} + u_{i,22} + u_{i,33} = \partial^2 u_i / \partial x^2 + \partial^2 u_i / \partial y^2 + \partial^2 u_i / \partial z^2$). We will assume this convention to be in place for these four indices unless otherwise stated throughout the remainder of the book.

In the case of a general nonsymmetric tensor, $\mathbf{A} = [A_{ij}]$, we use parentheses around the indices to denote its symmetric part and square brackets around the indices to denote its skew-symmetric part. Thus, $A_{ij} = A_{(ij)} + A_{[ij]}$ where

$$A_{(ij)} = A_{(ji)} \equiv \frac{A_{ij} + A_{ji}}{2}, \tag{4.1}$$

$$A_{[ij]} = -A_{[ji]} \equiv \frac{A_{ij} - A_{ji}}{2}, \tag{4.2}$$

which is known as the Euclidean decomposition. Note that if $\mathbf{A}$ is a nonsymmetric tensor, and $\mathbf{B} = [B_{ij}] = [B_{(ij)}]$ is a symmetric tensor, then

$$A_{ij} B_{ij} = A_{(ij)} B_{ij}, \tag{4.3}$$

which also has the corollary $A_{[ij]} B_{ij} = 0$. We will use these properties to reduce redundant computations below.

Isogeometric Analysis: Toward Integration of CAD and FEA   by J. A. Cottrell, T. J. R. Hughes, Y. Bazilevs
© 2009, John Wiley & Sons, Ltd

4.1 Formulating the equations of elastostatics

Let u_i denote the **displacement vector** and let $\boldsymbol{\sigma} = [\sigma_{ij}]$ denote the Cartesian components of the Cauchy **stress tensor**. The **infinitesimal strain tensor**, $\boldsymbol{\epsilon} = [\epsilon_{ij}]$ is defined to be the symmetric part of the displacement gradient,

$$\epsilon_{ij} = u_{(i,j)} \equiv \frac{u_{i,j} + u_{j,i}}{2}. \tag{4.4}$$

The constitutive law relating this strain tensor to the aforementioned stress tensor is the generalized Hooke's law, given by

$$\sigma_{ij} = c_{ijkl}\epsilon_{kl}, \tag{4.5}$$

where the c_{ijkl}'s are **elastic coefficients**, which are given functions of $\mathbf{x}$. If the c_{ijkl}'s are constant throughout the domain, the body is said to be "homogeneous." The elastic coefficients are assumed to satisfy several important properties. The first three,

$$c_{ijkl} = c_{klij}, \tag{4.6}$$

$$c_{ijkl} = c_{jikl}, \tag{4.7}$$

$$c_{ijkl} = c_{ijlk}, \tag{4.8}$$

relate to symmetry, with (4.6) referred to as "major symmetry," while (4.7) and (4.8) are referred to as "minor symmetries." The coefficients also satisfy positive-definiteness of the form

$$c_{ijkl}\psi_{ij}\psi_{kl} \geq 0, \tag{4.9}$$

$$c_{ijkl}\psi_{ij}\psi_{kl} = 0 \iff \psi_{ij} = 0, \tag{4.10}$$

for all symmetric $\boldsymbol{\psi}$ (i.e., $\psi_{ij} = \psi_{ji}$). These properties, combined with appropriate displacement boundary conditions, lead to the symmetry and positive-definiteness of the stiffness matrix $\mathbf{K}$. Furthermore, (4.7) implies the symmetry of the stress tensor $\boldsymbol{\sigma}$. From a physical point of view, the symmetry of the stress tensor derives from the conservation of angular momentum.

In the examples in this chapter, we will assume that the body is homogeneous. Additionally, we will assume it is **isotropic**. That is, the elastic coefficients have the form

$$c_{ijkl} = \lambda \delta_{ij}\delta_{kl} + \mu(\delta_{ik}\delta_{jl} + \delta_{il}\delta_{jk}), \tag{4.11}$$

where the **Kronecker delta** is defined by

$$\delta_{ij} = \begin{cases} 1 & i = j, \\ 0 & \text{otherwise.} \end{cases} \tag{4.12}$$

Constants λ and μ are the **Lamé parameters**. These are frequently expressed in terms of the **Young's modulus**, E, and **Poisson's ratio**, ν, as

$$\lambda = \frac{\nu E}{(1 + \nu)(1 - 2\nu)} \tag{4.13}$$

$$\nu = \frac{E}{2(1 + \nu)}. \tag{4.14}$$

4.1.1 Strong form

We can now formally state the strong form of the boundary value problem. Given $f_i : \Omega \to \mathbb{R}$, $g_i : \Gamma_{D_i} \to \mathbb{R}$, and $h_i : \Gamma_{N_i} \to \mathbb{R}$, find $u_i : \bar{\Omega} \to \mathbb{R}$ such that

$$\sigma_{ij,j} + f_i = 0 \quad \text{in } \Omega, \tag{4.15a}$$

$$u_i = g_i \quad \text{on } \Gamma_{D_i}, \tag{4.15b}$$

$$\sigma_{ij} n_j = h_i \quad \text{on } \Gamma_{N_i}, \tag{4.15c}$$

where σ_{ij} is defined in terms of u_i by (4.4) and (4.5).

Due to the fact that the unknown is a vector, observe that (4.15b) and (4.15c) represent generalizations of the Dirichlet and Neumann boundary conditions considered previously. We are applying these conditions in each direction independently and thus $\overline{\Gamma_{D_i} \cup \Gamma_{N_i}} = \Gamma$ and $\Gamma_{D_i} \cap \Gamma_{N_i} = \varnothing$ for $i = 1, \ldots, d$. In this context, g_i and h_i are referred to as "prescribed boundary displacements" and "tractions," respectively.

4.1.2 Weak form

Let us denote the trial solution space by $\mathcal{S}_i$ and the weighting space by $\mathcal{V}_i$. As before, each $u_i \in \mathcal{S}_i$ satisfies the Dirichlet condition $u_i = g_i$ on Γ_{D_i}, and each $w_i \in \mathcal{V}_i$ satisfies $w_i = 0$ on Γ_{D_i}. Proceeding as in Section 3.3.1 of Chapter 3, we multiply (4.15a) by a weighting function and integrate by parts to obtain a variational form of the problem: Given $f_i : \Omega \to \mathbb{R}$, $g_i : \Gamma_{D_i} \to \mathbb{R}$, and $h_i : \Gamma_{N_i} \to \mathbb{R}$, find $u_i \in \mathcal{S}_i$ such that for all $w_i \in \mathcal{V}_i$

$$\int_\Omega w_{(i,j)} \sigma_{ij} \, d\Omega = \int_\Omega w_i f_i \, d\Omega + \sum_{i=1}^d \left(\int_{\Gamma_{N_i}} w_i h_i \, d\Gamma \right). \tag{4.16}$$

Note that we have taken advantage of (4.3) in writing only the symmetric part of $w_{i,j}$ in the first term. In the last term, we have explicitly written the sum for clarity and to emphasize the fact that the domain of integration is actually changing for each of the d terms in the sum.

As we did in the previous chapter for the Laplace equation, we can rewrite (4.16) in a more concise form. Let $\mathcal{S} = \{\mathbf{u} | u_i \in \mathcal{S}_i\}$ and let $\mathcal{V} = \{\mathbf{w} | w_i \in \mathcal{V}_i\}$. The weak form of the problem becomes: Given $\mathbf{f} = \{f_i\}$, $\mathbf{g} = \{g_i\}$, and $\mathbf{h} = \{h_i\}$, find $\mathbf{u} \in \mathcal{S}$ such that for all $\mathbf{w} \in \mathcal{V}$

$$a(\mathbf{w}, \mathbf{u}) = L(\mathbf{w}), \tag{4.17}$$

where

$$a(\mathbf{w}, \mathbf{u}) = \int_{\Omega} w_{(i,j)} c_{ijkl} u_{(k,l)} \, d\Omega, \tag{4.18}$$

$$L(\mathbf{w}) = \int_{\Omega} w_i f_i \, d\Omega + \sum_{i=1}^{d} \left(\int_{\Gamma_{N_i}} w_i h_i \, d\Gamma \right). \tag{4.19}$$

4.1.3 Galerkin's method

To turn this weak statement of the problem into a system of algebraic equations, we again apply Galerkin's method and work in finite-dimensional subspaces $\mathcal{S}^h \subset \mathcal{S}$ and $\mathcal{V}^h \subset \mathcal{V}$. These subspaces are defined using the isoparametric NURBS basis, as before, but now with vector-valued control variables. Let us assume that we have determined a lifting $\mathbf{g}^h \in \mathcal{S}^h$, where $g_i^h|_{\Gamma_{D_i}} = g_i$, such that for all $\mathbf{u}^h \in \mathcal{S}^h$ we have the decomposition

$$\mathbf{u}^h = \mathbf{v}^h + \mathbf{g}^h, \tag{4.20}$$

where $\mathbf{v}^h \in \mathcal{V}^h$. The Galerkin approximation of (4.17) is given by: find $\mathbf{u}^h = \mathbf{v}^h + \mathbf{g}^h \in \mathcal{S}^h$ such that for all $\mathbf{w}^h \in \mathcal{V}^h$

$$a(\mathbf{w}^h, \mathbf{v}^h) = L(\mathbf{w}^h) - a(\mathbf{w}^h, \mathbf{g}^h). \tag{4.21}$$

We can make this more precise by defining $\boldsymbol{\eta} = \{1, \ldots, n_{np}\}$ to be the set containing the indices of all of the functions in the NURBS basis that defines the geometry. Similarly, let $\boldsymbol{\eta}_{g_i} \subset \boldsymbol{\eta}$ be the set containing the indices of all of the basis functions that are non-zero on Γ_{D_i}. Thus we can write the ith component of $\mathbf{u}^h \in \mathcal{S}^h$ as

$$u_i^h = \sum_{A \in \boldsymbol{\eta} - \boldsymbol{\eta}_{g_i}} N_A d_{iA} + \sum_{B \in \boldsymbol{\eta}_{g_i}} N_B g_{iB} = \sum_{A \in \boldsymbol{\eta} - \boldsymbol{\eta}_{g_i}} N_A d_{iA} + g_i^h, \tag{4.22}$$

where $\boldsymbol{\eta} - \boldsymbol{\eta}_{g_i}$ denotes set subtraction. Equation (4.22) is simply the vector-valued generalization of (3.28) from the previous chapter, with d_{iA} being the ith component of control variable $\mathbf{d}_A$. Similarly, the ith component of $\mathbf{w}^h \in \mathcal{V}^h$ is given by

$$w_i^h = \sum_{A \in \boldsymbol{\eta} - \boldsymbol{\eta}_{g_i}} N_A c_{iA}. \tag{4.23}$$

We can now represent $\mathbf{u}^h$ and $\mathbf{w}^h$ by

$$\mathbf{u}^h = u_i^h \mathbf{e}_i \quad \text{and} \quad \mathbf{w}^h = w_i^h \mathbf{e}_i, \tag{4.24}$$

where (in $\mathbb{R}^3$)

$$\mathbf{e}_1 = \begin{pmatrix} 1 \\ 0 \\ 0 \end{pmatrix}, \quad \mathbf{e}_2 = \begin{pmatrix} 0 \\ 1 \\ 0 \end{pmatrix}, \quad \text{and} \quad \mathbf{e}_3 = \begin{pmatrix} 0 \\ 0 \\ 1 \end{pmatrix}. \tag{4.25}$$

Our goal is to use (4.24) with (4.22) and (4.23) in (4.21) to obtain a matrix formulation of the problem. The extra complexity that we encounter due to the fact that the unknowns are vector-valued is that the number of equations we must solve, n_{eq}, is much larger than the number of functions, n_{np}. For a scalar problem, we used the connectivity array IEN to relate a local function number and an element number to a global basis function, and we had one equation for each such global function. Now, we have d equations for each global function. Thus, we introduce a second level of connectivity through the ID array, which relates a degree-of-freedom number $i = 1, \ldots, d$ and a global function number $A \in \boldsymbol{\eta} - \boldsymbol{\eta}_{g_i}$ and returns the equation number $P = \mathrm{ID}(i, A)$. Furthermore, to go from an element e with local shape function a and degree-of-freedom i, we can obtain the global equation number by composing ID and IEN to obtain $P = \mathrm{ID}(i, \mathrm{IEN}(a, e))$. Because we will use it so frequently, it is most convenient to define one final, three-dimensional connectivity array, LM, that incorporates both ID and IEN, such that $P = \mathrm{LM}(i, a, e) = \mathrm{ID}(i, \mathrm{IEN}(a, e))$.

We now build the matrix equation

$$\mathbf{Kd} = \mathbf{F}, \tag{4.26}$$

where

$$\mathbf{K} = [K_{PQ}], \tag{4.27}$$

$$\mathbf{d} = \{d_Q\}, \tag{4.28}$$

$$\mathbf{F} = \{F_P\}, \tag{4.29}$$

with

$$P = \mathrm{ID}(i, A) \quad \text{and} \quad Q = \mathrm{ID}(j, B), \tag{4.30}$$

such that

$$K_{PQ} = a(N_A \mathbf{e}_i, N_B \mathbf{e}_j), \tag{4.31}$$

$$F_P = L(N_A \mathbf{e}_i), \tag{4.32}$$

and

$$d_Q = d_{jB}. \tag{4.33}$$

4.1.4 Assembly

Recall that in the previous chapter we assembled the global system by looping – not through the basis functions as a glance at the formulation might imply – but through the elements, constructing local stiffness matrices and force vectors, which are then assembled into the global system by means of the connectivity arrays. We have presented sufficient information for such an approach to be implemented for linear elastostatics as well. In practice, however, the specific structure of the problem can be exploited to reduce the computational burden.

Let us define the **strain vector** for the case of $d = 3$ as

$$\epsilon(\mathbf{u}) = \left\{\begin{array}{c} u_{1,1} \\ u_{2,2} \\ u_{3,3} \\ u_{2,3} + u_{3,2} \\ u_{3,1} + u_{1,3} \\ u_{1,2} + u_{2,1} \end{array}\right\}. \tag{4.34}$$

Note that, according to our previous notational convention, $\epsilon = [\epsilon_{ij}]$ was a strain matrix. We will no longer need this matrix explicitly, and consequently reserve ϵ for the strain vector as in (4.34). Let us now condense the fourth rank tensor of elastic coefficients into the matrix

$$\mathbf{D} = \begin{bmatrix} D_{11} & D_{12} & D_{13} & D_{14} & D_{15} & D_{16} \\ & D_{22} & D_{23} & D_{24} & D_{25} & D_{26} \\ & & D_{33} & D_{34} & D_{35} & D_{36} \\ & & & D_{44} & D_{45} & D_{46} \\ & & & & D_{55} & D_{56} \\ & \text{symmetric} & & & & D_{66} \end{bmatrix}, \tag{4.35}$$

where

$$D_{IJ} = c_{ijkl}, \tag{4.36}$$

with the indices are related as indicated in Table 4.1. With ϵ and $\mathbf{D}$ in hand (committing the same notational crime as with the strain vector) we define the **stress vector** to be

$$\sigma = \left\{\begin{array}{c} \sigma_{11} \\ \sigma_{22} \\ \sigma_{33} \\ \sigma_{23} \\ \sigma_{31} \\ \sigma_{12} \end{array}\right\} = \mathbf{D}\epsilon(\mathbf{u}). \tag{4.37}$$

Table 4.1 For the elastic tensor, we collapse indices i and j into I, and we collapse k and l into J. If we are interested in the value of I in the first column, then we read i from the second column and j from the third. If, instead, we seek J from the first column, then the second column is to be read as k and the third as l

I/J	i/k	j/l
1	1	1
2	2	2
3	3	3
4	2	3
4	3	2
5	1	3
5	3	1
6	1	2
6	2	1

This is often referred to as the **Voight notation**. Note from (4.34) that we have removed the factors of one-half from the shearing components (the terms containing cross-derivatives). This has been accounted for in (4.37) as we now have one contribution from each shearing term without any one-half, as opposed to two identical contributions from terms containing the factor.

All of these new definitions allow us to rewrite (4.18) as

$$a(\mathbf{w}, \mathbf{u}) = \int_{\Omega} \epsilon(\mathbf{w})^T \mathbf{D}\epsilon(\mathbf{u}) \, d\Omega. \tag{4.38}$$

Furthermore, we can extend the notation by noting that

$$\epsilon(N_A \mathbf{e}_i) = \mathbf{B}_A \mathbf{e}_i, \tag{4.39}$$

where[1]

$$\mathbf{B}_A = \begin{bmatrix} N_{A,1} & 0 & 0 \\ 0 & N_{A,2} & 0 \\ 0 & 0 & N_{A,3} \\ 0 & N_{A,3} & N_{A,2} \\ N_{A,3} & 0 & N_{A,1} \\ N_{A,2} & N_{A,1} & 0 \end{bmatrix}. \tag{4.40}$$

Thus, we may rewrite the entries of the global stiffness matrix as

$$K_{PQ} = \mathbf{e}_i^T \int_{\Omega} \mathbf{B}_A^T \mathbf{D} \mathbf{B}_B \, d\Omega \, \mathbf{e}_j, \tag{4.41}$$

where the indices are related by (4.30).

We will build the sparse global stiffness matrix and force vector by looping through the elements and constructing dense local stiffness matrices and force vectors, which are then assembled into the global system. With d spatial dimensions (*i.e.*, degrees-of-freedom per control variable) and n_{en} local shape functions, we calculate the entries to the local stiffness matrix on element Ω^e as

$$k_{pq}^e = \mathbf{e}_i^T \int_{\Omega^e} \mathbf{B}_a^T \mathbf{D} \mathbf{B}_b \, d\Omega \, \mathbf{e}_j, \tag{4.42}$$

where

$$p = d(a-1) + i \quad \text{and} \quad q = d(b-1) + j. \tag{4.43}$$

Similarly, the elements of the local force vector are given by

$$f_p^e = \int_{\Omega^e} N_a f_i \, d\Omega + \int_{\Gamma_{h_i}^e} N_a h_i \, d\Omega - \sum_{q=1}^{d \cdot n_{en}} k_{pq}^e g_q^e, \tag{4.44}$$

where $\Gamma_{h_i}^e$ is the intersection of the boundary of the element with Γ_{h_i}, and $g_q^e = g_Q$ with $q = d(b-1) + j$ and $Q = \text{LM}(j, b, e)$. Of course, in practice we compute the integrals using Gaussian quadrature.

The proliferation of indices can be a bit confusing at first. We have already introduced IEN, which relates the local shape function numbers for a given element number to their corresponding global shape function numbers, ID, which relates the global shape function number and a degree-of-freedom number to a global equation number, and LM, which combines the previous two to relate the element number, degree-of-freedom number, and local shape function number to the appropriate global equation number. Without defining any new array, we have used (4.43) to connect the local shape function numbers with a local equation number. In practice, it is common to overload the LM array so that it can accommodate either two indices, the element number and local equation number, or three indices, the element number, local shape function number, and degree-of-freedom number, and return the local equation number in both cases, such that

$$\text{LM}(p, e) = \text{LM}(i, a, e), \tag{4.45}$$

with p as in (4.43). See Appendix A at the end of the book for a detailed discussion of these data structures, and see Appendix 4.C for an element assembly routine.

4.2 Infinite plate with circular hole under constant in-plane tension

As stated in the introduction to this chapter, NURBS are particularly well suited to linear elasticity. It is obvious that representing geometry accurately at all levels of discretization should lead to improved accuracy across all meshes as compared with less geometrically accurate methods. Furthermore, as we have seen, the standard formulation for linear elasticity uses the displacements as the unknown degrees-of-freedom. In practice, however, it is often the case that the quantity of interest is not the displacement but the stress. The stress is a function of the gradient of the displacement, and so any approach using elements that are only C^0 across element boundaries results in stress values being undefined at these element boundaries. Alternatively, a C^1 NURBS basis results in unambiguous, continuous stresses across element boundaries.

In this two-dimensional example, we present the NURBS-based isogeometric analysis of a problem in solid mechanics having an exact solution: an infinite plate with a circular hole under constant in-plane tension at infinity. We will systematically explore h- and k-refinement. The infinite plate is modeled by a finite quarter plate. The exact solution (Gould, 1999, pp. 120–123),

$$\sigma_{rr}(r, \theta) = \frac{T_x}{2}\left(1 - \frac{R^2}{r^2}\right) + \frac{T_x}{2}\left(1 - 4\frac{R^2}{r^2} + 3\frac{R^4}{r^4}\right)\cos 2\theta, \tag{4.46}$$

$$\sigma_{\theta\theta}(r, \theta) = \frac{T_x}{2}\left(1 + \frac{R^2}{r^2}\right) - \frac{T_x}{2}\left(1 + 3\frac{R^4}{r^4}\right)\cos 2\theta, \tag{4.47}$$

$$\sigma_{r\theta}(r, \theta) = -\frac{T_x}{2}\left(1 + 2\frac{R^2}{r^2} - 3\frac{R^4}{r^4}\right)\sin 2\theta, \tag{4.48}$$

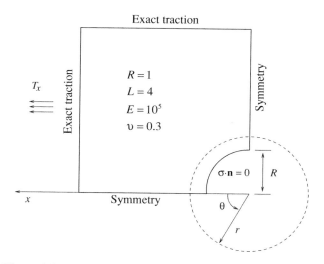

Figure 4.1 Elastic plate with a circular hole: problem definition.

where T_x is the magnitude of the applied stress for the infinite plate case, is applied as a Neumann boundary condition at the boundary of the finite quarter plate. The setup is illustrated in Figure 4.1. R is the radius of the hole, L is the edge length of the finite quarter plate, E is Young's modulus, and v is Poisson's ratio. A rational quadratic basis is the minimum order capable of representing a circular hole. The coarsest mesh, $\Xi \times \mathcal{H}$, is defined by the knot vectors

$$\Xi = \{0, 0, 0, 0.5, 1, 1, 1\}, \tag{4.49}$$

$$\mathcal{H} = \{0, 0, 0, 1, 1, 1\}, \tag{4.50}$$

The exact geometry is represented with only two elements, as shown in Figure 4.2a. The corresponding control net is shown in Figure 4.2b. A repeated control point is responsible for the upper left-hand corner.[2] This is not the only way to model the geometry, and may not even be the most preferable. This choice allows C^1-continuity across all interior element boundaries, but at the expense of having a singularity in the inverse of the geometrical mapping at the corner where the control points are repeated. This singularity does not create problems in the analysis, however, as we need never place a quadrature point at the location of the singularity. The fact that we still get excellent results demonstrates the overall robustness of this basis.

The first six meshes used in the analysis are shown in Figure 4.3. Contour plots of results obtained on meshes 1, 4, and 7 are presented in Figure 4.4. The applied stress is $T_x = 10$ and the contours show that the stress concentration of $\sigma_{xx} = 30$ at the edge of the hole (i.e., at $r = R, \theta = 3/2\pi$) is obtained as the mesh is refined.

Convergence results in the L^2-norm of stresses are shown in Figure 4.5. The cubic and quartic NURBS are obtained by order elevation of the quadratic NURBS on the coarsest mesh. Since the parameterization of the geometrical mapping does not change, the h-refinement algorithm (knot insertion) generates identical meshes for all polynomial orders. As a result, the continuity

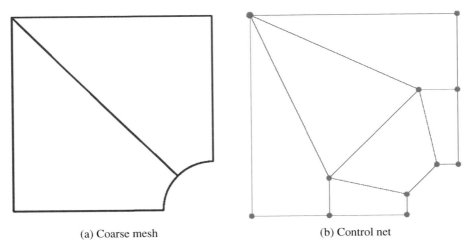

(a) Coarse mesh   (b) Control net

Figure 4.2   Mesh and control net for the elastic plate with circular hole. It is the knot values that define the elements, not the control net. Two control points at the same location create the upper-left corner. Coalescing control points is analogous to the degenerated element concept. See Hughes, 2000, chapter 3.

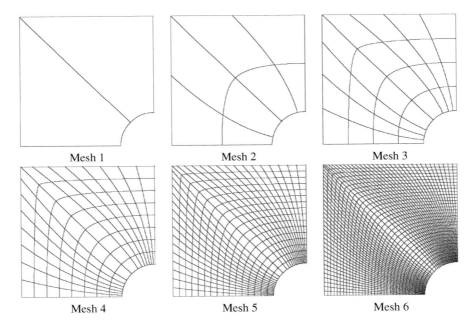

Mesh 1   Mesh 2   Mesh 3

Mesh 4   Mesh 5   Mesh 6

Figure 4.3   Elastic plate with circular hole. Meshes produced by *h*-refinement (knot insertion).

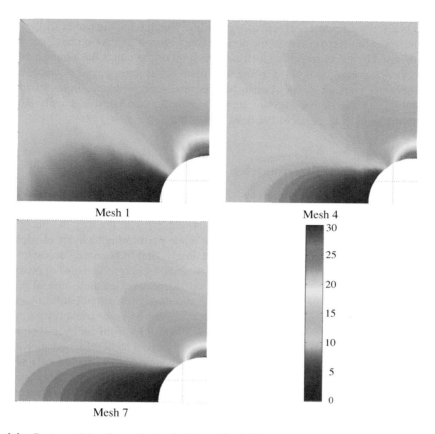

Figure 4.4 Contour plots of σ_{xx} obtained with quadratic NURBS. The applied stress is $T_x = 10$ and the stress concentration is $\sigma_{xx} = 30$ at $r = R, \theta = 3/2\pi$.

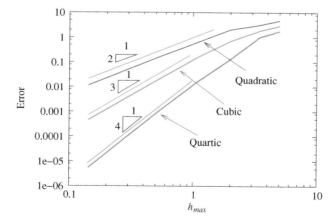

Figure 4.5 Error measured in the L^2-norm of stress versus the largest element diameter found in the mesh.

of the basis is C^{p-1} everywhere, except along the line which joins the center of the circular edge with the upper left-hand corner of the domain (see Figure 4.2a). Along this line, the basis is C^1 as is dictated by the coarsest mesh (from which all of the others have been generated). The mesh parameter, h_{max}, is defined as the maximum distance, in physical space, between diagonally opposite knot locations. As can be seen, the L^2-convergence rates of stress for quadratic, cubic, and quartic NURBS are approximately 2, 3, and 4, respectively. This is in keeping with the analytical results of Bazilevs *et al.*, 2006a.

4.3 Thin-walled structures modeled as solids

Analysis of thin-walled structures presents several challenges beyond those found in their more uniformly-dimensioned thick counterparts. They are notoriously sensitive to imperfections. Anyone who has crushed an aluminum can is familiar with the way in which a small dent in its side greatly reduces the buckling load. When performing analysis of such structures, accurate geometrical representations are absolutely vital to obtaining accurate results. Additionally, they are prone to boundary layer phenomena, locking, and a host of other obstacles to obtaining accurate numerical solutions. In this section we examine several problems involving thin-walled structures. We treat them as the three-dimensional solid objects that they are, despite the fact that classical finite element analysis of such structures traditionally employs surface-based shell elements. Our approach is in keeping with our overall philosophy of geometrical exactness at all levels of discretization (though a NURBS based shell analysis is a very promising approach as well – for an excellent comprehensive review of approaches to shell modeling, see Bischoff *et al.*, 2004). We see that, not only do NURBS accurately represent the geometry, but they resolve solution features quite well, even on surprisingly coarse discretizations. Convergence to analytical or benchmark solutions is observed in all cases.

4.3.1 Thin cylindrical shell with fixed ends subjected to constant internal pressure

The problem setup and a radial displacement profile for this problem are shown in Figure 4.6. The fixed ends create boundary layers which are difficult to accurately capture with low order finite element methods. The exact thin shell theory solution given in Timoshenko and Woinowsky-Krieger, 1959, pp. 476–477, is obtained under the assumption of plane stress. For the fully three-dimensional treatment of the problem with fixed-end conditions, we modify the Timoshenko and Woinowsky-Krieger, 1959 solution to obtain

$$u(x) = -\frac{PR^2}{\tilde{E}t}(1 - C_1 \sin \beta x \sinh \beta x - C_2 \cos \beta x \cosh \beta x) \quad (4.51)$$

$$x \in (-L/2, L/2),$$

$$C_1 = \frac{\sin \alpha \cosh \alpha - \cos \alpha \sinh \alpha}{\sinh \alpha \cosh \alpha + \sin \alpha \cos \alpha}, \quad (4.52)$$

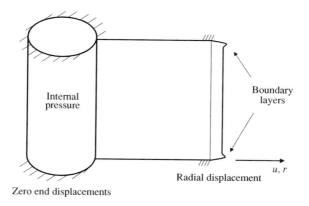

Figure 4.6   Thin cylindrical shell. Problem statement and displacement profile.

$$C_2 = \frac{\cos\alpha\,\sinh\alpha + \sin\alpha\,\cosh\alpha}{\sinh\alpha\,\cosh\alpha + \sin\alpha\,\cos\alpha}, \tag{4.53}$$

$$\beta = \left(\frac{\tilde{E}t}{4R^2 D}\right)^{1/4}, \qquad \alpha = \frac{\beta L}{2}, \qquad D = \frac{\tilde{E}t^3}{12(1-\tilde{\nu}^3)}, \tag{4.54}$$

where

$$\tilde{E} = \frac{E}{1-\nu^2}, \quad \text{and} \quad \tilde{\nu} = \frac{\nu}{1-\nu}. \tag{4.55}$$

The geometry of the shell is shown in Figure 4.7. Note that the radius to thickness ratio is 100, resulting in solid NURBS elements with a very high aspect ratio. A template for building such geometries is given in Appendix 4.B at the end of this chapter.

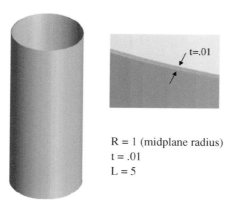

R = 1 (midplane radius)
t = .01
L = 5

Figure 4.7   Thin cylindrical shell geometry.

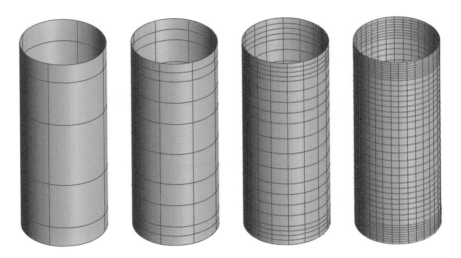

Figure 4.8   Thin cylindrical shell surface meshes. Meshes 1–4.

The meshes are depicted in Figures 4.8 and 4.9. In Figure 4.8, note that we have biased the mesh toward capturing the boundary layer by creating a coarse mesh with smaller elements near the fixed ends. Subsequent uniform refinement results in a high percentage of the total elements in the region containing the layer we wish to capture. This has been accomplished by defining the initial geometry using the knot vector $\mathcal{Z} = \{0, 0, 0, 1, 1, 1\}$, with the corresponding control points chosen such that the parameterization is linear in the axial direction. The knot values $\{1/10, 1/3, 2/3, 9/10\}$ were then inserted to define the elements of the coarsest mesh used for

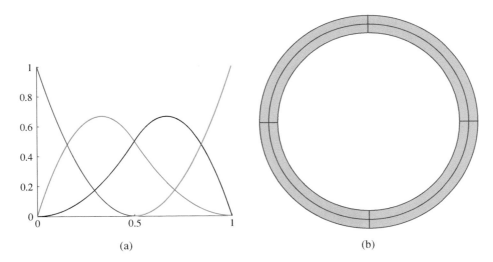

(a)   (b)

Figure 4.9   Thin cylindrical shell. (a) Quadratic basis functions through the thickness. (b) End view of the coarse mesh (not to scale).

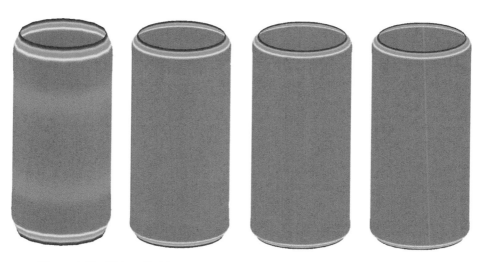

Figure 4.10   Thin cylindrical shell. Meshes 2–5. Radial displacement contour plots.

analysis. All further refinements were performed by bisecting the elements (in the surface, not through the thickness) via knot insertion.

The through-thickness mesh resolution for Mesh 1 is shown in Figure 4.9. Note that there are two elements in the radial direction and four in the circumferential direction. As the surface mesh is refined, the number of elements in the circumferential direction increases accordingly. However, the number of elements in the radial direction is fixed at two throughout the refinement process. The functions employed for this problem are quadratic NURBS. Their appearance in the radial, or through-thickness, direction is presented in Figure 4.9a. Radial displacement contours are presented in Figure 4.10 for Meshes 2–5. Note that, for all meshes, pointwise axisymmetric response is obtained. Note also the appearance of boundary layers. The convergence of the radial displacement profile is shown in Figure 4.11. Mesh 1 is too coarse to represent the boundary layers and the plateau between them. Mesh 3 picks up the plateau, but the boundary layers are still not accurately captured. The Mesh 5 solution is indistinguishable from the exact shell theory solution. In the detail on the right, the exact shell solution and Mesh 5 solution are seen to overlap in the boundary layer region.

4.3.2 The shell obstacle course

The so-called "shell obstacle course" consists of three problems: the Scordelis-Lo roof, the pinched hemisphere, and the pinched cylinder. These problems, and their relevance to the assessment of shell analysis procedures, have been discussed extensively in the literature. The problem descriptions presented in Figure 4.12 are adapted from Belytschko *et al.*, 1985. Two quadratic NURBS elements are employed in the through-thickness direction (see Figure 4.9), whereas *h*-refinement and *k*-refinement are utilized for surface meshing. Quadratic through quintic surface NURBS are employed in the convergence analysis of all

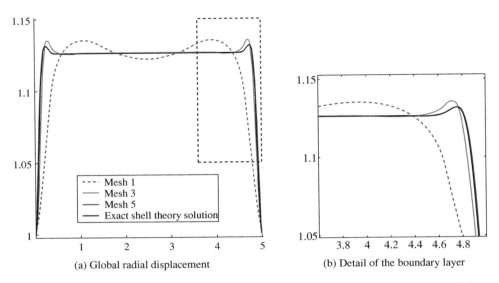

(a) Global radial displacement   (b) Detail of the boundary layer

Figure 4.11 Thin cylindrical shell. Convergence of the radial displacement to the exact shell theory solution. The Mesh 5 solution is indistinguishable from the exact solution.

cases. In one case, the pinched hemisphere, a one-element surface solution, starting with rational quadratics, the lowest-order NURBS capable of exactly representing spherical geometry, and culminating with tenth-order NURBS, is used to assess convergence. This analysis has the flavor of what are usually referred to as "spectral methods," which are higher-order accurate procedures, typically utilized for performing detailed studies of geometrically simple but physically complex phenomena, such as turbulence. (See Canuto *et al.*, 1988 and Moin, 2001 for detailed descriptions and applications of spectral methods.) Convergence is assessed by comparing the displacement of certain points in the shell with benchmark solutions presented in Belytschko *et al.*, 1985. Sample contour plots of the solutions for quadratic elements on the finest meshes studied are presented in Figure 4.13. Note that in each case the contours are very smooth.

4.3.2.1 Scordelis-Lo roof

The Scordelis-Lo roof is subjected to gravity loading. The ends are supported by fixed diaphragms and the side edges are free (see Figure 4.12a). The vertical displacement of the mid-point of the side edge is the quantity used to assess convergence. The second, fourth, and sixth meshes used in the study are shown in Figure 4.14. These meshes have 2, 8, and 32 surface elements per side, respectively. Due to symmetry, only one quadrant is meshed. Convergence of the displacement to the benchmark value is shown in Figure 4.15. In all cases, convergence is quite rapid. For the higher-order cases, namely, quartic and quintic, even one element provides a very accurate solution.

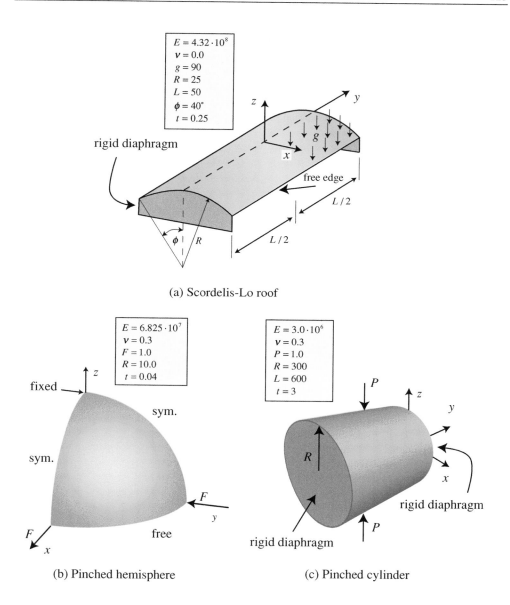

(a) Scordelis-Lo roof

(b) Pinched hemisphere

(c) Pinched cylinder

Figure 4.12 Shell obstacle course. Problem descriptions and data.

4.3.2.2 Pinched hemisphere

In the pinched hemisphere, equal and opposite concentrated forces are applied at antipodal points of the equator. The equator is otherwise considered to be free (see Figure 4.12b). The control points, knot vectors, and polynomial orders for the coarsest mesh are tabulated in Appendix 4.A at the end of this chapter. The second, fourth, and sixth meshes are shown in Figure 4.16. Due to symmetry, only one quadrant is meshed. Convergence of the displacement

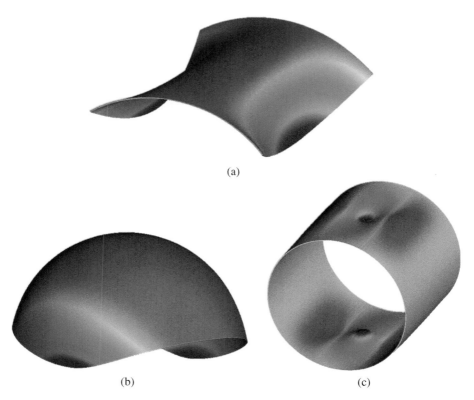

(a)

(b)   (c)

Figure 4.13 Shell obstacle course. (a) Scordelis-Lo roof in deformed configuration (scaling factor of 200 used). Contours of displacement in the direction of the gravity load. (b) Pinched hemisphere in deformed configuration (scaling factor of 33.3 used). Contours of displacement in the direction of the inward directed point load. (c) Pinched cylinder in deformed configuration (scaling factor of 3×10^6 used). Contours of displacement in the direction of the point load. Notice the highly localized displacement in the vicinity of the load.

at the location of the inward directed load is presented in Figure 4.17. The quadratic case converges very slowly, which is not surprising as quadratic, fully-integrated, solid C^0 finite elements are known to "lock" in shell analysis. Cubic solid C^0 finite elements also exhibit locking in similar circumstances but in the present case cubic NURBS behave reasonably well. Figure 4.18 presents convergence of the displacement for one surface element meshes. Notice that the lowest-order meshes lock but eventually accurate results are obtained. One tenth-order NURBS surface element is seen to provide an essentially exact result. To assess whether there is any tendency to oscillate, displacement in the direction of the inward directed point load is plotted for the single tenth-order NURBS surface element case in Figure 4.19. As is evident, the displacements are very smooth and monotone.

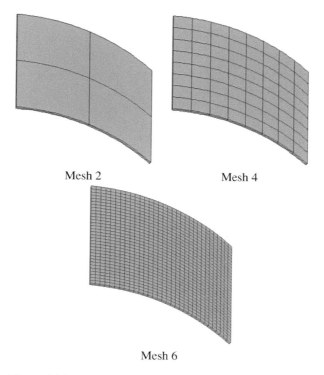

Mesh 2   Mesh 4

Mesh 6

Figure 4.14   Shell obstacle course. Scordelis-Lo roof meshes.

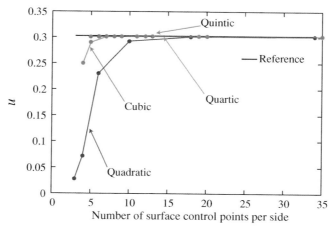

Figure 4.15   Shell obstacle course. Scordelis-Lo roof displacement convergence.

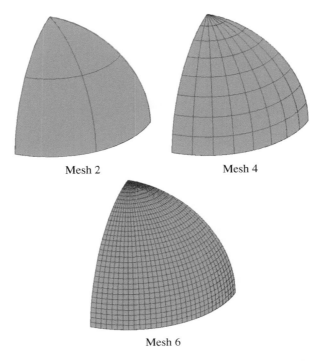

Mesh 2   Mesh 4

Mesh 6

Figure 4.16   Shell obstacle course. Pinched hemisphere meshes

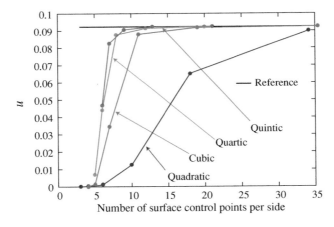

Figure 4.17   Shell obstacle course. Pinched hemisphere displacement convergence.

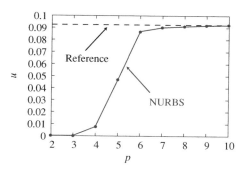

Figure 4.18 Shell obstacle course. Pinched hemisphere displacement convergence for one NURBS surface element.

4.3.2.3 Pinched cylinder

The pinched cylinder is subjected to equal and opposite concentrated forces at its midspan (see Figure 4.12c). The ends are supported by rigid diaphragms. This constraint results in highly localized deformation under the loads (see Figure 4.13c). Only one octant of the cylinder is used in the calculation due to symmetry. The second, fourth, and sixth meshes are shown in Figure 4.20. Convergence of the displacement under the load is presented in Figure 4.21. The NURBS elements converge to a very slightly softer solution than the benchmark solution. This may be due to transverse shear effects. It is well known that, as long as the characteristic surface element dimension is large compared with the thickness, formulations which permit transverse shear deformations typically closely approximate formulations which satisfy the Kirchhoff constraint (i.e., zero transverse shear strain). When this trend reverses, that is, as the surface element dimension approaches zero, holding the thickness constant, the displacement

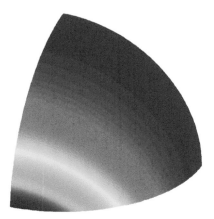

Figure 4.19 Shell obstacle course. Pinched hemisphere displacement in the direction of the inward directed point load for one surface element with $p = 10$.

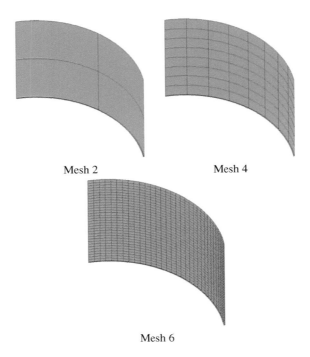

Mesh 2   Mesh 4

Mesh 6

Figure 4.20   Shell obstacle course. Pinched cylinder meshes.

under a concentrated load grows, and converges to infinity. (See Hughes and Franca, 1988 for elaboration.)

We believe that two quadratic NURBS through the thickness are unnecessary for typical thin shell analysis, unless warping of through-thickness sections is deemed important. See Bischoff *et al.*, 2004. We performed some tests with one quadratic NURBS through the thickness and the results were indistinguishable when compared with the two-quadratic-NURBS case. However,

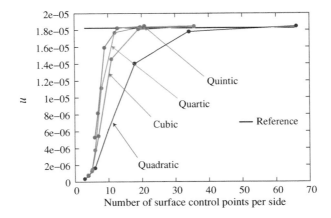

Figure 4.21   Shell obstacle course. Pinched cylinder displacement convergence.

when we reduced to linear variation through the thickness, convergence to correct solutions was not obtained. The classic shell theory hypothesis of invoking the plane stress condition in the through-thickness direction is sufficient to correct the deficiency of linear through-thickness displacement variation. Such a formulation may be competitive with traditional shell element formulations which employ displacement and rotation degrees-of-freedom at a reference surface. Using only displacement degrees-of-freedom (i.e., control variables) in the NURBS case considerably simplifies shell analysis, especially in nonlinear analysis wherein rotations are no longer vectorial and additive but require a multiplicative group structure. A further simplification is to use a NURBS surface and employ "rotationless" formulations, such as the one for plates described in Engel *et al.*, 2002. In this reference a discontinuous Galerkin formulation is proposed but all interface discontinuity (i.e., "jump") terms in it disappear if C^1, or higher, continuity is satisfied. This is simply attained with NURBS but very difficult to achieve in finite element analysis. Rotationless shell elements have recently gained popularity in computational mechanics (*e.g.*, Phaal and Calladine, 1992; Cirak *et al.*, 2000; Oñate and Zarate, 2000). The formulation of Engel *et al.*, 2002 has also been proposed as being an appropriate basis for so-called "strain gradient theories." These theories also require C^1-continuity, and NURBS would appear to be naturally suited to them. In Chapter 11, phase-field models are described and once again C^1-continuous basis functions are the natural choice.

4.3.3 Hyperboloidal shell

The hyperboloidal shell problem provides some insight into the role that the local control of continuity can play in analysis. This is still an active area of research, but some preliminary impressions emerge from examining this problem, as well as the stiffened shell problem of the next section.

The domain is the thin-walled solid seen in Figure 4.22, whose mid-surface is defined by

$$x^2 + z^2 - y^2 = 1, \quad y \in [-1, 1]. \tag{4.56}$$

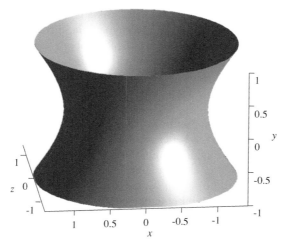

Figure 4.22 The geometry of the hyperboloidal shell.

The structure has a thickness of $t = 0.001$ in the direction normal to this mid-surface (all distances are in meters). The loading is a smoothly varying pressure normal to the surface,

$$p(\theta) = p_0 \cos(2\theta), \tag{4.57}$$

with $p_0 = 1.0$ MPa. The top and bottom of the structure are fixed.

4.3.3.1 Mesh generation and implementation

Only a quarter of the structure is modeled due to symmetry. The mid-surface is a conic section, namely a hyperbola, extruded in a path defined by another conic section, a circle. As rational quadratic NURBS are capable of representing all conic sections, this hyperboloidal surface of revolution can be represented exactly. However, the inner and outer surfaces of the structure are defined as offsets of the mid-surface, shifted by $\pm t/2$ in the normal direction, and are not conic sections. Moreover, they are not in the NURBS space, so the mesh will inherently be an approximate geometry. Note, however, that it is the offset of the hyperbola which is not represented exactly. In the radial direction, the offset of a circle is again a circle and therefore exists in the NURBS space. It is the *radii* of the circles denoting the inner and outer surfaces of the structure at a given height y that are not exact.

The decision was made to use two quadratic elements through the thickness of the structure. The knot value defining the boundary between the elements has a multiplicity equal to its polynomial order, 2, thereby making the geometrically exact mid-surface a discernible entity within the mesh – it is the boundary between the inner and outer layers of elements. Knots are then inserted into the appropriate knot vectors to define the elements in the mid-surface of the coarsest mesh. This mid-surface mesh is identical for all polynomial orders.

Once the coarse mid-surface mesh is fixed, so too is the number of basis functions in the axial direction. The offset curves that define the inner and outer surfaces of revolution must now be interpolated. The number of points along each curve that may be interpolated is equal to the number of basis functions in the axial direction. Due to the use of open knot vectors, the number of functions for a fixed mesh grows with p. Specifically, for the chosen mesh there are $14 + p$ basis functions in the axial (*i.e.*, y) direction. As a result, $14 + p$ points, equispaced in the parametric domain, are calculated along the hyperbola, then offset by $\pm t/2$ using the analytically computed normals to the curve. These offset points are then interpolated using C^{p-1} B-splines to create the approximate geometry. In this way, the quality of the overall geometric approximation improves as the polynomial order increases, though the mid-surface mesh is the same for all orders. The loading, however, does not differ as it is applied directly to the mid-surface itself.

To complete Mesh 1 for each polynomial order, knots are inserted near the fixed ends creating two rows of small elements in order to better resolve the boundary layer. The multiplicity of these knots is p, and so the basis functions are C^0 across these element boundaries, shown in red in Figure 4.23a. As discussed in Chapter 2, we could have introduced the *same number* of new degrees-of-freedom into the region by creating many small elements, each having $p - 1$ continuous derivatives across their boundaries. Instead we have chosen fewer elements with lower continuity. The motivation for introducing these C^0 mesh-lines is that previous experience has indicated that doing so helps to prevent the behavior in the layer from polluting results elsewhere in the domain. The main reason for this is that introducing a C^0 mesh-line results in localizing the support of the basis functions and thereby decreasing the coupling of

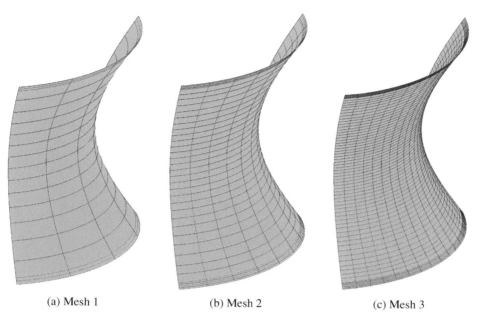

(a) Mesh 1   (b) Mesh 2   (c) Mesh 3

Figure 4.23 (a) Mesh 1. Basis functions have $p - 1$ continuous derivatives across blue element boundaries. They are only C^0 across red element boundaries. (b) Mesh 2. The second mesh is generated by uniform h-refinement of Mesh 1. The basis is C^{p-1} across the new element boundaries. (c) Mesh 3. The third mesh is generated by uniform h-refinement of Mesh 2. Again, the basis is C^{p-1} across the new element boundaries.

functions within the boundary layer to functions outside of the boundary layer (see Figure 4.24). The result is crisper layers and more compact representations of the global solution, particularly on coarse meshes.[3]

Meshes 2 and 3, seen in Figures 4.23b and 4.23c, respectively, are the result of subsequent uniform h-refinements. The basis is C^{p-1} across the element boundaries introduced through these refinements. The geometry and parameterization remain unchanged, and so Mesh 1 fixes the geometry for each polynomial order. In this way, the analysis is an h-method, repeated for several different polynomial orders, rather than a p-method repeated for several meshes. Recall, however, that the mid-surface meshes for Mesh 1, Mesh 2 and Mesh 3 are independent of the polynomial order. This was done in an effort to make the results from one polynomial order to the next as comparable as possible.

4.3.3.2 Results

The numbers of degrees-of-freedom and the numerically computed volume are reported in Table 4.2. The potential energy for each mesh, as well as the limit as $h \to 0$ estimated using Richardson extrapolation, is reported in Table 4.3. Plots of the deformed geometry as seen from two different angles are shown in Figure 4.25. The displacement has been amplified by a factor of 10 to make it more visible. Due to the sinusoidal character of the loading, the deformed structure has "compression lobes" and "expansion lobes." In both cases, the largest

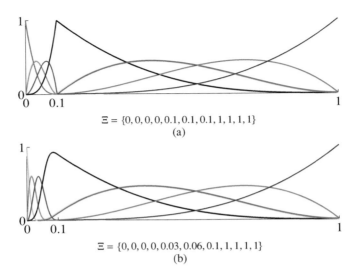

$$\Xi = \{0, 0, 0, 0, 0.1, 0.1, 0.1, 1, 1, 1, 1\}$$
(a)

$$\Xi = \{0, 0, 0, 0, 0.03, 0.06, 0.1, 1, 1, 1, 1\}$$
(b)

Figure 4.24 Example boundary layer meshes. Both meshes have the same number of basis functions. (a) Only one of the seven basis functions has support both inside the layer ($\xi < 0.1$) and outside of the layer ($\xi > 0.1$). (b) Three of the seven basis functions have support both inside and outside of the layer. This may not be visually apparent, but the red basis function is non-zero for $\xi < 0.1$.

gradients of the solution are contained in thin layers near the fixed ends of the structure. Plots of the radial displacement at a compression lobe are shown in Figure 4.26 and Figure 4.27. In these plots, results for each of the polynomial orders on Mesh 3 are shown. While the quadratics are far from converged, and cubics seem to be showing signs of the geometry error, the quartics and quintics lie practically on top of each other.

After this initial study was completed, a second study was performed using the maximum continuity possible. The shell geometries were identical to those presented above, as were the meshes outside of the boundary layer region. The width of the boundary layer portion of the mesh was kept the same as well, but instead of two rows of elements with C^0 boundaries, as in the initial study, many rows of elements with C^{p-1} boundaries were used (the number of rows depended on the polynomial order and was chosen so as to equate the number of degrees-of-freedom in each mesh of this study with the equivalent mesh in the previous

Table 4.2 Mesh data. Here p denotes the polynomial order in the plane of the surface, while q is the polynomial order through the thickness. The exact volume of the shell is $1.597530 * 10^{-2}$ m^3

p, q	Mesh 1 DOF	Mesh 2 DOF	Mesh 3 DOF	Volume of shell
2, 2	2160	6300	21060	$1.597535 * 10^{-2}$m^3
3, 2	3045	7755	23655	$1.597527 * 10^{-2}$m^3
4, 2	4080	9360	26400	$1.597530 * 10^{-2}$m^3
5, 2	5265	11115	29295	$1.597530 * 10^{-2}$m^3

Table 4.3   Potential energy. Estimated limit calculated using Richardson extrapolation

p, q	Mesh 1	Mesh 2	Mesh 3	Estimated limit
2, 2	−4.668902	−4.751145	−4.796779	−4.799821
3, 2	−4.783082	−4.794395	−4.801991	−4.802112
4, 2	−4.787948	−4.795994	−4.799878	−4.799893
5, 2	−4.791334	−4.798373	−4.799941	−4.799942
	$* 10^{-2}$MNm	$* 10^{-2}$MNm	$* 10^{-2}$MNm	$* 10^{-2}$MNm

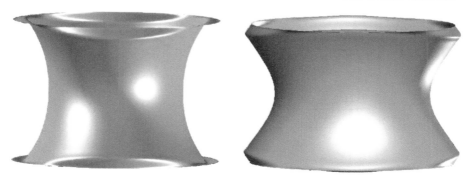

Figure 4.25   The deformed configuration viewed from two different angles. Compression lobes are visible where the loading is directed inward. Expansion lobes are visible where the loading is directed outward.

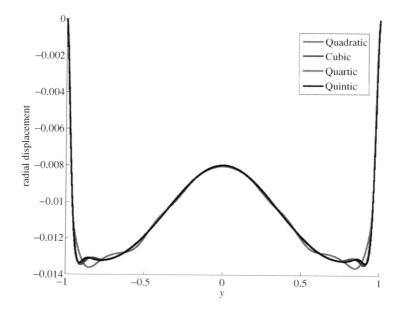

Figure 4.26   Compression lobe. Radial displacement versus height.

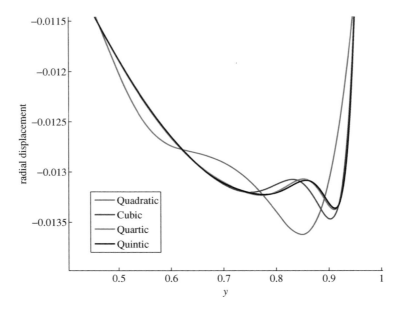

Figure 4.27 Compression lobe. Detail of radial displacement versus height.

study). As expected, the results on coarser meshes are always better for the case with decreased continuity. As the meshes are refined, this performance gap narrows, and eventually the trend reverses. See, for example, the results for quintic meshes in Table 4.4. As stated previously, we conclude that the use of functions with the maximum continuity possible at the given polynomial order will be more efficient asymptotically whenever the exact solution is smooth. On coarser meshes, however, reducing the continuity *and therefore decreasing the support of the basis functions* may result in a more accurate solution.

4.3.4 Hemispherical shell with a stiffener

The hemispherical shell with a stiffener problem (see Figure 4.28) was modeled with a single NURBS patch in Hughes *et al.*, 2005. As was shown in Section 2.3 of Chapter 2, use of a single patch leads to substantial distortion of the elements for this problem. While it speaks well of the overall robustness of the method that accurate results were still obtained, efficiency

Table 4.4 Comparison of the potential energy errors in the two approaches to boundary layer meshing, $p = 5$. Though the coarse mesh favors the C^0 boundary layer treatment, smooth functions prove to be more accurate once the meshes are sufficiently fine. The reference solution used in the error calculation is the estimated limit for the quintic case from the previous table

	Mesh 1	Mesh 2	Mesh 3
C^0 Boundary layer	0.1804%	0.0337%	0.0011%
C^{p-1} Boundary layer	0.3379%	0.0206%	0.0007%

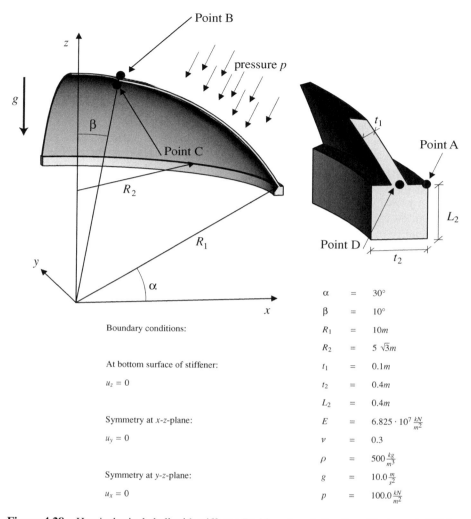

Figure 4.28 Hemispherical shell with stiffener. Problem description from Rank *et al.*, 2005.

clearly suffered. We wish to compare with the original results of Rank *et al.*, 2005, who used a trunk space *p*-refinement strategy. Such an approach does not use the full tensor product space of basis functions, but the much smaller *trunk* space, just large enough to ensure the optimal convergence rate at a given polynomial order (see Szabo *et al.*, 2004 for a discussion of the trunk space and the *p*-method in general). As NURBS necessarily have an underlying tensor product structure, at least on patches, an analogous isogeometric analysis approach exploiting the trunk space has not been attempted thus far.

Despite the tensor product structure of NURBS, *k*-refinement presents the possibility of improved efficiency. In fact, *k*-refinement, in conjunction with the use of multiple patches to create better quality meshes, and the use of local refinement as described in Section 3.5.1

of Chapter 3 to avoid placing functions in regions where they are not needed, enables the NURBS based approach to show an accuracy per degree-of-freedom comparable to the results presented in Rank *et al.*, 2005; see Figures 4.31–4.34.

As we have seen several times now, if a given mesh of higher-order and high continuity does not achieve the level of accuracy desired, one can add more degrees-of-freedom by inserting a knot in one of the parametric directions. The number of new degrees-of-freedom is *exactly* the same regardless of whether a new knot value is inserted (creating new elements by splitting existing ones), or whether an existing knot value is repeated (creating no new elements, but decreasing the continuity of the basis across the corresponding element boundaries). While a rigorous analysis of the two approaches has not yet been performed, in the present results it seems clear that in regions where the solution is very smooth (such as in the shell, a reasonable distance away from the stiffener), inserting a new knot, and thus more functions that maintain high continuity, was the more beneficial refinement. In the vicinity of a singularity (such as near the reentrant corner where the shell meets the stiffener and the stress is singular), it is more beneficial to repeat an existing knot value, decreasing the continuity of the basis and simultaneously decreasing the support of the basis functions in the physical space. Both of these effects help localize the singularity and prevent it from polluting the results elsewhere in the domain[4] (recall Figure 4.24).

The meshes for the multiple-patch treatment of the stiffened shell are shown in Figures 4.29 and 4.30. The locally refined, *k*-method meshes are seen in Figure 4.30a. In Figure 4.30b, we see the case where fewer elements are used. A *k*-type refinement is used everywhere except at the knot lines marked in red. The multiplicities of these knots were increased with the polynomial order such that the basis remained C^0 across them. The results for this mesh are labeled "Local *k**-ref" to indicate that the *k*-refinement paradigm was altered near the singularity. The displacements are plotted versus the number of degrees-of-freedom for points A and B in Figures 4.31 and 4.32. The calculated von Mises stresses are plotted versus the number of degrees-of-freedom in Figures 4.33 and 4.34. The trunk space *p*-method results

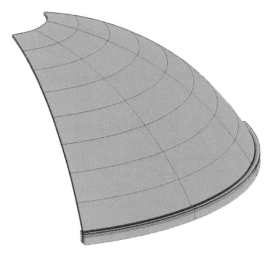

Figure 4.29   Hemispherical shell with stiffener. The coarse mesh may be refined in multiple ways.

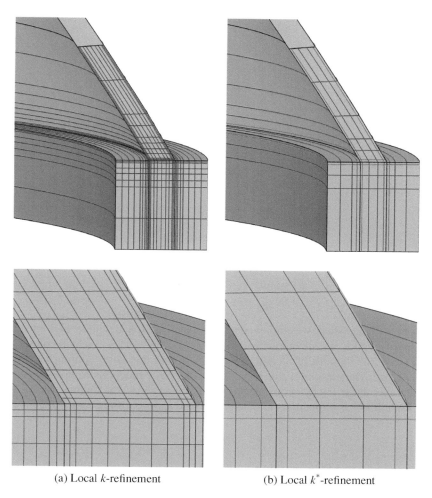

(a) Local k-refinement   (b) Local k^*-refinement

Figure 4.30   Hemispherical shell with stiffener. (a) A k-refinement approach with C^{p-1} continuity across element boundaries. Many small elements are used to get a well-resolved solution. (b) Functions are C^0 across the element boundaries in red, C^{p-1} elsewhere. Fewer elements are needed than in (a). In both cases, the basis is C^0 across patch boundaries, shown in black, and local refinement is implemented at the patch level.

from Rank *et al.*, 2005 are plotted for comparison. For displacements, the single patch results from Hughes *et al.*, 2005 are plotted as well.

In each of the problems in this section, we treated thin-walled geometries quite successfully with solid, trivariate discretizations. The last two problems, the hyperboloidal shell and the hemispherical shell with a ring stiffener, provided the opportunity to study the effects of smoothness of basis functions in the vicinity of singularities. For the hyperboloidal shell, we reduced smoothness locally in the vicinity of the boundary layer. In the case of coarse meshes, this improved accuracy, whereas for finer meshes the pure k-method was more accurate. In

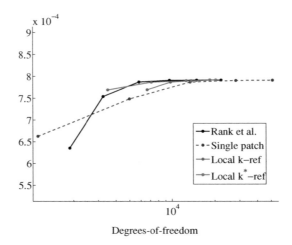

Figure 4.31 Hemispherical shell with stiffener. The displacement at point A is plotted versus the total number of degrees-of-freedom.

the case of the stiffened shell, we employed a multi-patch approach with local refinement and again compared smooth discretizations within the patches with ones in which continuity was reduced to C^0 in the vicinity of the singularity. We found this latter approach led to more rapid convergence. The reason for this seems to be that basis functions having support in the vicinity of the singularity tend to propagate information away from the singularity. The support of smooth k-method basis functions is greater than the support of the same order p-method

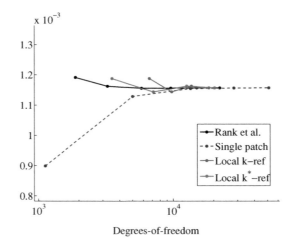

Figure 4.32 Hemispherical shell with stiffener. The displacement at point B is plotted versus the total number of degrees-of-freedom.

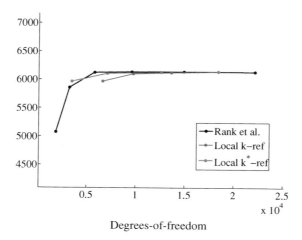

Figure 4.33 Hemispherical shell with stiffener. The von Mises stress at point A is plotted versus the total number of degrees-of-freedom.

functions when there are approximately the same numbers of degrees-of-freedom. As a result, the errors created by the singularities tend to propagate further for the smoother basis functions of the k-method. By judiciously locating a few surfaces of reduced continuity, the "pollution" created by the singularities seemed to be more locally confined. However, it does not seem to be a black and white issue, but rather to depend strongly on the nature of the exact solution. Further studies need to be performed to assess the trade-offs in a wider variety of problems.

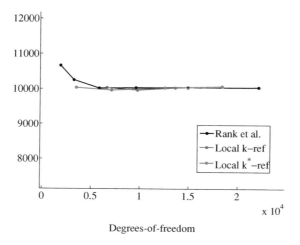

Figure 4.34 Hemispherical shell with stiffener. The von Mises stress at point B is plotted versus the total number of degrees-of-freedom.

Appendix 4.A: Geometrical data for the hemispherical shell

The mesh for the hemispherical shell is a quintessential example of a geometry that we can model quite simply with NURBS, but that is completely outside the space of piecewise polynomial geometries realizable in a classical finite element setting. Here we present the same geometrical data as used for the hemispherical shell in the shell obstacle course. We model a quarter of the domain, applying symmetry boundary conditions where appropriate (see Figure 4.12b). The coarsest possible mesh has only one element, which is rational quadratic in the two curved directions of the face and linear through the thickness. The ξ and η coordinates will map onto the latitude and longitude lines. We have $p = q = 2$,

$$\Xi = \{0, 0, 0, 1, 1, 1\}, \tag{4.A.1}$$

and

$$\mathcal{H} = \{0, 0, 0, 1, 1, 1\}. \tag{4.A.2}$$

Finally, through the thickness we have $r = 1$ and

$$\mathcal{Z} = \{0, 0, 1, 1\}. \tag{4.A.3}$$

The control points are tabulated in Table 4.A.1. Note that the radius of the mid-surface is $R = 10$ and the thickness is $t = 0.04$.

Before performing analysis, we performed k-refinement in the ζ-direction. First we order elevate to $r = 2$, and then insert a new knot at $\zeta = 0.5$. From there, further refinements were made and results compiled. See Section 4.3.2.

Appendix 4.B: Geometrical data for a cylindrical pipe

Another frequently occurring shape in engineering design is the cylinder. As with the other examples from the shell obstacle course, we modeled the cylindrical shell as a solid using

Table 4.A.1 Control points for the hemispherical shell

i	j	$B_{i,j,1}$	$B_{i,j,2}$	$w_{i,j,1}$	$w_{i,j,2}$
1	1	$(9.98, 0, 0)$	$(10.02, 0, 0)$	1	1
1	2	$(9.98, 0, 9.98)$	$(10.02, 0, 10.02)$	$1/\sqrt{2}$	$1/\sqrt{2}$
1	3	$(0, 0, 9.98)$	$(0, 0, 10.02)$	1	1
2	1	$(9.98, 9.98, 0)$	$(10.02, 10.02, 0)$	$1/\sqrt{2}$	$1/\sqrt{2}$
2	2	$(9.98, 9.98, 9.98)$	$(10.02, 10.02, 10.02)$	$1/2$	$1/2$
2	3	$(0, 0, 9.98)$	$(0, 0, 10.02)$	$1/\sqrt{2}$	$1/\sqrt{2}$
3	1	$(0, 9.98, 0)$	$(0, 10.02, 0)$	1	1
3	2	$(0, 9.98, 9.98)$	$(0, 10.02, 10.02)$	$1/\sqrt{2}$	$1/\sqrt{2}$
3	3	$(0, 0, 9.98)$	$(0, 0, 10.02)$	1	1

trivariate elements. Here, we will provide a template for such a geometry, but one with more uniform proportions. Transforming the resulting cylindrical pipe into one with any desired proportions is a simple task. Chapter 2, and the geometry tutorial of Section 2.4 in particular, should provide the necessary details.

We let the ξ-direction in the parameter space correspond to the θ-direction in cylindrical coordinates. The η-direction will align with the radial direction, while the ζ-direction will correspond to the z-axis. The corresponding polynomial orders will be $p = 2$, $q = 1$, and $r = 1$, respectively. For the knot vectors we have

$$\Xi = \{0, 0, 0, 1, 1, 2, 2, 3, 3, 4, 4, 4\}, \tag{4.B.1}$$

$$\mathcal{H} = \{0, 0, 1, 1\}, \tag{4.B.2}$$

$$\mathcal{Z} = \{0, 0, 1, 1\}. \tag{4.B.3}$$

Note that we specified the polynomial orders and knot vectors without specifying the dimensions of the object yet. This is natural as knot vectors in which the knots are integers are very simple to deal with. To determine the control points, however, we must know the dimensions of the cylindrical pipe. Let us consider a pipe with an inner radius of 1, an outer radius of 2, and a height of 5. The necessary control points are tabulated in Table 4.B.1.

Table 4.B.1 Control points for a cylinder

i	k	$\mathbf{B}_{i,1,k}$	$\mathbf{B}_{i,2,k}$	$w_{i,1,k}$	$w_{i,2,k}$
1	1	$(1, 0, 0)$	$(2, 0, 0)$	1	1
1	2	$(1, 0, 5)$	$(2, 0, 5)$	1	1
2	1	$(1, 1, 0)$	$(2, 2, 0)$	$1/\sqrt{2}$	$1/\sqrt{2}$
2	2	$(1, 1, 5)$	$(2, 2, 5)$	$1/\sqrt{2}$	$1/\sqrt{2}$
3	1	$(0, 1, 0)$	$(0, 2, 0)$	1	1
3	2	$(0, 1, 5)$	$(0, 2, 5)$	1	1
4	1	$(-1, 1, 0)$	$(-2, 2, 0)$	$1/\sqrt{2}$	$1/\sqrt{2}$
4	2	$(-1, 1, 5)$	$(-2, 2, 5)$	$1/\sqrt{2}$	$1/\sqrt{2}$
5	1	$(-1, 0, 0)$	$(-2, 0, 0)$	1	1
5	2	$(-1, 0, 5)$	$(-2, 0, 5)$	1	1
6	1	$(-1, -1, 0)$	$(-2, -2, 0)$	$1/\sqrt{2}$	$1/\sqrt{2}$
6	2	$(-1, -1, 5)$	$(-2, -2, 5)$	$1/\sqrt{2}$	$1/\sqrt{2}$
7	1	$(0, -1, 0)$	$(0, -2, 0)$	1	1
7	2	$(0, -1, 5)$	$(0, -2, 5)$	1	1
8	1	$(1, -1, 0)$	$(2, -2, 0)$	$1/\sqrt{2}$	$1/\sqrt{2}$
8	2	$(1, -1, 5)$	$(2, -2, 5)$	$1/\sqrt{2}$	$1/\sqrt{2}$
9	1	$(1, 0, 0)$	$(2, 0, 0)$	1	1
9	2	$(1, 0, 5)$	$(2, 0, 5)$	1	1

Appendix 4.C: Element assembly routine

In Appendix 3.A at the end of the previous chapter we presented pseudo-code for a shape function routine that would consume an element number and a quadrature point in the parent element, and would return the basis functions, their derivatives with respect to the physical coordinates, and the Jacobian determinant of the mapping, all evaluated at the image of the quadrature point in the physical domain. Though this was divided into three separate algorithms for the sake of clarity, let us assume that we have one single function, **Shape_function**, that performs the task in its entirety. In this appendix, we will use that function to build and assemble[5] the local stiffness matrices and load vectors required for the analysis of a homogeneous linear elastic solid.

The approach that we will take is to loop through the elements and within each element to loop through the quadrature points. At each point we will evaluate basis functions and their derivatives, and add the appropriate contributions to the local stiffness matrix and load vector. When all of the quadrature points have been traversed within a given element, these local contributions will be assembled into the global system by means of the **Assembly** function (not shown) which will utilize the IEN and ID (or, analogously, LM) arrays discussed in Appendix A at the end of this book. As the solution field is vector-valued for the case of three spatial dimensions, the local stiffness matrix is stored as a $n_{en} \times n_{en} \times 9$ array and the local load vector is stored as a $n_{en} \times 3$ array. **Assembly** will use the connectivity information to map the local degrees-of-freedom to the appropriate global equation numbers. We will assume that the treatment of Dirichlet boundary conditions will be handled appropriately therein.

Algorithm 4 contains the aforementioned loops through the elements and quadrature points. The contributions to the local stiffness matrix and local load vector are computed in **Build_K_local** and **Build_F_local**, shown in Algorithms 5 and 6, respectively. For simplicity, we will assume that the Young's modulus, Poisson ratio, and body load are uniform throughout the domain. We use the Young's modulus and Poisson ratio to construct the Lamé parameters as in (4.13) and (4.14). We also assume that the same set of quadrature points in the parent element are appropriate for all physical elements. Lastly, note that the definition that we have used for an element, the span between the knots, allows for the possibility of elements which are of zero measure in the parameter space. This happens whenever continuity has been decreased by knot replication. As a result, we will check to see if the element has positive measure before attempting to integrate over it.

Algorithm 4: Element assembly
Data: All of the geometrical, connectivity, material and load data must be input, as well as
the quadrature points and weights.
Result: The global stiffness matrix and global load vector will be assembled.

```
// Lamé parameters:
lambda = nu_P*E_Y / ((1.0+nu_P)*(1.0+2.0*E_Y));
mu = E_Y / (2.0*(1.0+nu_P));
```

```
// Element Loop:
```
for e = 1 to nel **do**
```
    // NURBS coordinates; convention consistent with Algorithm 7
    ni = INN[IEN[e][1]][1];
    nj = INN[IEN[e][1]][2];
    nk = INN[IEN[e][1]][3];

    // Check if element has zero measure
    if (KV_Xi[ni+1] == KV_Xi[ni])or(KV_Eta[nj+1] == KV_Eta[nj])or
       (KV_Zeta[nk+1] == KV_Zeta[nk])
    then
      | CYCLE;        // Jump to end of loop and proceed with next element
    end

    K_local[nen][nen][9] = 0.0;                   // Local stiffness matrix
    F_local[nen][3] = 0.0;                        // Local load vector

    for i = 1 to NQUAD do
        for j = 1 to NQUAD do
            for k = 1 to NQUAD do

                // gp is the vector of NQUAD quadrature points
                call Shape_function(gp[i],gp[j],gp[k],R,dR_dx,J);

                // Combine quadrature weights with Jacobian
                Jmod = J*gw[i]*gw[j]*gw[k];

                call Build_K_local(dR_dx,Jmod,lambda,mu,K_local);

                // Fb is the body load vector
                call Build_F_local(R,Jmod,Fb,F_local);

            end
        end
    end
    call Assembly(K_local,F_local);
```
end

Algorithm 5: `Build_K_local`

Data: The derivatives of the basis functions, modified Jacobian, material parameters λ and μ, and current local stiffness matrix must be given as inputs.

Result: The local stiffness matrix is updated.

```
for aa = 1 to nen do
    for bb = 1 to nen do
        // Contributions to the diagonal
        K_Local[aa][bb][1] = K_Local[aa][bb][1] + (
            (lambda + 2.0*mu)*dR_dx[aa][1]*dR_dx[bb][1] +
            mu*(dR_dx[aa][2]*dR_dx[bb][2] + dR_dx[aa][3]*dR_dx[bb][3])
            )*Jmod;

        K_Local[aa][bb][5] = K_Local[aa][bb][5] + (
            (lambda + 2.0*mu)*dR_dx[aa][2]*dR_dx[bb][2] +
            mu*(dR_dx[aa][1]*dR_dx[bb][1] + dR_dx[aa][3]*dR_dx[bb][3])
            )*Jmod;

        K_Local[aa][bb][9] = K_Local[aa][bb][9] + (
            (lambda + 2.0*mu)*dR_dx[aa][3]*dR_dx[bb][3] +
            mu*(dR_dx[aa][1]*dR_dx[bb][1] + dR_dx[aa][2]*dR_dx[bb][2])
            )*Jmod;

        // Contributions to the off-diagonal entries
        K_Local[aa][bb][2] = K_Local[aa][bb][2] + (
            lambda*dR_dx[aa][1]*dR_dx[bb][2] +
            mu*dR_dx[aa][2]*dR_dx[bb][1] )*Jmod;

        K_Local[aa][bb][3] = K_Local[aa][bb][3] + (
            lambda*dR_dx[aa][1]*dR_dx[bb][3] +
            mu*dR_dx[aa][3]*dR_dx[bb][1] )*Jmod;

        K_Local[aa][bb][4] = K_Local[aa][bb][4] + (
            lambda*dR_dx[aa][2]*dR_dx[bb][1] +
            mu*dR_dx[aa][1]*dR_dx[bb][2] )*Jmod;

        K_Local[aa][bb][6] = K_Local[aa][bb][6] + (
            lambda*dR_dx[aa][2]*dR_dx[bb][3] +
            mu*dR_dx[aa][3]*dR_dx[bb][2] )*Jmod;

        K_Local[aa][bb][7] = K_Local[aa][bb][7] + (
            lambda*dR_dx[aa][3]*dR_dx[bb][1] +
            mu*dR_dx[aa][1]*dR_dx[bb][3] )*Jmod;

        K_Local[aa][bb][8] = K_Local[aa][bb][8] + (
            lambda*dR_dx[aa][3]*dR_dx[bb][2] +
            mu*dR_dx[aa][2]*dR_dx[bb][3] )*Jmod;
    end
end
```

Algorithm 6: `Build_F_local`
Data: The basis functions, modified Jacobian, body load, and current local load vector must be given as inputs.
Result: The local load vector is updated.

for *aa = 1 to nen* **do**

```
// Fb is the body load vector
F_local[aa][1] = F_local[aa][1] + Fb[1]*R[aa]*Jmod;
F_local[aa][2] = F_local[aa][2] + Fb[2]*R[aa]*Jmod;
F_local[aa][3] = F_local[aa][3] + Fb[3]*R[aa]*Jmod;
```

end

Notes

1. Do not confuse the symbol $\mathbf{B}_A$ (*i.e.*, the strain-displacement matrix) with one of the control points used to define the geometry. There are but a finite number of letters in the alphabet, and the use of $\mathbf{B}$ in this context is the standard in the literature.
2. Coalescing adjacent control points reduces continuity by one order. Coalescing $p - 1$ adjacent control points for NURBS of order p reduces continuity to C^0.
3. Note that we would expect a very fine mesh (*e.g.*, one with all of the elements the size of those in the boundary layer) comprised of highly continuous functions to represent the solution more efficiently than a classical p-method on a per degree-of-freedom basis. On a coarse mesh, however, the efficiency of the method is degraded if a large percentage of the functions have support in both very small and very large elements. This issue is investigated in more detail in Section 4.3.4.
4. This is reminiscent of the heuristic notion that an *hp*-method should use large elements with higher-order in smooth regions and small elements of lower-order near singularities. Coupling this with control over the continuity across elements opens the door to the possibility of an *hpk*-method.
5. Here "build" refers to the act of performing the integrals necessary to compute the entries of the local stiffness matrix and load vector, while "assemble" refers to taking these local objects and adding their contributions to their global counterparts.

5

Vibrations and Wave Propagation*

The study of structural vibrations or, more specifically, of eigenvalue problems allows us to examine in more detail the approximation properties of the smooth NURBS functions independently of any geometrical considerations. In general, **spectrum analysis** is the term applied to the study of how numerically computed natural frequencies, ω_n^h, compare with the analytically computed natural frequencies, ω_n. We will see that, for a given number of degrees-of-freedom and bandwidth, the use of NURBS results in dramatically improved accuracy in spectral calculations over classical finite elements analysis.

5.1 Longitudinal vibrations of an elastic rod

Let us begin by considering one of the simplest vibrational model problems in one dimension: the longitudinal vibrations of an elastic rod. If we consider the domain $\Omega = (0, L) \subset \mathbb{R}$, there is no longer an issue of geometrical accuracy. We will begin by taking the mapping from the parameter space to the physical space to be the identity mapping (equivalently, we can think of this as simply working in the parameter space). As demonstrated by Cottrell et al., 2006, we will see that this is not necessarily the best choice. FEA basis functions and NURBS[1] are equally capable of representing this domain exactly, and so the quality of the results will depend entirely on the approximation properties of the basis.

5.1.1 Formulating the problem

To understand the formulation of the eigenproblem representing the longitudinal vibrations of a "fixed–fixed" elastic rod, let us begin by considering the elastodynamics equation from which it is derived (elastodynamics will be discussed in more detail in Chapter 6). The behavior of the rod, which is assumed to move only in the longitudinal direction, is governed by the equations of linear elasticity combined with Newton's second law, resulting in

$$(E u_{,x})_{,x} - \rho u_{,tt} = 0 \quad \text{in} \quad \Omega \times (0, T), \tag{5.1a}$$

$$u = 0 \quad \text{on} \quad \Gamma \times (0, T), \tag{5.1b}$$

---

* Many of the results in this chapter were originally obtained in Cottrell et al., 2006 and Hughes et al., 2008a

Isogeometric Analysis: Toward Integration of CAD and FEA by J. A. Cottrell, T. J. R. Hughes, Y. Bazilevs
© 2009, John Wiley & Sons, Ltd

where $\Omega = (0, L)$, $\rho : (0, L) \to \mathbb{R}$ is the density per unit length of the rod, $E : (0, L) \to \mathbb{R}$ is Young's modulus, and the "fixed–fixed" condition (5.1b) ensures that the ends of the rod do not move. For an actual dynamics problem, we would need to augment (5.1) with appropriate initial conditions of the form

$$u(x, 0) = u_0(x), \tag{5.2}$$

$$u_{,t}(x, 0) = v_0(x). \tag{5.3}$$

At present, however, we are not interested in the transient behavior of the rod. Instead, we are interested in the natural frequencies and modes in which the rod vibrates. We obtain these by separation of variables. In a slight abuse of notation, we assume $u(x, t)$ to have the form

$$u(x, t) = u(x)e^{i\omega t}, \tag{5.4}$$

where $u(x)$ is a function of only the spatial variable, x, while $i = \sqrt{-1}$, and ω is the natural frequency. Inserting (5.4) into (5.1a) and dividing by the common exponential term results in the eigenproblem we are seeking:

$$(Eu_{,x})_{,x} + \omega^2 \rho u = 0 \quad \text{in} \quad \Omega, \tag{5.5a}$$

$$u = 0 \quad \text{on} \quad \Gamma. \tag{5.5b}$$

Equation (5.5) constitutes an eigenproblem for the rod. The nontrivial solutions are countably infinite. That is, for $k = 1, 2, \ldots, \infty$, there is an eigenvalue $\lambda_k = (\omega_k)^2$ and corresponding eigenfunction $u_{(k)}$ satisfying (5.5). Furthermore, $0 < \lambda_1 \leq \lambda_2 \leq \ldots$, and the eigenfunctions are orthogonal. Though the eigenfunctions are only defined up to a multiplicative constant, we can remove the arbitrariness by augmenting the orthogonality condition to include normality. That is, we can demand that the eigenfunctions all obey the property

$$\int_0^L u_{(k)} \rho u_{(l)} \, dx = \delta_{kl}. \tag{5.6}$$

Following the now familiar process, we multiply (5.5a) by a test function w and integrate by parts to obtain the weak form of the equation: Find all eigenpairs $\{u, \lambda\}$, $u \in \mathcal{S}$, $\lambda = \omega^2 \in \mathbb{R}^+$, such that for all $w \in \mathcal{V}$

$$a(w, u) - \omega^2(w, \rho u) = 0, \tag{5.7}$$

where

$$a(w, u) = \int_0^L w_{,x} E u_{,x} \, dx, \tag{5.8}$$

$$(w, \rho u) = \int_0^L w \rho u \, dx. \tag{5.9}$$

Note that, due to the homogeneous boundary conditions, $S = V = H_0^1(0, L) = \{u \in H^1(0, L) | u(0) = u(L) = 0\}$.

The Galerkin formulation is obtained by restricting ourselves to finite-dimensional subspaces $S^h \subset S$ in the usual way. That is, w and u in (5.7) will be replaced by finite dimensional approximations w^h and u^h of the form

$$w^h = \sum_{A=1}^{n_{eq}} N_A d_A \quad \text{and} \quad u^h = \sum_{B=1}^{n_{eq}} N_B c_B, \tag{5.10}$$

respectively. The resulting eigenpairs will contain approximations of both natural modes $u_{(k)}^h$ and the natural frequencies ω_k^h. The problem becomes: Find all $\omega^h \in \mathbb{R}^+$ and $u^h \in S^h$ such that for all $w^h \in V^h$

$$a(w^h, u^h) - (\omega^h)^2 (w^h, \rho u^h) = 0. \tag{5.11}$$

Substituting the shape-function expansions for w^h and u^h in (5.11) gives rise to a matrix eigenvalue problem: Find natural frequency $\omega_k^h \in \mathbb{R}^+$ and eigenvector $\mathbf{\Psi}_k$, $k = 1, \ldots, n_{eq}$, such that

$$\left(\mathbf{K} - (\omega_k^h)^2 \mathbf{M}\right) \mathbf{\Psi}_k = \mathbf{0}, \tag{5.12}$$

where

$$\mathbf{K} = [K_{AB}], \tag{5.13}$$

$$\mathbf{M} = [M_{AB}], \tag{5.14}$$

with

$$K_{AB} = a(N_A, N_B), \tag{5.15}$$

$$M_{AB} = (N_A, \rho N_B), \tag{5.16}$$

and $\mathbf{\Psi}_k$ is the vector of control variables corresponding to $u_{(k)}^h$. The orthonormality condition, (5.6), can be expressed as

$$\mathbf{\Psi}_k^{\mathrm{T}} \mathbf{M} \mathbf{\Psi}_l = \delta_{kl}. \tag{5.17}$$

As in the previous chapter, we refer to $\mathbf{K}$ as the *stiffness matrix*. The new object, $\mathbf{M}$, is the *mass matrix*. Noting that $\rho > 0$, and that the NURBS basis functions are pointwise non-negative, we see from (5.9) that every entry in the mass matrix is also non-negative. This claim cannot be made for standard finite elements.

5.1.2 Results: NURBS vs. FEA

Let us consider the case where ρ, E, and L are each taken to be 1. Analytically, (5.5a) can be solved to obtain $\omega_n = n\pi$ for $n = 1, \ldots, \infty$. We can assess the quality of the numerical method by comparing the ratio of the computed modes, ω_n^h, with the analytical result. That is, $(\omega_n^h/\omega_n) = 1$ indicates that the numerical frequency is identical to the analytical result. In practice, the discrete frequencies will always obey the relationship

$$\omega_n \leq \omega_n^h \quad \text{for} \quad n = 1, \ldots, n_{eq}, \tag{5.18}$$

and so we expect the ratio (ω_n^h/ω_n) to be greater than 1 (see, *e.g.*, Strang and Fix, 1973), with larger values indicating decreased accuracy.

5.1.2.1 Initial results

Note that linear NURBS are identical to linear finite element functions, so let us begin the comparison with quadratic functions. We compare C^1-continuous quadratic NURBS functions with the classical C^0-continuous quadratic finite elements. The results are shown in Figure 5.1, where we have plotted the normalized frequency results, ω_n^h/ω_n, versus the mode number, n, normalized by the total number of degrees-of-freedom, $N \equiv n_{eq} = 999$.

Figure 5.1 illustrates the superior behavior of NURBS basis functions compared with finite elements. In this case, the finite element results depict a so-called acoustical branch for $n/N < 0.5$ and an optical branch for $n/N > 0.5$ (see Brillouin, 1953). This branching is due to the fact that there are two distinct types of difference equations for the finite elements: those corresponding to the end-point nodes at element boundaries, and those corresponding to mid-point nodes on element interiors; see Figure 5.2. The acoustical branch corresponds to modes in which the neighboring end- and mid-point nodes oscillate in phase with each other, and the optical branch modes are the modes in which they are out of phase[2]. Alternatively, the quadratic NURBS difference equations are all identical (recall Figure 2.3), and no such

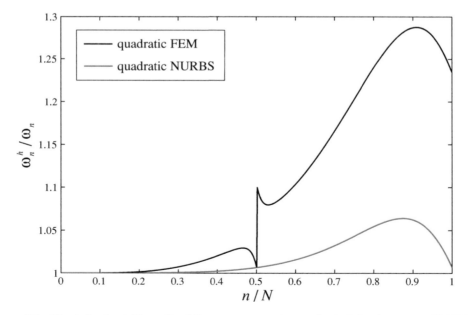

Figure 5.1 Fixed–fixed rod. Normalized discrete spectra using quadratic finite elements and NURBS.

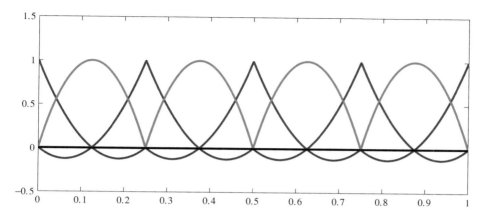

Figure 5.2 Nodal finite element basis functions for the quadratic *p*-method. Note the two distinct types of functions corresponding to end-nodes and mid-nodes. These lead to two distinct difference equations corresponding to the end-point nodes at element boundaries and the mid-point nodes in element interiors.

branching takes place. Observe that the NURBS are more accurate *throughout the entire spectrum*, not just in the upper half after the branching of the FEA spectrum.

The results of Figure 5.1 were obtained numerically. One could also have obtained these results by analytically solving the discrete equations, as will be seen in Section 5.1.3.

The same eigenvalue analysis can be performed using higher-order NURBS basis functions. The resulting spectra are presented in Figure 5.3; the analyses were carried out using $N = 1000$ degrees-of-freedom.

Increasing the order, p, of the basis functions, increases accuracy. Increasing p also results in the appearance of strange frequencies at the very end of the spectrum (see Figure 5.3),

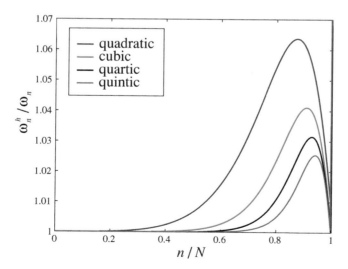

Figure 5.3 Fixed–fixed rod. Normalized discrete spectra using different order NURBS basis functions. Outliers appear in the very thin band on the right end of the spectrum.

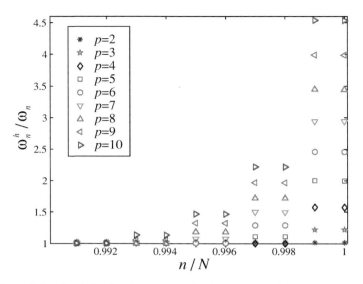

Figure 5.4   Fixed–fixed rod. Last normalized frequencies for $p = 2, \ldots, 10$.

referred to as "outlier frequencies," whose number and magnitude increase with p. In Figure 5.4, this behavior is highlighted by plotting the last computed frequencies for $p = 2, \ldots, 10$. To understand the outliers, recall the branching of the FEA spectrum seen in Figure 5.1. There are only two distinct equations in the discrete system, corresponding to element middle and end nodes, and this gives rise to the two branches. In the case of NURBS, as $N \to \infty$, all but a *finite* number of equations are the same, as seen in Figure 5.5. Those associated with the open knot vectors, at the ends of the domain, are different and are responsible for the outliers. The outliers constitute a discrete optical branch. The typical equation for a function

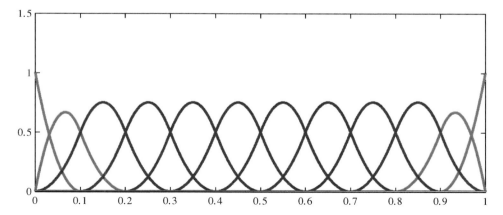

Figure 5.5   Basis functions for the quadratic NURBS. Note that the number of non-standard basis functions depends on the polynomial order, but not on the total number of elements in the domain.

on the interior of the domain gives rise to the continuous acoustic branch, as has been verified analytically; see Cottrell *et al.*, 2006. In finite element analysis, the frequencies associated with the optical branch are regarded as inaccurate and, obviously, the same is true for NURBS. In many applications, these frequencies are in a sense harmless. They can be ignored in vibration analysis and their participation in transient response can be suppressed through the use of dissipative, implicit time integration algorithms (see, e.g., Hilber *et al.*, 1977; Hilber and Hughes, 1978; Miranda *et al.*, 1989; Chung and Hulbert, 1993; Hughes, 2000). However, they would be detrimental in explicit transient analysis because the frequencies of the highest modes are grossly overestimated and stability would necessitate an unacceptably small time step. See Hughes, 2000, chapter 9. It will be shown in the next section how to completely eliminate the outliers by a reparameterization of the isogeometric mapping. Obviously, the greater the number of outliers, the less efficient is the discrete approximation, and in this situation FEA is at a severe disadvantage compared with NURBS.

5.1.2.2 Linear and nonlinear parameterizations

The most natural way to define a basis and geometrical mapping is to begin with a single linear element with one control point at each end. The parameterization of such a domain is, of course, linear. As the polynomial order is elevated and knots are inserted, the parameterization remains linear (i.e., constant Jacobian determinant), though the spacing of the control points will not be linear. In Cottrell *et al.*, 2006 it has been shown that when studying structural vibrations, a nonlinear parameterization such that the control points are uniformly spaced gives better results. In Figure 5.6, we show the one-dimensional distribution of 21 control points obtained for the two cases using cubic NURBS (top), along with plots of the corresponding parameterization $x = x(\xi)$ (middle) and Jacobian $J(\xi) = \frac{dx(\xi)}{d\xi}$ (bottom). Subsequently, we will refer to this choice, in which control points are uniformly distributed, as "nonlinear parameterization," in contrast with the linear parameterization.

5.1.2.3 Higher-order results

Let us proceed to higher orders, now using the nonlinear parameterization of the domain. Figure 5.7 shows a comparison of k- and p-method numerical spectra for $p = 1, \dots, 4$ (we recall that for $p = 1$ the two methods coincide). Here, the superiority of the isogeometric approach is evident, as one can see that optical branches of spectra *diverge* with p for classical C^0 finite elements. This negative result shows that even higher-order finite elements have no approximability for higher modes in vibration analysis, and possibly explains the fragility of higher-order finite element methods in nonlinear and dynamic applications in which higher modes necessarily participate. In contrast, the entire NURBS spectrum converges for all modes. This dramatic result is all the more compelling when we recall that the result is independent of the geometry in this one-dimensional setting. Results such as these can be understood from a more fundamental functional analysis perspective through the notion of Kolmogorov n-widths. See Appendix 5.A.

5.1.3 Analytically computing the discrete spectrum

Thus far, we have looked at numerical results. We can, however, compute spectra by analytically solving the discrete system. Beginning with the variational form (assuming unit coefficients),

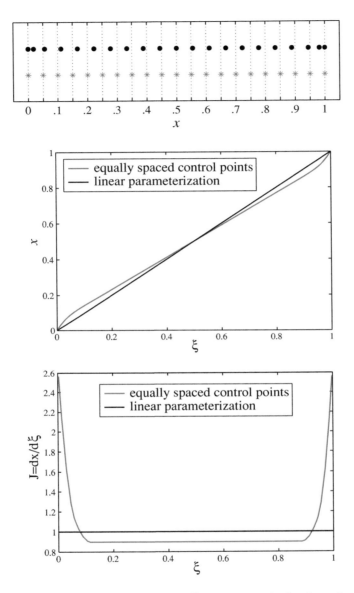

Figure 5.6 One-dimensional case: linear versus nonlinear parameterization determined by uniformly-spaced control points (cubic NURBS, 21 control points). Top: distribution of control points; black dots correspond to linear parameterization control points and red asterisks to uniformly-spaced control points. Middle: Plot of the parameterizations of the two cases. Bottom: Plot of the Jacobians for the two cases.

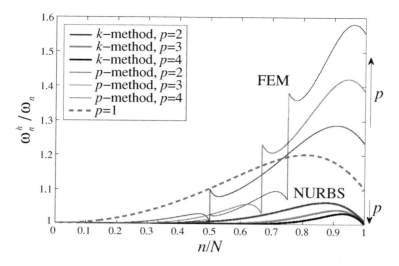

Figure 5.7 Longitudinal vibrations of an elastic rod. Comparison of k-method and p-method numerical spectra.

we have

$$\int_\Omega wu_{,tt}\, d\Omega + \int_\Omega w_{,x}u_{,x}\, d\Omega = 0. \tag{5.19}$$

Letting

$$w = N_A(x) \tag{5.20}$$

and

$$u = u_B(t)N_B(x), \tag{5.21}$$

with the N_A's being the quadratic B-splines used above, we can analytically perform the integration in (5.19) to obtain the **stencil**

$$\frac{h}{120}(\ddot{u}_{A-2} + 26\ddot{u}_{A-1} + 66\ddot{u}_A + 26\ddot{u}_{A+1} + \ddot{u}_{A+2})$$
$$-\frac{1}{6h}(u_{A-2} + 2u_{A-1} - 6u_A + 2u_{A+1} + u_{A+2}) = 0, \tag{5.22}$$

where we use the double-dot to denote differentiation of the coefficient with respect to time. We can compactly rewrite (5.22) as

$$\frac{h^2}{20}\alpha\ddot{u}_A - \beta u_A = 0, \tag{5.23}$$

where α and β are operators defined by

$$\alpha x_A = x_{A-2} + 26x_{A-1} + 66x_A + 26x_{A+1} + x_{A+2},$$
$$\beta x_A = x_{A-2} + 2x_{A-1} - 6x_A + 2x_{A+1} + x_{A+2}. \tag{5.24}$$

We assume the time varying control point to have the form

$$u_A(t) = \phi_A q(t), \tag{5.25}$$

where ϕ_A is a constant coefficient depending only on the control point, A, and $q(t)$ is a function of time that is the same for each control point. Substituting this expression into (5.23) and adding and subtracting $\dfrac{(\omega^h h)^2}{20}\alpha u_A$ leads to

$$(\ddot{q} + (\omega^h)^2 q)\frac{h^2}{20}\alpha\phi_A - (\frac{(\omega^h h)^2}{20}\alpha\phi_A + \beta\phi_A)q = 0. \tag{5.26}$$

The satisfaction of (5.26) is achieved by selecting ϕ_A and q such that

$$\ddot{q} + (\omega^h)^2 q = 0 \tag{5.27}$$

and

$$(\frac{(\omega^h h)^2}{20}\alpha + \beta)\phi_A = 0. \tag{5.28}$$

Assuming a solution for (5.28) of the form (for fixed–fixed boundary conditions)

$$\phi_A = C\sin(A\omega h), \qquad \omega = n\pi, \tag{5.29}$$

(5.28) can be rewritten as

$$(\frac{(\omega^h h)^2}{20}\alpha + \beta)\sin(A\omega h) = 0. \tag{5.30}$$

Recalling the definitions of α and β given in (5.24), and the trigonometric identity $\sin(a \pm b) = \sin(a)\cos(b) \pm \sin(b)\cos(a)$, it requires only a small amount of algebraic manipulation to obtain

$$\frac{(\omega^h h)^2}{20}(16 + 13\cos(\omega h) + \cos^2(\omega h)) - (2 - \cos(\omega h) - \cos^2(\omega h)) = 0, \tag{5.31}$$

which can be solved for $\dfrac{\omega^h}{\omega}$, giving:

$$\frac{\omega^h}{\omega} = \frac{1}{\omega h}\sqrt{\frac{20(2 - \cos(\omega h) - \cos^2(\omega h))}{16 + 13\cos(\omega h) + \cos^2(\omega h)}}. \tag{5.32}$$

Equation (5.32) is the analytical expression for the normalized discrete spectrum using quadratic NURBS basis functions. Analogous calculations can be performed for higher-order approximations. The expression for cubic NURBS is:

$$\frac{\omega^h}{\omega} = \frac{1}{\omega h}\sqrt{\frac{42(16 - 3\cos(\omega h) - 12\cos^2(\omega h) - \cos^3(\omega h))}{272 + 297\cos(\omega h) + 60\cos^2(\omega h) + \cos^3(\omega h)}}. \tag{5.33}$$

For large N, these curves are indistinguishable from the numerical results plotted in Figure 5.7.

5.1.4 Lumped mass approaches

The mass matrix as we have defined it in (5.16) leads to the so-called "consistent mass matrix" – an appropriate name as it is the definition that follows from the variational formulation and leads to optimal order estimates. In a world of unlimited computational resources, this is what we would normally use. There are circumstances, however, in which computational savings can be achieved by using a *diagonal* mass matrix that approximates the consistent mass. Such a **lumped mass matrix** can be inverted trivially by taking the reciprocal of each of its diagonal entries, leading to rapid equation solving. These savings can be even more compelling for time-dependent problems in which the system must be solved many times. In fact, lumped-mass, "explicit" procedures are the fundamental technology in large-scale automobile crash simulation programs and many metal forming applications.

The name "lumped mass" refers to the fact that all of the mass ends up in one spot: the diagonal of the matrix. There are several techniques for obtaining lumped masses. One of the most common approaches is the row-sum technique in which the elements of each row are summed together and lumped on the diagonal. That is

$$\tilde{M}_{AB} = \begin{cases} \int_\Omega \rho N_A \, d\Omega & A = B, \\ 0 & A \neq B. \end{cases} \tag{5.34}$$

To see that this is indeed a row-sum, note that

$$\sum_{B=1}^{n_{eq}} \int_\Omega \rho N_A N_B \, d\Omega = \int_\Omega \rho N_A \left(\sum_{B=1}^{n_{eq}} N_B\right) d\Omega = \int_\Omega \rho N_A \, d\Omega, \tag{5.35}$$

where the last equality follows from the partition of unity property of the basis.

For NURBS, the pointwise positivity of the basis functions ensures that all entries in a row-sum lumped mass matrix will be positive. This is clearly desirable as the consistent mass matrix is positive-definite, and so we would like the lumped mass matrix to be as well. On the other hand, higher-order C^0 finite elements can produce zero or negative lumped masses, which is unacceptable in engineering analysis (see Hughes, 2000, chapter 7).

5.1.4.1 Order of accuracy using row-sum lumped mass

In a manner identical to the approach of Section 5.1.3, the order of accuracy of lumped mass can be obtained analytically. Employing linearly parameterized quadratic NURBS, we derive

$$\frac{\omega^h}{\omega} = \frac{1}{\omega h}\sqrt{\frac{2}{3}(2 - \cos(\omega h) - \cos^2(\omega h))}, \tag{5.36}$$

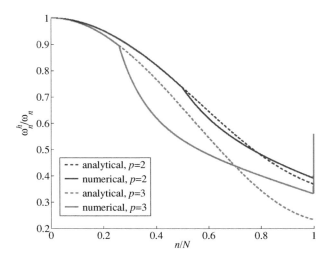

Figure 5.8 Rod problem and row-sum lumped mass. Analytical versus numerical discrete spectra computed using quadratic and cubic NURBS.

while with cubic NURBS we obtain

$$\frac{\omega^h}{\omega} = \frac{1}{\omega h}\sqrt{\frac{1}{15}(16 - 3\cos(\omega h) - 12\cos^2(\omega h) - \cos^3(\omega h))}. \tag{5.37}$$

In these cases, the analytical expressions do not reproduce the behavior of the numerical spectra (which displays some branching), but the two approaches do correspond in the low-frequency part of the spectrum before the slope discontinuity, as shown in Figure 5.8. This is the relevant part as far as order of accuracy is concerned. So, by means of Taylor expansions, we obtain, for quadratic NURBS,

$$\frac{\omega^h}{\omega} \sim 1 - \frac{(\omega h)^2}{8}, \tag{5.38}$$

while for cubic NURBS we get

$$\frac{\omega^h}{\omega} \sim 1 - \frac{(\omega h)^2}{6}. \tag{5.39}$$

As is evident from Figure 5.8, by increasing the order p, higher-order accuracy is not achieved. For row-sum lumped mass, it is always equal to 2. Finally, Figures 5.9 and 5.10 confirm the validity of expressions (5.38) and (5.39), respectively, for low frequencies.

5.1.4.2 A Petrov–Galerkin approach to mass lumping

There is an alternative approach to mass lumping which has some theoretical appeal, but has not yet been investigated thoroughly. Using the techniques described in Schumaker, 2007, we

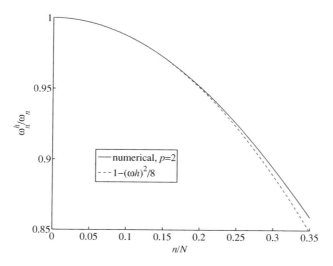

Figure 5.9 Rod problem and row-sum lumped mass. Normalized discrete spectrum using quadratic NURBS compared with $1 - (\omega h)^2/8$ for low frequencies.

can construct a locally supported dual basis, $\{N_A^*\}$, to a given B-spline basis, $\{N_A\}$, such that

$$\int_\Omega N_A^* N_B \, d\Omega = \delta_{AB}. \qquad (5.40)$$

In general, constructing a basis such that (5.40) holds is not difficult at all. In fact, it can be done from a linear combination of the spline functions themselves just by setting up and

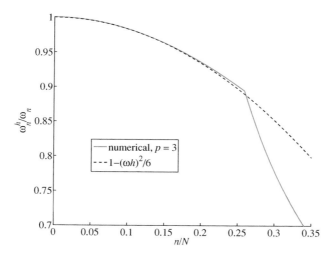

Figure 5.10 Rod problem and row-sum lumped mass. Normalized discrete spectrum using cubic NURBS compared with $1 - (\omega h)^2/6$ for low frequencies.

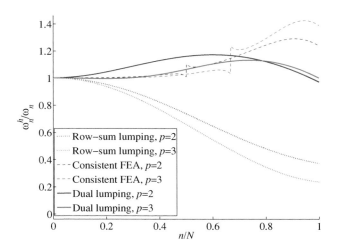

Figure 5.11   Fixed–fixed rod spectra computed with dual mass lumping and row-sum mass lumping, for quadratic and cubic NURBS elements. Consistent classical finite element results are shown for comparison.

solving a simple linear system. Such an approach, however, leads to a globally supported dual basis. The power of the construction of Schumaker, 2007 is that N_A^* and N_A have *exactly* the same support.

A locally supported dual basis opens the possibility of a Petrov–Galerkin method where the solution space is spanned by the spline basis, but the weighting space is spanned by the dual basis. This would lead to a consistent mass matrix that is naturally diagonal. Figure 5.11, shows the analytically computed spectra for this **dual lumping** approach, as well as the analytical results for row-sum lumping as a reference. Not only are the dual-lumping results dramatically better, they also improve as the order increases. Notably, the dual lumped cubic NURBS are competitive with both consistent quadratic and cubic classical finite element approaches.

The use of a fully-integrated stiffness matrix might lead one to expect the order of convergence to increase with p, rather than reaching a plateau as with the row-sum technique. Unfortunately, this is not the case. The dual basis we have constructed is incomplete (not even constant functions can be represented) and so standard error estimates do not apply. In fact, it is not clear that the word "convergence" is really appropriate in this setting. It can be said, however, that all four of the curves in Figure 5.11 have second-order accuracy. The improved results for dual lumping as the order increases are a result of reduced coefficients in the Taylor-expansion of the solution, not higher-order accuracy. In particular, for quadratics we have

$$\frac{\omega^h}{\omega} \sim 1 + \frac{79}{648}(\omega h)^2, \tag{5.41}$$

while for cubics we have

$$\frac{\omega^h}{\omega} \sim 1 + (3\sqrt{2} - \frac{205}{48})(\omega h)^2. \tag{5.42}$$

That is, the second order coefficient for quadratics is approximately 0.1204, while for cubics it is approximately -0.0282.

Aside from the lack of completeness of the dual basis, another downside of this approach is that the dual basis is very difficult to construct. The expressions are non-intuitive and do not lend themselves easily to recursive algorithmic definitions as is the case for B-splines. Also, the dual basis is only piecewise continuous, containing jump discontinuities. Worse still, these discontinuities do not align with the element boundaries (*i.e.*, the knots), meaning that standard quadrature procedures would be inappropriate. Figure 5.12a shows the function N_A^*

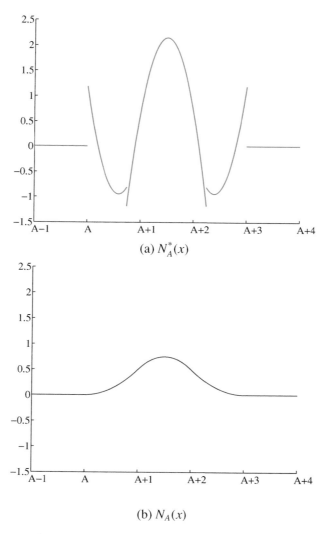

(a) $N_A^*(x)$

(b) $N_A(x)$

Figure 5.12   (a) The C^{-1} dual function $N_A^*(x)$. Note that the discontinuities do not occur at the knots. (b) The associated C^1 quadratic B-spline function $N_A(x)$.

corresponding to the quadratic spline function N_A shown in Figure 5.12b. For further details of the construction of dual bases, see Schumaker, 2007.

We believe that the subject of lumped mass matrices and NURBS deserves further attention. It still may be possible to develop methods that have higher-order accuracy.

An additional consequence of the weighting functions being discontinuous is that we cannot integrate by parts. That is, the stiffness matrix for the rod problem is given by

$$K_{AB} = \int_\Omega N_A^* \frac{\partial^2 N_B}{\partial x^2} \, d\Omega. \tag{5.43}$$

Interestingly, the resulting stiffness matrix remains symmetric in the cases considered.

5.2 Rotation-free analysis of the transverse vibrations of a Bernoulli–Euler beam

We now consider the transverse vibrations of a simply-supported, unit length Bernoulli–Euler beam. Such a beam is the one-dimensional analogue of the Poisson–Kirchhoff plate in that the formulation assumes that there are zero transverse shear strains during bending of the beam (see Section 5.4).

For the Bernoulli–Euler beam, the natural frequencies and modes, assuming unit material and cross-sectional parameters, are governed by:

$$u_{,xxxx} - \omega^2 u = 0 \quad \text{in} \quad \Omega,$$
$$u = 0 \quad \text{on} \quad \Gamma, \tag{5.44}$$
$$u_{,xx} = 0 \quad \text{on} \quad \Gamma,$$

where $\Omega = (0, 1)$, $\Gamma = \partial\Omega = \{0, 1\}$. The analytical frequencies are given by

$$\omega_n = (n\pi)^2, \text{ with } n = 1, 2, 3, \ldots \tag{5.45}$$

The numerical experiments and results for the Bernoulli–Euler beam problem are analogous to the ones reported for the rod. The nonlinear parameterization described earlier is utilized. Note that the classical beam finite element employed to solve problem (5.44) is a two-node Hermite cubic element (see Figure 5.13). Figure 5.14 presents the discrete spectra obtained using different order finite element and NURBS basis functions. The Hermite FEA functions are always C^1, while the NURBS are C^{p-1}. The degrees-of-freedom for the Hermite elements are displacement and slope (i.e., rotation) at each node, whereas the control variables for the NURBS elements are simply displacements. These latter elements are "rotation-free." Again, k-refinement results are dramatically better on a per degree-of-freedom basis.

In the case of a rotation-free beam with zero slope boundary condition, one needs to use Lagrange multiplier and/or penalty methods to enforce the boundary condition. See Hughes et al., 2008a for details.

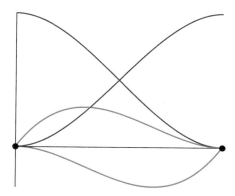

Figure 5.13 Classical Hermite cubic functions. The blue functions are interpolatory at the nodes, at which they always have a slope of zero. Conversely, the functions in red always take a value of zero at the nodes, but their slopes are interpolatory there. This means that Dirichlet conditions relating to slope may be easily implemented strongly. Unfortunately, having two types of functions still leads to a branching of the spectrum, as with the rod.

5.3 Transverse vibrations of an elastic membrane

We present some numerical experiments for the transverse vibrations of an elastic membrane. First, we consider the two-dimensional counterpart of the linear versus nonlinear parameterization discussion of Section 5.1.

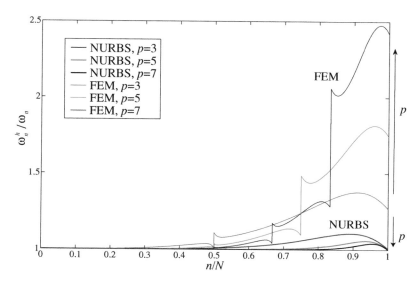

Figure 5.14 Simply-supported beam. Normalized discrete spectra for higher-order finite elements and NURBS.

5.3.1 Linear and nonlinear parameterizations revisited

As we have seen previously, when higher-order functions are used with open knot vectors, a linear, even spacing of the control points does *not* result in a linear parameterization of the domain. Again, the reason for this is that the functions near the beginning and end of the domain are not identical to those in the interior and so they result in adjustments to the control point positions when a linearly parameterized domain is being order-elevated. We can see this for two dimensions in Figure 5.15 for the case of a square domain obtained using $p = q = 4$, where p and q are the orders of the basis functions in the ξ and η (*i.e.*, x and y) directions, and 11×11 control points. We show the control points and mesh for both the linear and nonlinear parameterizations. As before, numerical experiments indicate that the nonlinear parameterization gives rise to better spectral properties by eliminating the outlier frequencies. The reader with a background in finite elements may find these notions alien. However, it is important to deal with them in order to gain an understanding of B-splines and NURBS methodologies.

5.3.2 Formulation and results

The membrane under consideration is an idealization of the elastic behavior of a material with zero thickness. Mathematically, the equations are similar to the one-dimensional rod in Section 5.1. The equation for the axial displacement of the rod is the same as the equation for the transverse displacement of a string with the string tension replacing the Young's modulus. The string is the one-dimensional analogue of the membrane. The natural frequencies and modes, assuming unit tension, density and edge length, are governed by:

$$\Delta u(x, y) + \omega^2 u(x, y) = 0 \quad \text{in} \quad \Omega = (0, 1) \times (0, 1),$$
$$u(x, y) = 0 \quad \text{on} \quad \Gamma = \partial \Omega. \tag{5.46}$$

The exact natural frequencies are (see, e.g., Meirovitch, 1967):

$$\omega_{mn} = \pi \sqrt{m^2 + n^2}, m, n = 1, 2, 3 \dots \tag{5.47}$$

The conversion of (5.46) into its weak, finite-dimensional counterpart proceeds in the usual way, as does the subsequent assembly into a system of algebraic equations. As we have seen for second-order systems, $H^1(\Omega)$ is the appropriate setting for the problem, and so the finite-dimensional subspaces $\mathcal{S}^h = \mathcal{V}^h$ do not require higher-order continuity; C^0 is sufficient. This means that a straight–forward application of Galerkin's method applies to both the isogeometric approach and the FEA approach. Any difference in the quality of the results is again due to the properties of the basis.

Figure 5.16 reports the numerical spectra obtained using 70x70 degrees-of-freedom. The results exhibit similarities to the 1D cases and the superiority of the isogeometric approach is also clear. Again, for higher frequencies, finite element spectra seem to diverge with p.

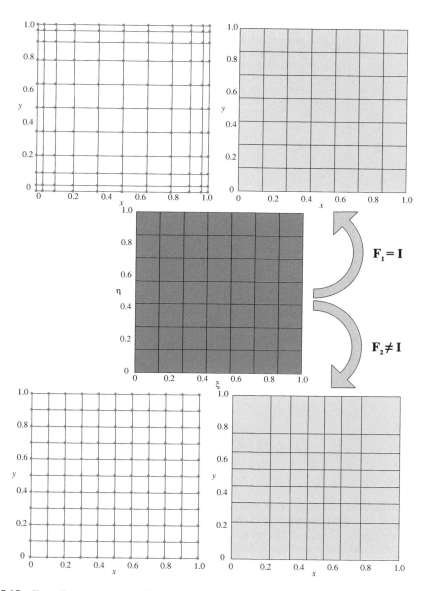

Figure 5.15 Two-dimensional case: linear versus nonlinear parameterization determined by uniformly-spaced control points ($p = q = 4$, 11×11 control points). Top: control net (left) and mesh (right) obtained employing the linear parameterization, both plotted on the *physical domain*. Middle: uniform mesh on the *parametric domain*. Bottom: control net (left) and mesh (right) obtained employing the nonlinear parameterization, both plotted on the *physical domain*.

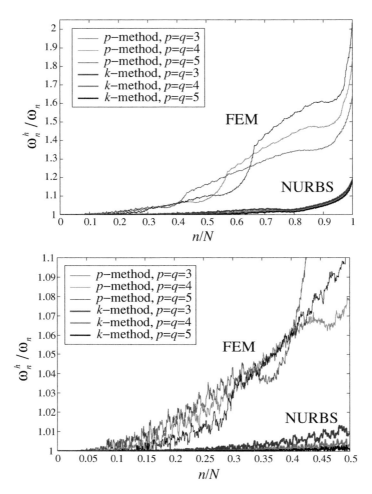

Figure 5.16 Transverse vibrations of an elastic membrane. Comparison of k-method and p-method numerical spectra. Top: entire spectrum. Bottom: detail of the first half of the spectrum. The oscillations in the spectra are due to the mismatches between the numerically obtained frequencies and analytical frequencies. If the exact correspondence is found, the discrete spectra becomes smooth, but it is very difficult to find the correct correspondence. The details of the problem are discussed in Hughes *et al.*, 2008a, along with a two-dimensional example in which the correct correspondence is known.

5.4 Rotation-free analysis of the transverse vibrations of a Poisson–Kirchhoff plate

We consider the transverse vibrations of a simply-supported, square plate governed by Poisson–Kirchhoff plate theory, a two-dimensional analogue of the Bernoulli–Euler beam problem. We do not show any FEA results for this problem, largely because C^1 discretizations in more than one dimension are difficult to construct in classical ways. Alternatively, the smooth NURBS bases are easy to construct, and they perform quite well.

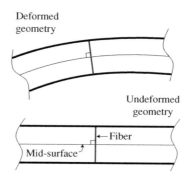

Figure 5.17 The Kirchhoff hypothesis states that straight fibers normal to the mid-surface prior to deformation remain straight *and* remain normal to the mid-surface after deformation.

Let $\Omega \subset \mathbb{R}^2$ denote the mid-surface area of a plate of thickness t. The kinematic behavior of the plate is defined by the transverse displacement of the mid-surface, $u = u(\mathbf{x})$ for $\mathbf{x} \in \Omega$. We consider the rotation of fibers normal to the mid-surface to be equal to the slope of the mid-surface. Thus, *normal fibers remain normal throughout the deformation of the plate*, resulting in zero transverse shear strain; see Figure 5.17. This is the Kirchhoff hypothesis. The transverse displacement is the dependent variable of the theory and all other quantities are derived from it.

The full derivation of the Poisson–Kirchhoff theory is outside the scope of this book, but it can be shown that vibration analysis of such a plate reduces to a biharmonic problem (hence the need for continuous derivatives). The natural frequencies and modes, assuming unit flexural stiffness, density and edge length, are governed by:

$$\Delta\left(\Delta u(x, y)\right) - \omega^2 u(x, y) = 0 \quad \text{in} \quad \Omega = (0, 1) \times (0, 1),$$
$$u(x, y) = 0 \quad \text{on} \quad \Gamma = \partial\Omega, \tag{5.48}$$

for which the exact natural frequencies (see, e.g., Meirovitch, 1967) are:

$$\omega_{mn} = \pi^2(m^2 + n^2), \, m, n = 1, 2, 3 \ldots \tag{5.49}$$

As we have noted, the NURBS formulation results in a rotation-free approach, as was the case for the Bernoulli–Euler beam. The numerical results are similar to the ones obtained for the elastic membrane. Figure 5.18 shows the spectra obtained employing a uniformly-spaced control net.

5.5 Vibrations of a clamped thin circular plate using three-dimensional solid elements

It has been shown in Chapter 4 that higher-order, three-dimensional NURBS elements could be effectively utilized in the analysis of thin structures. In this section we consider the vibrations of a clamped, thin circular plate modeled as a three-dimensional solid. A coarse mesh, but one capable of exactly representing the geometry, is utilized and the order of the basis functions is increased by way of k-refinement.

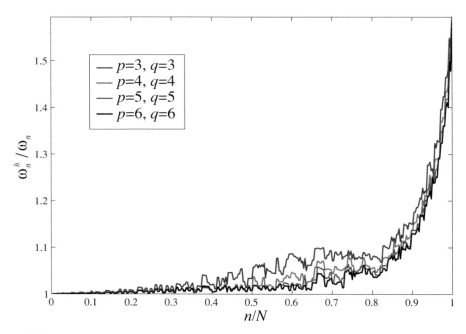

Figure 5.18   Poisson–Kirchhoff plate. Normalized discrete spectra using a uniformly-spaced 90×90 control net corresponding to the nonlinear parameterization of Figure 5.15.

5.5.1 Formulating the problem

The exact Poisson–Kirchhoff solution for this problem, given, for example, in Meirovitch, 1967, is

$$\omega_{mn} = C_{mn}^2 \frac{\pi^2}{R^2} \sqrt{\frac{D}{\rho t}} \; [rad/s], \tag{5.50}$$

where R is the radius of the plate, t is the thickness, $D = \dfrac{Et^3}{12(1 - v^2)}$ is the flexural stiffness (E and v are Young's modulus and Poisson's ratio, respectively) and ρ is the density (mass per unit volume). For the first three frequencies, the values of the coefficients C_{mn} are $C_{01} = 1.015$, $C_{11} = 1.468$ and $C_{02} = 2.007$. The data for the problem are presented in Table 5.1. Note that, because the radius to thickness ratio is 100, the plate may be considered thin, and the results of Poisson–Kirchhoff theory may be considered valid.

In practice, however, we model the plate as a three-dimensional solid. We begin with the unforced equations of linear elastodynamics (these will be discussed in more detail in Chapter 6)

$$\rho u_{j,tt} = \sigma_{jk,k} \quad \text{in } \Omega \times (0, T), \tag{5.51a}$$

$$u_j = g_j \quad \text{on } \Gamma_{D_j} \times (0, T), \tag{5.51b}$$

$$\sigma_{jk} n_k = h_j \quad \text{on } \Gamma_{N_j} \times (0, T). \tag{5.51c}$$

Table 5.1 Clamped circular
plate. Geometric and material
parameters

R	$2\ [m]$
t	$.02\ [m]$
E	$30{\cdot}10^6\ [KN/m^2]$
ν	0.2
ρ	$2.320\ [KNs/m^4]$

As was the case with the rod, if we were interested in actually solving (5.51) for the transient behavior, we would need to augment these equations with initial conditions

$$u(\mathbf{x}, 0) = u_0(\mathbf{x}) \quad \text{in} \quad \Omega, \tag{5.52}$$

$$u_{,t}(\mathbf{x}, 0) = v_0(\mathbf{x}) \quad \text{in} \quad \Omega. \tag{5.53}$$

At present, however, our goal is to pursue the same path for the one-dimensional problems and obtain an eigenproblem. We separate variables, assuming $\mathbf{u}$ to have the form (committing the same notational crime as with the rod)

$$u_j(\mathbf{x}, t) = u_j(\mathbf{x})e^{i\omega t}, \tag{5.54}$$

where we refer to individual components of $\mathbf{u}$ using the letter j, reserving i in this context for the imaginary unit $i = \sqrt{1}$.

Inserting (5.54) into (5.51), multiplying by weighting function $\mathbf{w}$, and integrating by parts leads to the weak form

$$a(\mathbf{w}, \mathbf{u}) - \omega^2\,(\mathbf{w}, \rho\mathbf{u}) = 0, \tag{5.55}$$

where

$$(\mathbf{w}, \rho\mathbf{u}) = \int_\Omega \rho\mathbf{w} \cdot \mathbf{u}\,d\mathbf{x}, \tag{5.56}$$

and $a(\cdot, \cdot)$ is as in (4.18).

We proceed, via Galerkin's method, to obtain matrix equations by inserting the shape-function expansions for $\mathbf{w}^h$ and $\mathbf{u}^h$ into (5.55) to obtain the matrix eigenvalue problem: Find natural frequency $\omega_k^h \in \mathbb{R}^+$ and eigenvector $\mathbf{\Psi}_k$ such that

$$\left(\mathbf{K} - (\omega_k^h)^2\mathbf{M}\right)\mathbf{\Psi}_k = \mathbf{0}, \tag{5.57}$$

where

$$\mathbf{K} = [K_{PQ}], \tag{5.58}$$

$$\mathbf{M} = [M_{PQ}], \tag{5.59}$$

and

$$K_{PQ} = a(N_A \mathbf{e}_i, N_B \mathbf{e}_j),$$ (5.60)

and

$$M_{PQ} = \delta_{ij}(N_A, \rho N_B),$$ (5.61)

with $P = \text{ID}(i, A)$ and $Q = \text{ID}(j, B)$.

5.5.2 Results

For this geometry we no longer have the luxury of evenly spacing the control points in order to tweak the parameterization. The mesh used is topologically identical to that of the solid cylinder of Figure 2.33. Though the dimensions and number of elements are different, they both have a degeneracy along the axis of the cylindrical geometry (note that the thin circular plate is indeed a solid cylinder with a very short "length," that is, thickness) due to the coalescence of many control points to the same point in physical space. This relationship is enforced during assembly by mapping the jth degree-of-freedom for each of the corresponding control variables to a single equation number.

The initial control net consists of $9 \times 4 \times 3$ control points in the θ, r, and z directions, respectively, and quadratic approximations in all three parametric directions. Figure 5.19 shows the mesh, consisting of eight elements within a single patch. The numerical results are compared with the exact solution in Table 5.2, where p, q, and r are the orders of the basis functions in the circumferential, radial, and vertical directions, respectively. Figure 5.20 shows the first three eigenmodes (computed using $p = 4$, $q = 5$, $r = 2$), which are in qualitative agreement with the ones depicted in Meirovitch, 1967. The relative errors (i.e., $(\omega^h - \omega)/\omega$) for these cases are, respectively, 0.0054, 0.00027, and 0.0012. It is important to note that the first and third modes exhibit pointwise radial symmetry.

5.6 The NASA aluminum testbed cylinder

As a final example of isogeometric analysis applied to structural vibrations we consider the NASA aluminum testbed cylinder (ATC). The ATC is a structure inspired by the features of an airplane fuselage which is used by NASA to validate many of the modeling tools involved in the analysis and prediction of interior aircraft noise. It represents an application of isogeometric analysis to a "real world" geometry found in the aerospace industry, demonstrating the

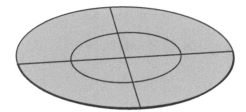

Figure 5.19   Clamped circular plate. Mesh of eight solid elements.

Table 5.2 Clamped circular plate. Numerical results compared with the exact solution

p	q	r	$\omega_{01}[rad/s]$	$\omega_{11}[rad/s]$	$\omega_{02}[rad/s]$
2	2	2	138.133	1648.800	2052.440
2	3	2	56.702	267.765	276.684
3	3	2	56.051	126.684	232.788
3	4	2	54.284	124.417	212.451
4	4	2	54.284	113.209	212.451
4	5	2	54.153	112.700	210.840
exact			53.863	112.670	210.597

feasibility of constructing exact geometrical models of complicated objects, as well as the usage of NURBS on large-scale problems. More importantly, it demonstrates the profound increase in geometrical modeling capability in simply going from linear or quadratic polynomials to *quadratic* NURBS. While higher-order basis functions may be very interesting in analysis, for the geometry it seems to be a fork in the road. It is one of the major accomplishments of isogeometric research up to this point. A thorough discussion of this example is contained in Cottrell *et al.*, 2006.

The ATC is shown in Figures 5.21a and 5.22a. An isogeometric model (see Figures 5.21b and 5.22b) was constructed from design drawings. There are three distinct members composing the frame: nine identical main ribs; twenty-four identical prismatic stringers, and two end ribs. Every geometrical feature of the design drawings is *exactly* represented in the model. Figures 5.23–5.25 show some of the geometrical details. These features are exactly preserved through all levels of refinement.

The downside to the complicated geometry is that, as in the previous example of the clamped plate, we can no longer easily control the parameterization in an effort to eliminate outlier frequencies. Nevertheless, the results obtained compare favorably with the experimental values measured in the laboratory (as reported in Grosveld *et al.*, 2002 and Buehrle *et al.*, 2001). The formulation for the problem is exactly that of Section 5.5.

Numerical results for the frame and skin assembly are presented in Figure 5.26, along with the experimental data. The mesh consisted of 228,936 rational quadratic elements and 2,219,184 degrees-of-freedom. One could reduce the number of degrees-of-freedom significantly by exploiting rotational symmetry and modeling only 1/24 of the frame assembly (as others have done, see Couchman *et al.*, 2003), but part of the goal of this work was to demonstrate the feasibility of modeling an entire real structure of engineering interest using isoparametric NURBS elements, and so no such simplifications were employed.

5.7 Wave propagation

The classical equation governing wave propagation is

$$\Delta u - \frac{1}{c}\frac{d^2u}{dt^2} = 0, \tag{5.62}$$

Eigenmode corresponding to ω_{01}

Eigenmode corresponding to ω_{11}

Eigenmode corresponding to ω_{02}

Figure 5.20   The lowest three computed eigenmodes of the clamped circular plate.

where c is the wave propagation speed. Particular solutions of (5.62) are plane waves of frequency ω traveling in the direction $\mathbf{n}$ at speed c, which can be expressed as the time-harmonic wave train

$$u(\mathbf{x}, t) = Re\left(Ae^{i(k\mathbf{n}\cdot\mathbf{x}-\omega t)}\right), \qquad\qquad (5.63)$$

where $k = \omega/c$ is the wave-number, ω is the angular frequency, and A is a complex amplitude. The wavelength (with units of length) is defined by $\lambda = 2\pi/k$, while the dual measure of period (with units of time) is defined by $T = 2\pi/\omega$.

(a)

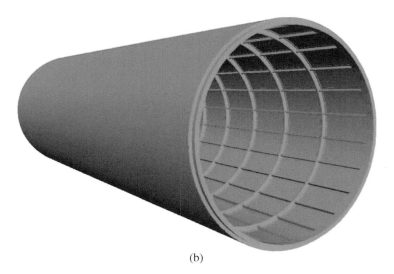

(b)

Figure 5.21   NASA aluminum testbed cylinder (ATC). (a) Frame and skin assembly of the actual ATC. Note that the exterior pegs and rings visible on the skin are part of the measuring apparatus and not part of the ATC itself. (b) The isogeometric model of the frame and skin assembly.

Assuming time-harmonic solutions, as we did for vibrations, we insert

$$u(\mathbf{x}, t) = e^{i\omega t} u(\mathbf{x}), \tag{5.64}$$

into (5.62) reducing the linear wave equation to the Helmholtz equation

$$\Delta u + k^2 u = 0, \tag{5.65}$$

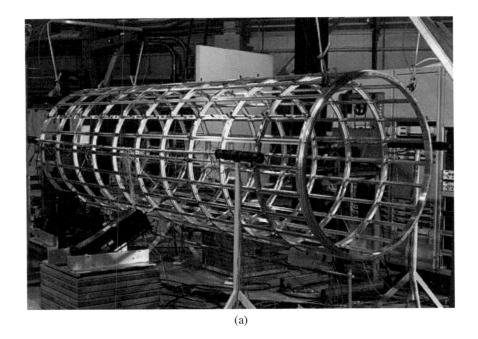

(a)

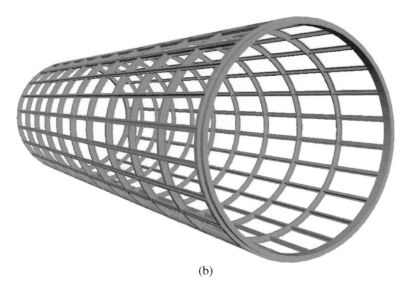

(b)

Figure 5.22 NASA ATC. (a) Frame assembly of actual ATC. (b) The isogeometric model of the frame assembly.

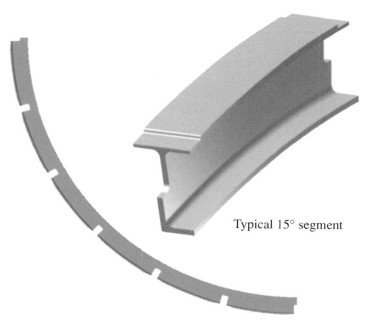

Typical 15° segment

Figure 5.23   NASA ATC. Isogeometric model of the main rib.

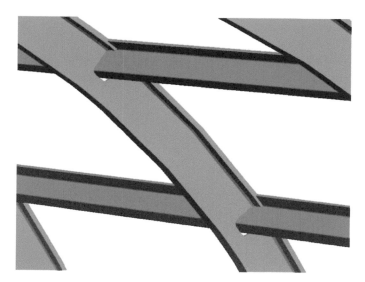

Figure 5.24   NASA ATC. Stringer–main-rib junction. The gaps between the stringers and main ribs, visible in the figure, are a feature of the exact geometry.

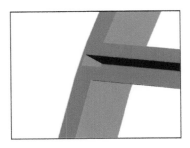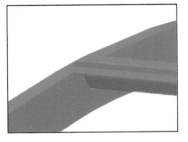

Figure 5.25   NASA ATC. Stringer–end-rib junction.

whose solutions in $\mathbb{R}^n$ are linear combinations of plane waves in space

$$u(\mathbf{x}) = e^{ik\mathbf{n}\cdot\mathbf{x}}. \tag{5.66}$$

The difference between the Helmholtz equation and the equation of free vibration (e.g., equation (5.46)) is that in the former case k^2 is known, whereas in the latter case ω^2 is unknown and must be determined as part of the solution of the eigenproblem.

5.7.1 Dispersion analysis

We proceed to apply Galerkin's method to the Helmholtz equation. After discretization, (5.65) gives rise to the matrix equation

$$\left(\mathbf{K} - k^2\mathbf{M}\right)\mathbf{d} = 0, \tag{5.67}$$

where $\mathbf{d} = \{d_A\}$ is the vector of coefficients defining discrete solution u^h via

$$u^h(\mathbf{x}) = \sum N_A(x)d_A. \tag{5.68}$$

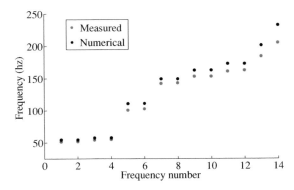

Figure 5.26   NASA ATC. Comparison of numerical and experimental frequency results for the frame and skin assembly.

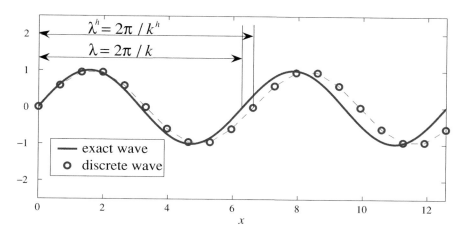

Figure 5.27   Different exact and numerical wave-numbers produce waves with different wavelengths.

The numerical solution of (5.67) is a linear combination of plane waves having numerical wave-number k^h, where, in general, $k^h \neq k$.

Thus, discrete and exact waves have different wavelengths, $2\pi/k^h$ and $2\pi/k$ (see Figure 5.27).

The fundamental issue, which is addressed by ***dispersion analysis***, is to determine how close the discrete wave-number k^h is to its continuous counterpart k.

5.7.2 Duality principle

There is a symmetry between the vibration problems considered previously and the current wave propagation problem (see Hughes *et al.*, 2008a). Compare (5.67) with (5.57) and notice the form of the equations. They are identical under the substitution $\omega^h \leftrightarrow k$. Furthermore, if we exchange ω and k^h as well, we achieve a duality between spectrum analysis and dispersion analysis in the domain where k^h is real.

If we consider the longitudinal vibrations of an elastic rod with unit material properties, as in Section 5.1, modeled using quadratic NURBS, we get the results seen in Figure 5.28 where we have plotted ω^h/ω versus the mode number n normalized by the total number of modes N (this is the same curve plotted for quadratic NURBS in Figure 5.7). Similarly, if we consider the one-dimensional Helmholtz equation using the same quadratic NURBS basis and plot k/k^h versus n/N, we again obtain *exactly* the curve seen in Figure 5.28. This is no accident. The duality principal is independent of the choice of basis used. We can reinterpret spectrum analysis as dispersion analysis, and vice-versa. The conclusion to be drawn is that the excellent behavior of the NURBS basis in vibration analysis carries over to wave propagation, where superior results are again obtained.

Despite the similarities, there are subtle differences between spectrum and dispersion analysis that are of note, particularly for higher-order elements. The first is the existence of "outlier frequencies" in spectrum analysis such as we saw in Section 5.1.2. The second is the existence of complex wave-numbers, which lead to spurious evanescent waves in dispersion analysis.

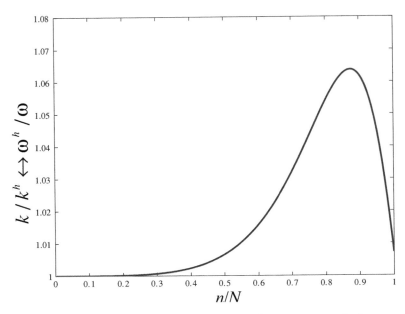

Figure 5.28   Unified dispersion and spectrum analysis for C^1-continuous quadratic approximation.

The non-zero imaginary part of $k^h h$ produces an amplitude modulation of the discrete solutions which is an unphysical feature of the numerical solution. See Hughes *et al.*, 2008a for a thorough discussion of the topic.

Appendix 5.A: Kolmogorov *n*-widths

The approximation result (3.B.8) is a basic tool for proving convergence of NURBS to the solution of partial differential equations with *h*-refined meshes (see Bazilevs *et al.*, 2006a for examples). Note that the continuity of the basis functions does not explicitly appear in (3.B.8). Consequently, the order of convergence in (3.B.8) depends only on the order of the basis functions employed. However, the results of eigenvalue calculations indicate that there is a dramatic difference between C^0- and C^1-continuous p^{th}-order basis functions (see, *e.g.*, Figures 5.7 and 5.14). In Figures 5.7 and 5.14, as p is increased, the upper part of the spectrum *diverges* for C^0-continuous classical finite elements whereas it *converges* for C^{p-1}-continuous NURBS (*i.e.*, B-splines in this case). This phenomenon is not revealed by standard approximation theory results of the form (3.B.8). Consequently, we much conclude that there is a lot of information hiding in the so-called "constant" C in (3.B.8).

     It would be desirable to develop a mathematical framework that revealed behavior like that seen in Figures 5.7 and 5.14 from the outset. The concept of Kolmogorov *n*-widths seems to hold the potential to do so. A sketch of some of the main ideas follows: Let X be a normed, linear space, equipped with norm $\| \cdot \|_X$. In the cases of primary interest here, X would be a Sobolev space. Let X_n be an *n*-dimensional subspace of X. Assume we wish to approximate a given $x \in A \subset X$, where A is a subset of X, with a member $x_n \in X_n$. We define the ***distance***

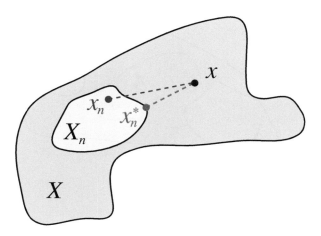

Figure 5.A.1   The point x_n^* is the closest approximation in X_n to x with respect to the norm $\| \cdot \|_X$.

between x and X_n as

$$E(x, X_n; X) = \inf_{x_n \in X_n} \|x - x_n\|_X, \tag{5.A.1}$$

where inf stands for infimum (see Figure 5.A.1). If there exists an x_n^* such that

$$\|x - x_n^*\|_X = E(x, X_n; X) \tag{5.A.2}$$

then x_n^* is called the ***best approximation*** of x (Figure 5.A.1).

Now we assume we are interested in approximating all $x \in A$. For each $x \in A$, the best we can do is expressed by (5.A.2). The question we wish to have answered is, for which $x \in A$ do we get the *worst* best-approximation? In other words, for which $x \in A$ is $\inf_{x_n \in X_n} \|x - x_n\|_X$ the largest? The idea is to anticipate situations such as those depicted in Figures 5.7 and 5.14. The worst best-approximation is obtained by computing the supremum of (5.A.2) over all $x \in A$; we define the ***deviation***, or "sup-inf," as

$$E(A, X_n; X) = \sup_{x \in A} \inf_{x_n \in X_n} \|x - x_n\|_X. \tag{5.A.3}$$

See Figure 5.A.2 for a schematic illustration. Sup-inf's are useful for comparing the approximation quality of different finite element subspaces, such as C^0 and C^{p-1} splines, but prior to that we might ask what is the best n-dimensional subspace for approximating A? This is given by the ***Kolmogorov n-width***, or "inf-sup-inf," namely,

$$d_n(A, X) = \inf_{\substack{X_n \subset X \\ \dim X_n = n}} \sup_{x \in A} \inf_{x_n \in X_n} \|x - x_n\|_X \tag{5.A.4}$$

$$= \inf_{\substack{X_n \subset X \\ \dim X_n = n}} E(A, X_n; X). \tag{5.A.5}$$

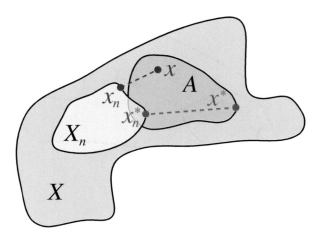

Figure 5.A.2 The distance between subspaces X_n and A is determined by the "worst-case scenario." That is, if the distance between point $x^* \in A$ and its best approximation $x_n^* \in X_n$ is the supremum over all such best-fit pairs, then $\|x^* - x_n^*\|_X$ defines the distance between X_n and A.

If there exists an $\tilde{X}_n$ such that

$$E(A, \tilde{X}_n; X) = d_n(A, X), \tag{5.A.6}$$

then $\tilde{X}_n$ is called an ***optimal n-dimensional subspace***. In this case, we can define the ***optimality ratio***, that is, the sup-inf divided by the inf-sup-inf, for a given X_n:

$$\Lambda(A, X_n; X) = \frac{E(A, X_n; X)}{d_n(A, X)}. \tag{5.A.7}$$

To illustrate how one might use this measure for comparing spaces, consider the following example of a uniform mesh on the unit interval $[0, 1]$. Let $X = H^1(0, 1)$ denote the Sobolev space of (Lebesgue) square-integrable functions with square-integrable derivatives. Let

$$A = B^5(0, 1) = \{x | x \in H^5(0, 1), \|x\|_X \le 1\}, \tag{5.A.8}$$

where $H^5(0, 1)$ is the Sobolev space of functions having five square-integrable derivatives. $B^5(0, 1)$ is referred to as the unit ball in $H^5(0, 1)$ in the $H^1(0, 1)$-topology. A comparison of optimality ratios for quartic C^0 and C^3 splines is shown in Figure 5.A.3. Note that as n increases, the optimality ratio of the C^3 case approaches 1. Apparently, the C^3 case is converging toward an optimal subspace. In contrast, in the C^0 case, the optimality ratio converges to approximately 5.5, indicating that for each n there is at least one member of $B^5(0, 1)$ that is much more poorly approximated by C^0 splines than C^3 splines. This result seems to be qualitatively consistent with what we saw in Figures 5.7 and 5.14. Smooth spline bases, that is the k-method, exhibit better behavior than classical C^0 elements. For further results and methodology used to compute them, see Evans *et al.*, 2009.

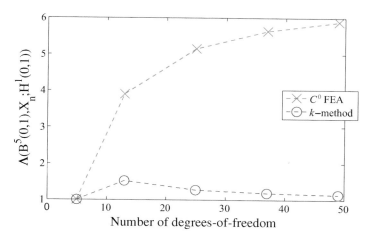

Figure 5.A.3   The optimality ratio for approximating the H^5 unit ball in H^1 using quartic ($p = 4$) elements. As the number of degrees-of-freedom increases, the optimality ratio of C^0 FEA functions diverges, while the optimality ratio of C^3-continuous splines converges toward 1.

Remark

Andrei Kolmogorov (1903–1987) was one of the most important mathematicians of the 20th century. See Figure 5.A.4. In addition to n-widths, he made fundamental contributions to turbulence (*e.g.*, the Kolmogorov inertial spectrum), classical mechanics (*e.g.*, the Kolmogorov–Arnold–Moser theorem), and many other areas. He is perhaps best known for his work in probability theory (*e.g.*, the Kolmogorov axioms), laying the foundation for the modern treatment of the subject.

Figure 5.A.4   Andrei Kolmogorov

Notes

1. In this simple domain, the NURBS reduce to the special case of B-splines.
2. The names "acoustical" branch and "optical" branch originate from the study of a one-dimensional lattice of sodium and chlorine ions vibrating longitudinally. The theoretical values of the natural frequencies when the sodium oscillates in phase with the chlorine are near the range of audible frequencies, and so were dubbed "acoustical," while the theoretical frequencies for the two elements to vibrate out of phase are much higher, near the spectrum of visible light, hence the name "optical."

6

Time-Dependent Problems

As we move from static and eigenvalue problems and begin to approximate the transient behavior of systems, we must employ some type of time integration procedure. The literature is replete with different approaches, many of which have been designed with a specific application in mind. Most of these techniques can either be described as semi-discrete methods or space–time methods. We introduce simple versions of both in this chapter, using the application of elastodynamics as our motivating example.

6.1 Elastodynamics

Elastodynamics is the study of the transient behavior of elastic solids. The developments in this chapter generalize those of Chapter 4 by augmenting the static linear elasticity equations. Whereas in the static case we sought an equilibrium solution in which all forces were balanced, we now consider the case where the forces are imbalanced, and this drives the acceleration of the object by Newton's second law. For elaboration on the notation and constitutive laws, see Chapter 4.

The strong form of the initial/boundary-value problem is

$$\rho u_{i,tt} = \sigma_{ij,j} + f_i \quad \text{in} \quad \Omega \times (0, T) \tag{6.1a}$$

$$u_i = g_i \quad \text{on} \quad \Gamma_{D_i} \times (0, T) \tag{6.1b}$$

$$\sigma_{ij} n_j = h_i \quad \text{on} \quad \Gamma_{N_i} \times (0, T) \tag{6.1c}$$

$$u_i(\mathbf{x}, 0) = u_{0i}(\mathbf{x}) \quad \mathbf{x} \in \Omega \tag{6.1d}$$

$$u_{i,t}(\mathbf{x}, 0) = \dot{u}_{0i}(\mathbf{x}) \quad \mathbf{x} \in \Omega \tag{6.1e}$$

$$\tag{6.1f}$$

where

$$f_i : \Omega \times (0, T) \to \mathbb{R} \tag{6.2}$$

$$g_i : \Gamma_{D_i} \times (0, T) \to \mathbb{R} \tag{6.3}$$

$$h_i : \Gamma_{N_i} \times (0, T) \to \mathbb{R} \tag{6.4}$$

Isogeometric Analysis: Toward Integration of CAD and FEA   by J. A. Cottrell, T. J. R. Hughes, Y. Bazilevs
© 2009, John Wiley & Sons, Ltd

are prescribed data, the given initial displacement and velocity are given by

$$u_{0i} : \Omega \to \mathbb{R}, \tag{6.5}$$

and

$$\dot{u}_{0i} : \Omega \to \mathbb{R}, \tag{6.6}$$

respectively, and the density $\rho : \Omega \to \mathbb{R}^+$ is also specified.

6.2 Semi-discrete methods

The term *semi-discrete* refers to the fact we discretize space using a Galerkin finite element scheme, and formulate the problem as though the time were continuous. In particular, we will represent the solution as a linear combination of basis functions that depend only on space and coefficients that depend upon time. We begin by constructing the weak form. Let us define the space of trial solutions as

$$S_t = \{\mathbf{u}(\cdot, t) | u_i(\mathbf{x}, t) = g_i(\mathbf{x}, t), \mathbf{x} \in \Gamma_{D_i}, \mathbf{u}(\cdot, t) \in H^1(\Omega)\}. \tag{6.7}$$

This definition varies as a function of time because the boundary conditions can evolve in time. The weighting space $\mathcal{V}$ has no time dependence at all.

The weak formulation is now obtained exactly as in the static case. We multiply by a test function and integrate by parts. We assume the Dirichlet data are built directly into the trial solution space, while the Neumann data are incorporated naturally; see Chapter 3. The problem statement is: Given $\mathbf{f}$, $\mathbf{g}$, $\mathbf{h}$, $\mathbf{u}_0$, and $\dot{\mathbf{u}}_0$, find $\mathbf{u}(t) \in S_t$ such that for all $\mathbf{w} \in \mathcal{V}$

$$(\mathbf{w}, \rho\ddot{\mathbf{u}}) + a(\mathbf{w}, \mathbf{u}) = L(\mathbf{w}), \tag{6.8}$$

$$(\mathbf{w}, \rho\mathbf{u}(0)) = (\mathbf{w}, \rho\mathbf{u}_0), \tag{6.9}$$

$$(\mathbf{w}, \rho\dot{\mathbf{u}}(0)) = (\mathbf{w}, \rho\dot{\mathbf{u}}_0), \tag{6.10}$$

where $a(\cdot, \cdot)$ and $L(\cdot)$ are defined as in (4.18) and (4.19), respectively. Note that, in keeping with the semi-discrete approach, we have only integrated with respect to space, leaving time untouched.

6.2.1 Matrix formulation

Proceeding to discretize in space, leaving time continuous, we follow the familiar Galerkin approach. Defining finite-dimensional subspaces $S_t^h \subset S_t$ and $\mathcal{V}^h \subset \mathcal{V}$ that are spanned by the isoparametric basis, we seek a solution of the form $\mathbf{u}^h = \mathbf{v}^h + \mathbf{g}^h \in S_t^h$, with $\mathbf{v}^h \in \mathcal{V}^h$ and $\mathbf{g}^h \in S_t^h$, such that for all $\mathbf{w}^h \in \mathcal{V}^h$

$$(\mathbf{w}^h, \rho\ddot{\mathbf{v}}^h) + a(\mathbf{w}^h, \mathbf{v}^h) = L(\mathbf{w}^h) - a(\mathbf{w}^h, \mathbf{g}^h) - (\mathbf{w}^h, \rho\ddot{\mathbf{g}}^h), \tag{6.11}$$

$$\left(\mathbf{w}^h, \rho\mathbf{v}^h(0)\right) = \left(\mathbf{w}^h, \rho\mathbf{u}_0\right) - \left(\mathbf{w}^h, \rho\mathbf{g}^h(0)\right), \tag{6.12}$$

$$\left(\mathbf{w}^h, \rho\dot{\mathbf{v}}^h(0)\right) = \left(\mathbf{w}^h, \rho\dot{\mathbf{u}}_0\right) - \left(\mathbf{w}^h, \rho\dot{\mathbf{g}}^h(0)\right). \tag{6.13}$$

Representing $\mathbf{v}^h$ by

$$v_i^h = \sum_{A \in \eta - \eta_g} N_A(\mathbf{x}) d_{iA}(t) \tag{6.14}$$

allows us to apply the usual arguments and arrive at a matrix problem. Let

$$\mathbf{M} = [M_{PQ}], \tag{6.15}$$

$$\mathbf{K} = [K_{PQ}], \tag{6.16}$$

$$\mathbf{F} = \{F_P(t)\}, \tag{6.17}$$

$$\mathbf{d}(t) = \{d_Q(t)\}, \tag{6.18}$$

$$\mathbf{d}_0 = \{d_{0Q}\}, \tag{6.19}$$

$$\dot{\mathbf{d}}_0 = \{\dot{d}_{0Q}\}, \tag{6.20}$$

where, with $P = \text{ID}(i, A)$ and $Q = \text{ID}(j, B)$, we have defined

$$M_{PQ} = (N_A \mathbf{e}_i, \rho N_B \mathbf{e}_j) = \delta_{ij} \int_\Omega N_A \rho N_B \, d\Omega, \tag{6.21}$$

$$K_{PQ} = a(N_A \mathbf{e}_i, N_B \mathbf{e}_j), \tag{6.22}$$

$$F_P = L(N_A \mathbf{e}_i) - a(N_A \mathbf{e}_i, \mathbf{g}^h) - (N_A \mathbf{e}_i, \rho \ddot{\mathbf{g}}^h). \tag{6.23}$$

Let us define the intermediate vectors

$$\tilde{d}_{0P} = \left(N_A \mathbf{e}_i, \rho(\mathbf{u}_0 - \mathbf{g}^h)\right), \tag{6.24}$$

$$\dot{\tilde{d}}_{0P} = \left(N_A \mathbf{e}_i, \rho(\dot{\mathbf{u}}_0 - \dot{\mathbf{g}}^h)\right), \tag{6.25}$$

and denote the Q, P entry of the inverse of the mass matrix, $\mathbf{M}^{-1}$, by M_{QP}^{-1}, and thus define

$$d_{0Q} = M_{QP}^{-1} \tilde{d}_{0P}, \tag{6.26}$$

$$\dot{d}_{0Q} = M_{QP}^{-1} \dot{\tilde{d}}_{0P}. \tag{6.27}$$

We can then rewrite (6.11)–(6.13) as a matrix problem,

$$\mathbf{M}\ddot{\mathbf{d}}(t) + \mathbf{K}\mathbf{d}(t) = \mathbf{F}(t), \quad t \in (0, T) \tag{6.28}$$

$$\mathbf{d}(0) = \mathbf{d}_0 \tag{6.29}$$

$$\dot{\mathbf{d}}(0) = \dot{\mathbf{d}}_0. \tag{6.30}$$

This is a system of ordinary differential equations (ODE) for the coefficients $d_P(t)$.

6.2.2 Viscous damping

In structural dynamics we often work with systems of the form

$$\mathbf{M}\ddot{\mathbf{d}} + \mathbf{C}\dot{\mathbf{d}} + \mathbf{K}\mathbf{d} = \mathbf{F}, \tag{6.31}$$

where $\mathbf{C}$ is the ***viscous damping matrix***. It is often convenient or appropriate to assume that the damping has one part that is proportional to the mass and another that is proportional to the stiffness. If we augment (6.1a) to read

$$\rho u_{i,tt} + a\rho u_{i,t} = \sigma_{ij,j} + f_i, \tag{6.32}$$

and modify the generalized Hooke's law to

$$\sigma_{ij} = c_{ijkl}(u_{(k,l)} + b\dot{u}_{(k,l)}), \tag{6.33}$$

where a and b are parameters, then the form of $\mathbf{C}$ in (6.31) is the ***Rayleigh damping matrix***:

$$\mathbf{C} = a\mathbf{M} + b\mathbf{K}. \tag{6.34}$$

The effect of the viscous damping matrix is also felt in modifying the forces due to prescribed displacement boundary conditions. Specifically, we replace the forcing vector by

$$F_P = \text{right-hand side of (6.23)} - \left(N_A\mathbf{e}_i, a\dot{\mathbf{g}}^h\right) - a\left(N_A\mathbf{e}_i, b\dot{\mathbf{g}}^h\right), \tag{6.35}$$

where $P = \text{ID}(i, A)$.

6.2.3 Predictor/multicorrector Newmark algorithms

There are many techniques for solving semi-discrete systems of the form (6.31). Many of them are discussed in Hughes, 2000. Here we focus on an important class of methods known as the ***predictor/multicorrector Newmark algorithm***. We define a step size, Δt and iterate within each time step, n, in order to find $\mathbf{a}_{n+1} \approx \ddot{\mathbf{d}}(t_{n+1})$, $\mathbf{v}_{n+1} \approx \dot{\mathbf{d}}(t_{n+1})$, and $\mathbf{d}_{n+1} \approx \mathbf{d}(t_{n+1})$. At the beginning of each time step, we set the iteration counter to $i = 0$ and enter a ***predictor phase*** where we initialize the approximations as

$$\mathbf{d}^i_{n+1} = \tilde{\mathbf{d}}_{n+1}, \tag{6.36}$$

$$\mathbf{v}^i_{n+1} = \tilde{\mathbf{v}}_{n+1}, \tag{6.37}$$

$$\mathbf{a}^i_{n+1} = \tilde{\mathbf{a}}_{n+1}, \tag{6.38}$$

where $\tilde{\mathbf{d}}_{n+1}$, $\tilde{\mathbf{v}}_{n+1}$, and $\tilde{\mathbf{a}}_{n+1}$ can be chosen in a variety of ways, but they must be consistent with the ***Newmark formulas***, that is,

$$\tilde{\mathbf{d}}_{n+1} = \mathbf{d}_n + \Delta t \mathbf{v}_n + \frac{(\Delta t)^2}{2}\left((1 - 2\beta)\mathbf{a}_n + 2\beta\tilde{\mathbf{a}}_{n+1}\right), \tag{6.39}$$

$$\tilde{\mathbf{v}}_{n+1} = \mathbf{v}_n + \Delta t\left((1 - \gamma)\mathbf{a}_n + \gamma\tilde{\mathbf{a}}_{n+1}\right), \tag{6.40}$$

where Δt is the time step, and β and γ are parameters. Typical choices of the predictors used in practice are as follows:

Constant displacement predictor

$$\tilde{\mathbf{d}}_{n+1} = \mathbf{d}_n \tag{6.41}$$

$$\tilde{\mathbf{a}}_{n+1} = -\frac{1}{\beta \Delta t}\mathbf{v}_n - \frac{(1-2\beta)}{2\beta}\mathbf{a}_n \tag{6.42}$$

$$\tilde{\mathbf{v}}_{n+1} \text{ defined by (6.40)} \tag{6.43}$$

Constant velocity predictor

$$\tilde{\mathbf{v}}_{n+1} = \mathbf{v}_n \tag{6.44}$$

$$\tilde{\mathbf{a}}_{n+1} = -\frac{(1-\gamma)}{\gamma}\mathbf{a}_n \tag{6.45}$$

$$\tilde{\mathbf{d}}_{n+1} \text{ defined by (6.39)} \tag{6.46}$$

Zero acceleration predictor

$$\tilde{\mathbf{a}}_{n+1} = 0 \tag{6.47}$$

$$\tilde{\mathbf{v}}_{n+1} = \mathbf{v}_n + \Delta t(1-\gamma)\mathbf{a}_n \tag{6.48}$$

$$\tilde{\mathbf{d}}_{n+1} = \mathbf{d}_n + \Delta t \mathbf{v}_n + \frac{(\Delta t)^2}{2}(1-2\beta)\mathbf{a}_n \tag{6.49}$$

The constant displacement predictor is often preferred in nonlinear solid mechanics, especially in problems involving large deformations and contact. The constant velocity predictor is usually preferred in problems of fluid mechanics and fluid–structure interaction. The zero acceleration predictor is often used in linear analysis, the situation under consideration here. Obviously, there are many other possibilities.

We use these values to compute a residual as

$$\Delta \mathbf{F}_{n+1}^i = \mathbf{F}_{n+1} - \mathbf{M}_{n+1}\mathbf{a}_{n+1}^i - \mathbf{C}_{n+1}\mathbf{v}_{n+1}^i - \mathbf{K}_{n+1}\mathbf{d}_{n+1}^i. \tag{6.50}$$

We then use this residual to calculate a correction to the acceleration term by solving

$$\mathbf{M}^*\Delta \mathbf{a} = \Delta \mathbf{F}_{n+1}^i, \tag{6.51}$$

where $\mathbf{M}^*$ depends on the exact method being used. For example, we frequently take $\mathbf{M}^*$ to have the form

$$\mathbf{M}^* = \mathbf{M} + \gamma \Delta t \mathbf{C} + \beta(\Delta t)^2 \mathbf{K}. \tag{6.52}$$

The particular choices of the parameters β and γ determine the properties of the method; for example, $2\beta \geq \gamma \geq 1/2$ achieves unconditional stability and $\gamma = 1/2$ attains second-order accuracy.

Another popular choice is $\mathbf{M}^* = \tilde{\mathbf{M}}$, where $\tilde{\mathbf{M}}$ is the lumped mass matrix (see Chapter 5, Section 5.1.4). In this case, the solution of (6.51) is trivial. This is referred to as an *explicit* algorithm because no coupled system of equations needs to be solved to advance the solution.

All other cases are referred to as *implicit*. See Hughes, 2000 for a discussion and stability analysis of implicit and explicit algorithms.

Regardless of the choice of $\mathbf{M}^*$, once we have used it to calculate $\Delta\mathbf{a}$ we enter a *corrector phase* in which we update the solution, *viz.*,

$$\mathbf{a}_{n+1}^{i+1} = \mathbf{a}_{n+1}^i + \Delta\mathbf{a}, \tag{6.53}$$

$$\mathbf{v}_{n+1}^{i+1} = \mathbf{v}_{n+1}^i + \gamma\Delta t\Delta\mathbf{a}, \tag{6.54}$$

$$\mathbf{d}_{n+1}^{i+1} = \mathbf{d}_{n+1}^i + \beta(\Delta t)^2\Delta\mathbf{a}. \tag{6.55}$$

We then check the residual for convergence. If $\|\Delta\mathbf{F}_{n+1}^i\| \le \epsilon\|\Delta\mathbf{F}_{n+1}^0\|$ for some predetermined tolerance ϵ, we move on to the next time step. If not, we increment the iteration counter i by 1 and return to (6.50) and repeat the process. In linear analysis, in exact precision, with (6.52) used to define $\mathbf{M}^*$, $\Delta\mathbf{F}_{n+1}^i \equiv \mathbf{0}$, $i = 1$. However, if the precision of the solution of the linear algebraic system in (6.51) is only approximate, as is often the case in practice, additional iterations may need to be performed. Of course, iteration is the rule in nonlinear analysis, which we will discuss in the next chapter.

The derivation of (6.50)–(6.55) follows from requiring that the correctors satisfy (6.31) and the Newmark formulas:

$$\mathbf{M}\mathbf{a}_{n+1}^{i+1} + \mathbf{C}\mathbf{v}_{n+1}^{i+1} + \mathbf{K}\mathbf{d}_{n+1}^{i+1} = \mathbf{F}_{n+1}, \tag{6.56}$$

$$\mathbf{d}_{n+1}^{i+1} = \mathbf{d}_n + \Delta t\mathbf{v}_n + \frac{(\Delta t)^2}{2}\left((1 - 2\beta)\mathbf{a}_n + 2\beta\mathbf{a}_{n+1}^{i+1}\right), \tag{6.57}$$

$$\mathbf{v}_{n+1}^{i+1} = \mathbf{v}_n + \Delta t\left((1 - \gamma)\mathbf{a}_n + \gamma\mathbf{a}_{n+1}^{i+1}\right). \tag{6.58}$$

We assume (6.57) and (6.58) hold for the ith iterates as well:

$$\mathbf{d}_{n+1}^i = \mathbf{d}_n + \Delta t\mathbf{v}_n + \frac{(\Delta t)^2}{2}\left((1 - 2\beta)\mathbf{a}_n + 2\beta\mathbf{a}_{n+1}^i\right), \tag{6.59}$$

$$\mathbf{v}_{n+1}^i = \mathbf{v}_n + \Delta t\left((1 - \gamma)\mathbf{a}_n + \gamma\mathbf{a}_{n+1}^i\right). \tag{6.60}$$

Subtracting (6.59) and (6.60) from (6.57) and (6.58), respectively, yields (6.53)–(6.55); (6.50)–(6.52) are obtained by substituting (6.53)–(6.55) into (6.56).

The predictor/multicorrector Newmark algorithm is summarized in the flow chart of Figure 6.1.

6.2.3.1 Remarks

1. The classical Newmark algorithm (Newmark, 1959) consists of the Newmark formulas, (6.39) and (6.40), and

$$\mathbf{M}\mathbf{a}_{n+1} + \mathbf{C}\mathbf{v}_{n+1} + \mathbf{K}\mathbf{d}_{n+1} = \mathbf{F}_{n+1}. \tag{6.61}$$

It is a special case of the predictor/multicorrector version in which $\mathbf{M}^*$ is defined by (6.52) and only one iteration is used to obtain $\mathbf{d}_{n+1}$, $\mathbf{v}_{n+1}$, and $\mathbf{a}_{n+1}$. The predictor/multicorrector

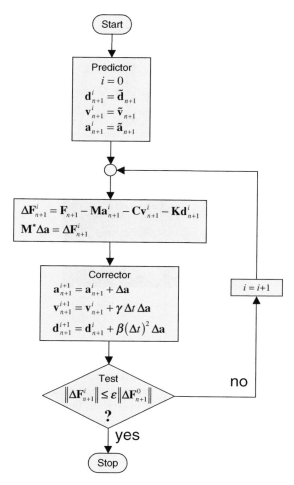

Figure 6.1   Flow chart for the predictor/multicorrector Newmark algorithm. This process takes place within each time step n.

Newmark algorithm was introduced by Hughes *et al.*, 1979 in order to unify the treatment of implicit and explicit algorithms, develop a second-order accurate explicit procedure for fluid dynamic applications, and serve as a framework for implicit–explicit mesh partitions. See Hughes *et al.*, 1979 and Hughes, 2000 for further details.

2. Nathan Newmark (1910–1981) was a world famous applied mechanician and earthquake engineer. He was also a pioneer in electronic computation. The computing laboratory at the University of Illinois, Urbana-Champaign, is named in his honor. See Figure 6.2.

6.3 Space–time finite elements

Unlike the semi-discrete approach, in a ***space–time finite element method*** we discretize both space *and* time. We create a basis in space–time by taking a tensor product of the basis with

Figure 6.2   Nathan Newmark. (Courtesy of the University of Illinois archives.)

which we have described the geometry and a one-dimensional basis in the "time-direction." Thus, we define a space–time domain $Q = \Omega \times (0, T)$ with boundary $\partial Q = P \cup \Omega \times \{0\} \cup \Omega \times \{T\}$, where $P = \Gamma \times (0, T)$ is referred to as the lateral boundary (see Figure 6.3). Clearly, adding another dimension to the mesh greatly increases the amount of data that will need to be stored, as well as the number of floating point operations that will be required to solve the problem. We attempt to minimize this inflated problem size by partitioning the domain into a sequence of *space–time slabs* $Q_n = \Omega \times (t_n, t_{n+1})$, as seen in Figure 6.3, where the basis is discontinuous across the slab boundaries in the time-direction. This allows us to solve only one slab at a time, frequently with each slab having only one element in the time-direction. We take the initial conditions on each slab to be the result at the end of the previous slab, which we enforce weakly as in a discontinuous Galerkin (DG) method. This is equivalent to weakly enforcing continuity of the solution across each slab boundary. Though this approach results in solutions that are discontinuous across the slab boundaries, the cost of solving for the solution on each slab Q_n one at a time, and repeating this N times, is much less than solving just one time on the entire space–time domain Q.

In the semi-discrete case we assumed the basis functions only depended on space, and the coefficients depended upon time. In the space–time setting, the coefficients are constants and the basis functions depend on both space and time. That is

$$w_i^h(\mathbf{x}, t) = \sum_A N_A(\mathbf{x}, t)c_{iA}, \tag{6.62}$$

and

$$u_j^h(\mathbf{x}, t) = \sum_B N_B(\mathbf{x}, t)d_{jB}. \tag{6.63}$$

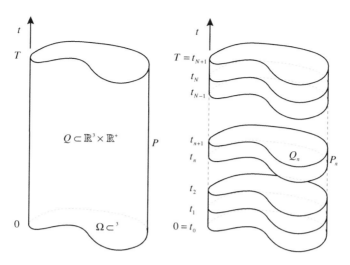

Figure 6.3   Space–time domain (left) and slicing into space–time slabs (right).

The basis functions N_A may be standard finite element or NURBS basis functions defined on the slab. In d space dimensions, the basis functions are $d + 1$-dimensional.

As the solution is now discontinuous, at each slab interface we must distinguish between the solution coming from the "lower slab" and that coming from the "upper slab". We do this with superscripts "+" and "−" such that $\mathbf{u}^h(t_n^-)$ is the solution at time t_n corresponding to slab $Q_{n-1} = \Omega \times (t_{n-1}, t_n)$, and $\mathbf{u}^h(t_n^+)$ is the solution at time t_n corresponding to slab $Q_n = \Omega \times (t_n, t_{n+1})$. We obtain a weak form of (6.1) by multiplying by a weighting function and integrating over the entire slab Q_n. The integration in time gives us the opportunity to impose the initial condition on the bottom of the slab naturally by replacing the unknown $\mathbf{u}^h(t_n^+)$ with the known condition that we would like to impose, namely, $\mathbf{u}^h(t_n^-)$. This is completely analogous to the imposition of Neumann data on the boundary, which should be familiar by now. The resulting Galerkin problem is: find $\mathbf{u}^h \in \mathcal{S}$ such that for all $\mathbf{w}^h \in \mathcal{V}$

$$a_n(\mathbf{w}^h, \mathbf{u}^h) = L_n(\mathbf{w}^h), \tag{6.64}$$

where

$$
\begin{aligned}
a_n(\mathbf{w}^h, \mathbf{u}^h) = {} & \int_{Q_n} w_i^h \rho \ddot{u}_i^h \, dQ + \int_{Q_n} w_{(i,j)}^h \sigma_{ij}\left(\mathbf{u}^h\right) dQ \\
& + \int_{\Omega} w_i^h(t_n^+) \rho \dot{u}_i^h(t_n^+) \, d\Omega \\
& + \int_{\Omega} w_{(i,j)}^h(t_n^+) \sigma_{ij}(\mathbf{u}^h(t_n^+)) \, d\Omega,
\end{aligned} \tag{6.65}
$$

and

$$L_n(\mathbf{w}^h) = \int_{Q_n} \dot{w}_i^h f_i \, dQ + \int_{Z_n} \dot{w}_i^h h_i \, dP$$

$$+ \int_{\Omega} \dot{w}_i^h(t_n^+) \rho \dot{u}_i^h(t_n^-) \, d\Omega$$

$$+ \int_{\Omega} w_{(i,j)}^h(t_n^+) \sigma_{ij}(\mathbf{u}^h(t_n^-)) \, d\Omega, \tag{6.66}$$

with

$$\sigma_{ij}(\mathbf{u}^h) = c_{ijkl} \epsilon_{kl}(\mathbf{u}^h)$$

$$= c_{ijkl} u_{(k,l)}. \tag{6.67}$$

The Neumann boundary of the slab is denoted $Z_n = \Gamma_N \times (t_n, t_{n+1})$. See Chapter 4 for further details.

Assuming the N_A are C^1-continuous on the slab, the Euler–Lagrange form of (6.64) is

$$0 = \int_{Q_n} \dot{w}_i^h \left(\rho \ddot{u}_i^h - \sigma_{ij,j}(\mathbf{u}^h) - f_i \right) dQ$$

$$- \int_{Z_n} \dot{w}_i^h \left(h_i - \sigma_{ij}(\mathbf{u}^h) n_j \right) dP$$

$$+ \int_{\Omega} \dot{w}_i^h(t_n^+) \rho \left[\!\left[\dot{u}_i^h(t_n) \right]\!\right] d\Omega$$

$$+ \int_{\Omega} w_i^h(t_n^+) \left[\!\left[\sigma_{ij,j}(\mathbf{u}^h(t_n)) \right]\!\right] d\Omega, \tag{6.68}$$

where the **temporal jump operator** is defined by

$$\left[\!\left[w(t_n) \right]\!\right] = w(t_n^+) - w(t_n^-), \tag{6.69}$$

and n_j is the jth component of the outward unit normal vector to Γ_N. The first term in (6.68) weakly enforces the differential equation within slab Q_n. The second term weakly enforces the Neumann boundary condition. The third and fourth terms are responsible for the continuity of the velocity and stress divergence, respectively, across the slab boundary at t_n. The weak form treats the values at t_n emanating from slab Q_{n-1} as known data and weakly enforces them as initial conditions on the problem for slab Q_n. The effect is to "penalize" the discontinuous approximations of these continuous functions by adding terms to the residual that are proportional to the jumps in the numerical solution. Such an approach is *consistent* in the sense that the exact solution, $\mathbf{u}$, that satisfies (6.1) will also satisfy (6.68).

On each slab, we use (6.64) to form a matrix problem, which we can solve for the solution within the slab. That solution provides the initial conditions we need for the next slab, and the process continues. This is a very intellectually satisfying approach, and potentially very accurate, but it can be costly as well. The best approach to time integration, semi-discrete or space–time, depends on the demands of the application under consideration. Both methods are firmly established in the literature, with semi-discrete methods more common in practice – for historical reasons if no other. For additional details, see Hughes and Hulbert, 1988 and Hulbert and Hughes, 1990. Explicit discontinuous Galerkin space–time methods, in which calculations can proceed on an element-by-element basis, have been developed by Abedi *et al.*, 2005. See also French 1993, 1998; and French and Peterson, 1996.

7

Nonlinear Isogeometric Analysis

In Chapter 3 we discussed many approaches to the isogeometric analysis of linear differential equations. In particular, we discussed the major components of a NURBS based Galerkin finite element approach. In this chapter, we will expand our investigations to the nonlinear regime. We begin with a brief discussion of the Newton–Raphson method for solving non-linear algebraic equations. Subsequently, guided by the simple but illustrative example of nonlinear heat conduction in one dimension, we will discuss the isogeometric analysis of nonlinear differential equations in a Galerkin setting.

7.1 The Newton–Raphson method

The simplest and best known technique for solving nonlinear algebraic equations is the Newton–Raphson method. It is an iterative approach to finding roots based on Taylor's theorem. In one dimension, suppose that we are trying to find a root x^* of a differentiable, nonlinear function $F(x)$. Beginning with $i = 0$, we will construct a sequence of approximations x^i such that $x^i \to x^*$ and $F(x^*) = 0$. At any step i, we know that $x^* = x^i + \Delta x^i$ for some unknown Δx^i. Though we cannot find Δx^i exactly, we can use a Taylor expansion of F to find an approximation. We have that

$$
0 = F(x^*) = F(x^i + \Delta x^i)
$$

$$
= F(x^i) + \frac{d F(x^i)}{dx} \Delta x^i + O((\Delta x^i)^2)
$$

$$
\approx F(x^i) + \frac{d F(x^i)}{dx} \Delta x^i. \tag{7.1}
$$

Rearranging, we get

$$
\Delta x^i \approx - \left(\frac{d F(x^i)}{dx} \right)^{-1} F(x^i). \tag{7.2}
$$

Isogeometric Analysis: Toward Integration of CAD and FEA   by J. A. Cottrell, T. J. R. Hughes, Y. Bazilevs
© 2009, John Wiley & Sons, Ltd

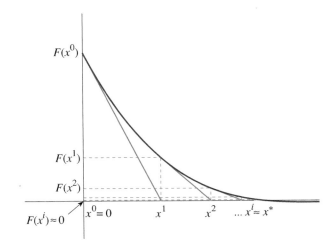

Figure 7.1   An example iteration path with the Newton–Raphson method.

If F were linear, this equation would be exact and we would have $x^* = x^i + \Delta x^i$. For the nonlinear case, we take

$$x^{i+1} = x^i + \Delta x^i \tag{7.3}$$

and repeat the process. We will continue to iterate until $F(x^i) \leq \epsilon |F(x^0)|$ for some predetermined tolerance ϵ. When that threshold is reached, we take $x^i \approx x^*$.

This process is depicted in Figure 7.1. The derivative $dF(x^i)/dx$ is the slope of the line that is tangent to the curve at x^i. We determine x^{i+1} by following this tangent line until it crosses the x-axis.

7.2 Isogeometric analysis of nonlinear differential equations

In the previous chapters, we used Galerkin's method to turn each linear differential equation into a system of linear algebraic equations. Now we will use the same approach to turn nonlinear differential equations into a system of nonlinear algebraic equations. This system can then be solved using the Newton–Raphson approach. In order to make these developments concrete, let us consider an example.

7.2.1 Nonlinear heat conduction

Consider the one-dimensional domain $\Omega = (0, 1)$. Assuming the boundary conditions are comprised of a fixed temperature imposed at $x = 1$ and a heat flux boundary condition specified at $x = 0$, the strong form of the nonlinear heat conduction problem is written as

$$q_{,x} = f \quad \text{in} \quad \Omega, \tag{7.4a}$$

$$u(1) = g, \tag{7.4b}$$

$$-(-1)q(0) = h, \tag{7.4c}$$

$$q(u, u_{,x}, x) = -\kappa(u, x)u_{,x}. \tag{7.4d}$$

The second negative sign in (7.4c) reflects the fact that the outward unit normal at $x = 0$ is $\mathbf{n} = -1$. The constitutive equation (7.4d) contains the nonlinearity that makes this problem fundamentally different to those considered thus far. The thermal conductivity κ depends not only on space, but on the temperature u as well. This constitutive equation is the nonlinear Fourier law of heat conduction.

Let us multiply (7.4a) by a test function and integrate by parts to obtain a weak form: Find $u \in \mathcal{S} = \{u | u(1) = g\}$ such that for all $w \in \mathcal{V} = \{w | w(1) = 0\}$

$$\eta(w; u) = L(w), \tag{7.5}$$

where

$$\eta(w; u) = \int_0^1 w_{,x} \kappa(u, x) u_{,x} \, dx \tag{7.6}$$

and

$$L(w) = \int_0^1 wf \, dx + w(0)h. \tag{7.7}$$

Note that the treatment of both the Dirichlet condition (7.4b) and the Neumann condition (7.4c) is identical to the way such conditions were treated in the linear case.

7.2.2 Applying the Newton–Raphson method

Let us proceed formally and attempt to apply the Newton–Raphson method to solve (7.5). Recall that Newton–Raphson is a root-finding method. For this example, the function that we want to drive to zero is the residual of the equation, defined by

$$R(w; u) = \eta(w; u) - L(w). \tag{7.8}$$

Our goal is to find u^* such that

$$R(w; u^*) = 0 \quad \forall w \in \mathcal{V}. \tag{7.9}$$

As in Section 7.1, we will proceed iteratively, beginning with some initial trial solution u^1, and constructing a sequence u^i such that $u^i \to u^*$. In analogy with (7.2), we seek to solve an equation of the form

$$
\text{``}\Delta u^i = -\left(\frac{dR(w;u^i)}{du}\right)^{-1} R(w;u^i),\text{''}
\tag{7.10}
$$

so that we might update the solution as

$$
u^{i+1} = u^i + \Delta u^i,
\tag{7.11}
$$

and check to see if $\|R(w;u^{i+1})\| < \epsilon\|R(w;u^1)\|$.

At a glance this approach might appear straightforward enough, but things are not as simple as they seem. Note that u is not just a scalar as was x in Section 7.1; it is a function. Thus, we must be very careful about the meaning of terms such as "$(dR/du)^{-1}$." Also, we are dealing with a weak formulation. We must respect the fact that (7.9) is to hold for all $w \in \mathcal{V}$, not necessarily any conceivable w. These issues are beyond the scope of this book. However, the use of the Galerkin finite element formulation allows us to work in a finite-dimensional setting where everything takes on a tangible meaning.

7.2.3 Nonlinear finite element analysis

Returning to an isoparametric setting, we assume that

$$
u^h = \sum_{B=1}^{n_{eq}} N_B d_B,
\tag{7.12}
$$

where N_B refers to the basis functions used to represent the geometry. To simplify the notation, let us drop the superscript h from the trial and weighting functions.

As u is uniquely defined by specifying each of its n_{eq} coefficients, the independent variables are the coefficients d_B, and it is on them that we must iterate. We begin by choosing some initial vector of coefficients $\mathbf{d}^0$. There are various techniques for doing so, but in many cases taking $\mathbf{d}^0 = \mathbf{0}$ is sufficient. Thus at any iteration i we have trial solution u^i that we obtain from inserting $\mathbf{d}^i$ into (7.12). As we want $R(w;u^*) = 0$ for all $w \in \mathcal{V}$, we must be able to choose the weighting function to be any one of the basis functions. Thus, in analogy with (7.1), Taylor's theorem leads us to

$$
\begin{aligned}
0 &= R(N_A; d_C^* N_C) \\
&= R(N_A; d_C^i N_C + \Delta d_C^i N_C) \\
&\approx R(N_A; d_C^i N_C) + \frac{\partial R(N_A; d_C^i N_C)}{\partial d_B} \Delta d_B^i \\
&= R(N_A; u^i) + \left(\frac{\partial R(N_A; u^i)}{\partial d_B}\right) \Delta d_B^i,
\end{aligned}
\tag{7.13}
$$

where we sum on repeated indices, as usual, and we have dropped the h superscript on u^i, that is, u^i is the ith iterate of u^h. Solving for Δd_B^i we have

$$\Delta d_B^i = -\left(\frac{\partial R(N_A; u^i)}{\partial d_B}\right)^{-1} R(N_A; u^i). \qquad (7.14)$$

We define

$$K_{AB}^i = \frac{\partial R(N_A; u^i)}{\partial d_B}$$

$$= \frac{\partial \eta(N_A; u^i)}{\partial d_B}$$

$$= \int_0^1 N_{A,x} \frac{\partial \kappa(u^i, x)}{\partial u} N_B\, u_{,x}^i\, dx$$

$$+ \int_0^1 N_{A,x} \kappa(u^i, x) N_{B,x}\, dx, \qquad (7.15)$$

and

$$R_A^i = R(N_A; u^i) = \eta(N_A; u^i) - L(N_A). \qquad (7.16)$$

Thus, with

$$\mathbf{K}^i = [K_{AB}^i], \qquad (7.17)$$

$$\mathbf{R}^i = \{R_A^i\}, \qquad (7.18)$$

$$\Delta \mathbf{d}^i = \{\Delta d_B^i\}, \qquad (7.19)$$

we can rewrite (7.14) as a matrix equation,

$$\Delta \mathbf{d}^i = -\left(\mathbf{K}^i\right)^{-1} \mathbf{R}^i. \qquad (7.20)$$

We call $\mathbf{K}^i$ the *consistent tangent matrix* because it plays the same role in multiple dimensions that was played by the slope dF/dx in the one-dimensional case of (7.2). We assemble $\mathbf{K}^i$ and the *residual vector* $\mathbf{R}^i$ by looping through the elements and using quadrature to build local tangent matrices and local residual vectors, which we then assemble into a global system just as we did for linear problems. The only difference is that we must repeat this process multiple times as we converge toward a solution, rather than being able to reach it in a single step.

Figure 7.2 shows a flow chart for the Newton–Raphson algorithm. Note that if we are solving a time dependent problem, then this whole process must be performed within each time step. There are many variants of this basic method, most of which are aimed at improving

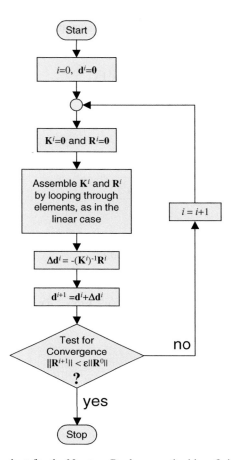

Figure 7.2   Flow chart for the Newton–Raphson method in a finite element setting.

the computational efficiency of the approach. One may choose not to recompute the tangent at every step, or to approximate the tangent in some way.[1] Other variations on the basic approach include choosing $\mathbf{d}^{i+1} = \mathbf{d}^i + \alpha \Delta \mathbf{d}^i$ where $\alpha \neq 1$. It may also be prudent to initialize to some value other than $\mathbf{d}^0 = \mathbf{0}$. For instance, in a time dependent problem, it might make more sense to initialize to the solution at the previous time step, as described in the previous chapter. The basic structure, however, will follow Figure 7.2.

7.3 Nonlinear time integration: The generalized-α method

In this section we present a time integration algorithm for semi-discrete nonlinear equations. The method is the generalized-α method proposed by Chung and Hulbert, 1993 for the equations of structural dynamics, and extended to the equations of fluid mechanics by Jansen et al., 1999. This is the approach used in all of the nonlinear time-dependent calculations in this book.

Consider an abstract, time dependent, nonlinear problem for the vector-valued function $\mathbf{u}(\mathbf{x}, t)$. We can write the variational form of the problem as

$$\eta(\mathbf{w}; \mathbf{u}, \dot{\mathbf{u}}, \ddot{\mathbf{u}}) = L(\mathbf{w}), \tag{7.21}$$

where we have made it clear the solution, $\mathbf{u}$, and its first and second time derivatives, $\dot{\mathbf{u}}$ and $\ddot{\mathbf{u}}$, respectively, can appear explicitly in the semilinear form η. We may think of (7.21) as the weak form of the nonlinear equations of motion and $\mathbf{u}$ the displacement, for definiteness. However, there are other possible interpretations. Following a semi-discrete approach as in Chapter 6, let the vector of control variables, $\mathbf{U}$, depend on time such that

$$\mathbf{u}^h(\mathbf{x}, t) = \mathbf{U}_B(t) N_B(\mathbf{x}). \tag{7.22}$$

Let $\dot{\mathbf{U}}$ and $\ddot{\mathbf{U}}$ denote the first and second time derivatives, respectively, of $\mathbf{U}$. Thus, we can define the residual vector as

$$\mathbf{R}(\mathbf{U}, \dot{\mathbf{U}}, \ddot{\mathbf{U}}) = \{R_P\}, \tag{7.23}$$

where $P = \text{ID}(A, i)$ and

$$R_P = \eta(N_A \mathbf{e}_i; \mathbf{u}^h, \dot{\mathbf{u}}^h, \ddot{\mathbf{u}}^h) - L(N_A \mathbf{e}_i). \tag{7.24}$$

In the time-discrete case, we will adopt the notation of Chapter 6 and replace $\mathbf{U}$, $\dot{\mathbf{U}}$, and $\ddot{\mathbf{U}}$ with $\mathbf{d}$, $\mathbf{v}$, and $\mathbf{a}$, respectively.

The generalized-α time integration algorithm is stated as follows: given $(\mathbf{d}_n, \mathbf{v}_n, \mathbf{a}_n)$, find $(\mathbf{d}_{n+1}, \mathbf{v}_{n+1}, \mathbf{a}_{n+1}, \mathbf{d}_{n+\alpha_f}, \mathbf{v}_{n+\alpha_f}, \mathbf{a}_{n+\alpha_m})$, such that

$$\mathbf{R}(\mathbf{d}_{n+\alpha_f}, \mathbf{v}_{n+\alpha_f}, \mathbf{a}_{n+\alpha_m}) = 0, \tag{7.25}$$

$$\mathbf{d}_{n+\alpha_f} = \mathbf{d}_n + \alpha_f(\mathbf{d}_{n+1} - \mathbf{d}_n), \tag{7.26}$$

$$\mathbf{v}_{n+\alpha_f} = \mathbf{v}_n + \alpha_f(\mathbf{v}_{n+1} - \mathbf{v}_n), \tag{7.27}$$

$$\mathbf{a}_{n+\alpha_m} = \mathbf{a}_n + \alpha_m(\mathbf{a}_{n+1} - \mathbf{a}_n), \tag{7.28}$$

$$\mathbf{v}_{n+1} = \mathbf{v}_n + \Delta t((1 - \gamma)\mathbf{a}_n + \gamma \mathbf{a}_{n+1}), \tag{7.29}$$

$$\mathbf{d}_{n+1} = \mathbf{d}_n + \Delta t \mathbf{v}_n + \frac{(\Delta t)^2}{2}((1 - 2\beta)\mathbf{a}_n + 2\beta \mathbf{a}_{n+1}), \tag{7.30}$$

where $\Delta t = t_{n+1} - t_n$ is the time step, α_f, α_m, γ, and β are real-valued parameters that define the method and are selected to ensure second-order accuracy and unconditional stability. For a second-order linear ordinary differential equation system with constant coefficients, Chung and Hulbert, 1993 showed that second-order accuracy is attained if

$$\gamma = \frac{1}{2} - \alpha_f + \alpha_m, \tag{7.31}$$

and

$$\beta = \frac{1}{4}(1 - \alpha_f + \alpha_m)^2, \tag{7.32}$$

while unconditional stability requires

$$\alpha_m \geq \alpha_f \geq \frac{1}{2}. \tag{7.33}$$

Results (7.31) and (7.33) were also shown by Jansen *et al.*, 1999 to hold true for a first-order linear ordinary differential equation system with constant coefficients. Condition (7.32) only pertains to the second-order case. In the case of a first-order system, **d** plays no role. In order to have strict control over high-frequency damping, α_m and α_f are parameterized by ρ_∞, the spectral radius of the amplification matrix at infinitely large time step. Optimal high-frequency damping occurs when all the eigenvalues of the amplification matrix take on the same value, namely ρ_∞. In this case, for the second-order system, Chung and Hulbert, 1993 derive

$$\alpha_m^c = \frac{2 - \rho_\infty^c}{1 + \rho_\infty^c}, \tag{7.34}$$

$$\alpha_f^c = \frac{1}{1 + \rho_\infty^c},$$

while for the first-order system Jansen *et al.*, 1999 give

$$\alpha_m^j = \frac{1}{2}\left(\frac{3 - \rho_\infty^j}{1 + \rho_\infty^j}\right), \tag{7.35}$$

$$\alpha_f^j = \frac{1}{1 + \rho_\infty^j},$$

where superscripts distinguish the quantities coming from two different methods. The above equations show that for the same values of ρ_∞ (that is, $\rho_\infty^c = \rho_\infty^j$) there is a mismatch between α_m^c and α_f^c. This inconsistency may be eliminated by setting $\rho_\infty^c = \rho_\infty^j = 1$, the case of zero high-frequency damping corresponding to the mid-point rule, but this is not sufficiently robust for nonlinear calculations. In the examples considered in the following chapters, the expressions (7.35) have been used, making the fluid part of the problem optimally damped, and thus the eigenvalues of the amplification matrix for a second-order linear ordinary differential equation system at infinitely large time step are given by an expression obtained in Chung and Hulbert, 1993:

$$\lim_{\Delta t \to \infty} \lambda = \left\{ \frac{-1 + (\alpha_m^j - \alpha_f^j)}{1 + (\alpha_m^j - \alpha_f^j)}, \frac{-1 + (\alpha_m^j - \alpha_f^j)}{1 + (\alpha_m^j - \alpha_f^j)}, 1 - \frac{1}{\alpha_f^j} \right\}. \tag{7.36}$$

Substituting (7.35) into (7.36), we obtain

$$\lim_{\Delta t \to \infty} \lambda = \left\{ \frac{-1 - 3\rho_\infty^j}{3 + \rho_\infty^j}, \frac{-1 - 3\rho_\infty^j}{3 + \rho_\infty^j}, -\rho_\infty^j \right\}. \tag{7.37}$$

The first two eigenvalues are different from $-\rho_\infty^j$, but it is a simple matter to show that they are monotone decreasing functions of ρ_∞^j and

$$\frac{1}{3} \leq \left| \frac{-1 - 3\rho_\infty^j}{3 + \rho_\infty^j} \right| \leq 1 \quad \forall |\rho_\infty^j| \leq 1. \tag{7.38}$$

This, in turn, implies that the spectral radius of the amplification matrix never exceeds unity in magnitude and no instabilities are incurred for a second-order system. Note that this choice of parameters maintains second-order accuracy and unconditional stability because conditions (7.31)–(7.33) still hold true.

To solve the nonlinear system of equations (7.25)–(7.30), we employ a Newton–Raphson method, which can be viewed as a two-phase predictor–multicorrector algorithm. As in the case considered in the previous chapter, there are various possibilities for the predictors. We shall assume the case of the **constant velocity predictor**:

Predictor phase. Set

$$\mathbf{v}_{n+1}^0 = \mathbf{v}_n, \tag{7.39}$$

$$\mathbf{a}_{n+1}^0 = \frac{(\gamma - 1)}{\gamma} \mathbf{a}_n, \tag{7.40}$$

$$\mathbf{d}_{n+1}^0 = \mathbf{d}_n + \Delta t \mathbf{v}_n + \frac{\Delta t^2}{2}((1 - 2\beta)\mathbf{a}_n + 2\beta \mathbf{a}_{n+1}^0). \tag{7.41}$$

The superscript 0 is the iteration index. Note that the predictor is consistent with the generalized-α equations (7.29)–(7.30), which are identical to the Newmark formulas.

Multicorrector phase. Repeat the following steps for $i = 0, 1, 2, \ldots, i_{max}$, or until convergence is achieved.

1. Evaluate iterates at the intermediate time levels as

$$\mathbf{d}_{n+\alpha_f}^i = \mathbf{d}_n + \alpha_f(\mathbf{d}_{n+1}^i - \mathbf{d}_n) \tag{7.42}$$

$$\mathbf{v}_{n+\alpha_f}^i = \mathbf{v}_n + \alpha_f(\mathbf{v}_{n+1}^i - \mathbf{v}_n) \tag{7.43}$$

$$\mathbf{a}_{n+\alpha_m}^i = \mathbf{a}_n + \alpha_m(\mathbf{a}_{n+1}^i - \mathbf{a}_n) \tag{7.44}$$

2. Use the intermediate solutions to assemble the residuals of the continuity and momentum equations and the corresponding matrices in the linear system

$$\frac{d\mathbf{R}^i}{d\mathbf{a}_{n+1}} \Delta \mathbf{a} = -\mathbf{R}_{n+1}^i, \tag{7.45}$$

where

$$\mathbf{R}_{n+1}^i = \mathbf{R}(\mathbf{d}_{n+\alpha_f}^i, \mathbf{v}_{n+\alpha_f}^i, \mathbf{a}_{n+\alpha_m}^i), \tag{7.46}$$

and

$$\frac{d\mathbf{R}^i}{d\mathbf{a}_{n+1}} = \frac{d\mathbf{R}}{d\mathbf{a}_{n+1}}(\mathbf{d}^i_{n+\alpha_f}, \mathbf{v}^i_{n+\alpha_f}, \mathbf{a}^i_{n+\alpha_m}). \tag{7.47}$$

The explicit calculation of the total derivative $d\mathbf{R}/d\mathbf{a}_{n+1}$ is described below. Solve this linear system to a specified tolerance. Various direct and iterative strategies can be employed. In nonlinear applications the preconditioned GMRES algorithm has enjoyed widespread use (see Saad and Schultz, 1986).

3. Having solved the linear system, update the iterates as

$$\mathbf{a}^{i+1}_{n+1} = \mathbf{a}^i_{n+1} + \Delta\mathbf{a} \tag{7.48}$$

$$\mathbf{v}^{i+1}_{n+1} = \mathbf{v}^i_{n+1} + \gamma\Delta t\,\Delta\mathbf{a} \tag{7.49}$$

$$\mathbf{d}^{i+1}_{n+1} = \mathbf{d}^i_{n+1} + \beta(\Delta t)^2\Delta\mathbf{a} \tag{7.50}$$

This process is summarized in Figure 7.3.

Note that (7.45) is the nonlinear analogue of (6.51) in the previous chapter. Defining

$$\mathbf{M}^* \equiv \frac{d\mathbf{R}^i}{d\mathbf{a}_{n+1}}, \tag{7.51}$$

(7.45) may be rewritten as

$$\mathbf{M}^*\Delta\mathbf{a} = -\mathbf{R}^i_{n+1}. \tag{7.52}$$

Dropping the indices and recalling (7.46), repeated application of the chain rule yields

$$\frac{d\mathbf{R}}{d\mathbf{a}_{n+1}} = \frac{\partial\mathbf{R}}{\partial\mathbf{a}_{n+\alpha_m}}\frac{\partial\mathbf{a}_{n+\alpha_m}}{\partial\mathbf{a}_{n+1}}$$

$$+ \frac{\partial\mathbf{R}}{\partial\mathbf{v}_{n+\alpha_f}}\frac{\partial\mathbf{v}_{n+\alpha_f}}{\partial\mathbf{v}_{n+1}}\frac{\partial\mathbf{v}_{n+1}}{\partial\mathbf{a}_{n+1}}$$

$$+ \frac{\partial\mathbf{R}}{\partial\mathbf{d}_{n+\alpha_f}}\frac{\partial\mathbf{d}_{n+\alpha_f}}{\partial\mathbf{d}_{n+1}}\frac{\partial\mathbf{d}_{n+1}}{\partial\mathbf{a}_{n+1}}. \tag{7.53}$$

Using (7.26)–(7.30), we obtain

$$\mathbf{M}^* = \alpha_m\mathbf{M} + \alpha_f\gamma\Delta t\mathbf{C} + \alpha_f\beta(\Delta t)^2\mathbf{K}, \tag{7.54}$$

where

$$\mathbf{M} \equiv \frac{\partial\mathbf{R}}{\partial\mathbf{a}_{n+\alpha_m}} \tag{7.55}$$

$$\mathbf{C} \equiv \frac{\partial\mathbf{R}}{\partial\mathbf{v}_{n+\alpha_f}} \tag{7.56}$$

$$\mathbf{K} \equiv \frac{\partial\mathbf{R}}{\partial\mathbf{d}_{n+\alpha_f}}. \tag{7.57}$$

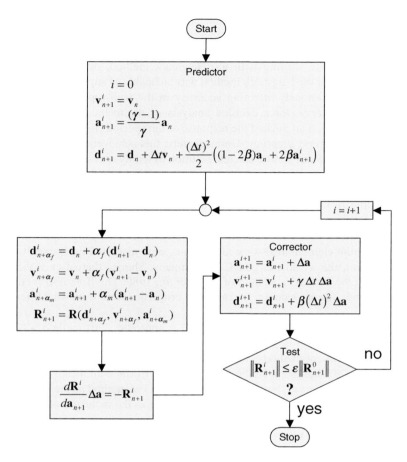

Figure 7.3 Flow chart of the algorithm for the generalized-α method. This process takes place within each time step.

In the special case of $\alpha_m = \alpha_f = 1$, this is the analogue of definition (6.52) for the linear case, and the generalized-α method becomes simply a nonlinear Newmark algorithm.

Remark

The progenitor of the generalized-α method was the HHT-α method (Hilber *et al.*, 1977). The problem facing time integration of finite element models in the 1970's was the fact that the higher-frequency modes were totally inaccurate, representing artifacts of discretization rather than accurate modal behavior of the partial differential equations being discretized. It became apparent that the higher-frequency modal components needed to be suppressed because they were completely spurious. This was accepted at the time as a by-product of discretization and something that had to be lived with. (The results presented in Figures 5.7 and 5.14 illustrate that there are now other and better discretization procedures than the classical C^0 p-method finite elements.) Given the presence of these undesirable modes, it was attempted to develop time integration algorithms which eliminated their participation.

The first individual to recognize this was E. L. Wilson, a professor at the University of California at Berkeley who, starting with the Newmark method, developed the Wilson-θ method. It was realized that the Newmark method was incapable of suppressing higher-modal components without deleteriously affecting accuracy in the well-represented lower modes (see Hughes, 2000, chapter 9, for a detailed analysis of the technical problems). Wilson developed a procedure which attenuated the response of the spurious higher modes, and at the same time retained quite good accuracy. Unfortunately, this achievement was marred by an unforeseen problem, a tendency for the discrete solution to wildly overshoot the exact solution when the loading was very abrupt, such as, for example, in the cases of blast loading and impact between deformable bodies.

The effort leading to the HHT-α algorithm was similarly inspired, but it also became a conscious design requirement that the "overshoot problem" would be mitigated. This was the achievement of HHT-α and it was immediately adopted as a default algorithm in the commercial nonlinear finite element analysis code Abaqus, now marketed by Dassault Systemes (http://www.3ds.com), the world's largest CAD company. Shortly thereafter, a variation on the theme appeared; the Bossak-α method of Wood *et al.*, 1980. The Bossak-α method never caught on, perhaps because its accuracy and stability properties were somewhat inferior to the HHT-α method.

A number of years elapsed before the generalized-α appeared. This method was a simple combination of the HHT-α and Bossak-α methods, but the important result was that superior properties to both constituent methods could be attained through their combination. This was the main contribution of Chung and Hulbert, 1993, who performed an incisive analysis. It all seems rather simple in retrospect, but one can view the HHT-α method as the inclusion of the parameter α_f in the generalized-α algorithm and Bossak-α as the inclusion of the α_m.

As things stand right now, the generalized-α method is enjoying increased usage in software. Up until this time, HHT-α has received the bulk of attention. The original paper of Hilber *et al.*, 1977 is presently the most cited paper in the history of the international journal *Earthquake*

Figure 7.4   David Hilbert.   **Figure 7.5**   Hans-Martin Hilber.

Engineering and Structural Dynamics. A subsequent paper refined understanding of the behavior of the method (Hilber and Hughes, 1978). Nonlinear aspects and a predictor/multicorrector generalization were described in Miranda *et al.*, 1989.

Despite the numerous citations the original HHT-α paper has received, it seemed to vanish from sight for many years. It turned out that the citations were being mistakenly attributed in the Citation Index to a senior author named "Hilbert," not Hilber. This error is perhaps understandable because Hilbert, that is David Hilbert (see Figure 7.4), is one of the most famous names in the history of mathematics. Eventually the error was corrected and the attribution was made to the actual senior author, Hans-Martin Hilber (see Figure 7.5). Hilber, after receiving his Ph.D. from the University of California at Berkeley, pursued a professional career at the engineering firm RIB in southern Germany.

Note

1. While there is some flexibility in the definition of the tangent matrix, the residual vector must be accurate at each iteration for convergence of the method to be achieved.

8

Nearly Incompressible Solids

The problems of linear elastostatics and elastodynamics that we encountered in Chapters 4 and 6, respectively, were implicitly assuming compressible materials. Such problems can be relatively simple to formulate and solve. For parameter values corresponding to many common metals, a straightforward application of Galerkin's method with your element of choice can yield a reasonable solution (though accuracy may depend on an accurate description of the geometry, as we discussed). For other materials, such as rubber, the problems can become much more difficult. Rubber is highly deformable when sheared, but relatively stiff with respect to pressure forces. As a result, rubber is described as a ***nearly incompressible*** material. As we discuss in the next section, the reason this presents such a difficulty is that in the limit of incompressibility, the Poisson's ratio, v, goes to $1/2$, and thus the Lamé parameter

$$\lambda = \frac{vE}{(1+v)(1-2v)} \tag{8.1}$$

appearing in the standard formulation tends toward infinity, and the problem becomes ill-posed in the limit. Even before the incompressible limit is reached, the discrete system resulting from a finite element approach can become quite ill-conditioned. "Locking" occurs and standard elements have difficulty or fail entirely. This is also a challenge in nonlinear problems such as modeling the elastic-plastic response of metals or undrained soils, where nearly incompressible behavior is prevalent. A full discussion of the early experiences and development of effective methodologies for these problems in the linear setting is presented in Hughes, 2000, chapter 4.

It would seem at this stage of the development of finite element technology that higher-order approaches would play an important role in nonlinear structural mechanics, but this is not the case. Nonlinear finite element structural analysis is dominated by the use of low-order "displacement" elements that are specially designed to avoid volumetric or incompressible locking. The only higher-order approach that claims any success is the p-method, in which the polynomial order within elements is increased on a fixed mesh (see Szabo and Babuska, 1991 and Szabo *et al.*, 2004). Though is seems that for standard higher-order C^0-continuous finite elements, volumetric locking is alleviated as the element polynomial order is increased, there is evidence that the accuracy of the solution at any fixed polynomial order is far from optimal. In addition, numerical experience indicates standard higher-order elements are much more

Isogeometric Analysis: Toward Integration of CAD and FEA   by J. A. Cottrell, T. J. R. Hughes, Y. Bazilevs
© 2009, John Wiley & Sons, Ltd

"fragile" than low-order elements. This lack of robustness is particularly apparent in nonlinear dynamic analysis of structures involving contact and impact, and subject to high wave number inputs, such as blast waves. We have already seen some evidence of this in the context of structural vibrations in Chapter 5 where it was revealed that the higher modes produced by the p-method diverge with p. That is, whereas formal accuracy is increased, the improvement is confined to lower modes, while at the same time the higher modes get worse as p increases. This may help to explain why robustness decreases with p in the classical setting.

As we have seen, isogeometric analysis offers a promising alternative to the pitfalls of classical p-refinement. It has been shown that k-refined meshes behave entirely differently than standard finite element methods with respect to higher modal components. In fact, in some cases all discrete modes converge to exact ones and nearly spectral accuracy is achieved. To us, this suggests that robust and higher-order accurate finite element methods applicable to nonlinear structural analysis may be a possibility. Initiatory steps have been taken in this direction by Elguedj *et al.*, 2008.

We also must deal with the locking problem in both small- and finite-deformation regimes. The most general and practically useful approach is by way of a pure displacement formulation (*i.e.*, no pressure degrees-of-freedom) as is generally employed in large-scale structural analysis programs (see, *e.g.*, Livermore Software Technology Corporation, 2007 and Maker, 1995). In order to achieve good behavior, it is imperative to use some form of "projection" to reduce the number of volumetric constraints. This is absolutely essential for lower-order elements, and very important for higher-order elements as well (as we will see). The $\bar{B}$ scheme (see Hughes, 1977) is a formalism that utilizes projection, and here we discuss a family of higher-order $\bar{B}$ schemes as developed in the work of Elguedj *et al.*, 2008.

The large-deformation counterpart of such $\bar{B}$ approaches involve projection of the deformation gradient, a so-called $\bar{F}$ scheme, involving a product decomposition into volumetric and deviatoric factors. Again following Elguedj *et al.*, 2008, we will present the $\bar{F}$ formulation based on a modified minimum potential energy principle. The basics of the method are presented, with the interested reader encouraged to seek the details in the original paper. We also present several numerical calculations on quasi-static, small- and large-deformation test problems.

Remark
Note that other approaches to the locking problem are possible, even some rooted in isogeometric analysis. In particular, the stream function approach of Auricchio *et al.*, 2007 takes advantage of the higher-order continuity of NURBS-based isogeometric analysis to obtain a divergence free (*i.e.*, incompressible) displacement field from the differentiation of a potential.

8.1 $\bar{B}$ formulation for linear elasticity using NURBS

As we saw in Chapter 4, the boundary value problem of compressible elasticity for a body Ω is given by:

Given $\mathbf{f} : \Omega \to \mathbb{R}^3$, $\mathbf{g} : \Gamma_g \to \mathbb{R}^3$, and $\mathbf{h} : \Gamma_h \to \mathbb{R}^3$, find $\mathbf{u} : \overline{\Omega} \to \mathbb{R}^3$ such that:

$$\mathrm{div}\boldsymbol{\sigma} + \mathbf{f} = 0 \qquad \text{in } \Omega, \tag{8.2}$$

$$\mathbf{u} = \mathbf{g} \qquad \text{on } \Gamma_g, \tag{8.3}$$

$$\boldsymbol{\sigma} \cdot \mathbf{n} = \mathbf{h} \qquad \text{on } \Gamma_h, \tag{8.4}$$

where $\mathbf{n}$ is the exterior unit normal on Γ, the boundary of Ω, $\mathbf{g}$ is the prescribed displacement on Γ_g and $\mathbf{h}$ is the prescribed traction on Γ_h, which together form the boundary $\Gamma = \overline{\Gamma_h \cup \Gamma_g}$ of Ω, and $\mathbf{f}$ is the body force. The stress tensor $\boldsymbol{\sigma}$ is defined in terms of the strain tensor $\boldsymbol{\varepsilon}$ by the generalized Hooke's law:

$$\boldsymbol{\varepsilon} = \nabla^s \mathbf{u} = \frac{1}{2}(\nabla \mathbf{u} + \nabla \mathbf{u}^T) \text{ or } \varepsilon_{ij} = \frac{1}{2}\left(\frac{\partial u_i}{\partial x_j} + \frac{\partial u_j}{\partial x_i}\right), \tag{8.5}$$

$$\boldsymbol{\sigma} = \mathbf{c} : \boldsymbol{\varepsilon} \text{ or } \sigma_{ij} = c_{ijkl}\varepsilon_{kl} \tag{8.6}$$

In the compressible isotropic linear elastic case, Hooke's law can be expressed in terms of the Lamé parameters λ and μ by:

$$c_{ijkl} = \lambda\delta_{ij}\delta_{kl} + \mu(\delta_{ik}\delta_{jl} + \delta_{il}\delta_{jk}), \tag{8.7}$$

$$\sigma_{ij} = \lambda u_{k,k}\delta_{ij} + 2\mu\varepsilon_{ij}, \tag{8.8}$$

where

$$\lambda = \frac{2\nu\mu}{(1-2\nu)}, \quad \mu = \frac{E}{2(1+\nu)}, \tag{8.9}$$

and ν is Poisson's ratio and E is Young's modulus, as in Chapter 4.

As mentioned in the introduction to this chapter, as $\nu \to \frac{1}{2}$, λ approaches infinity, that is, there is a singular limit in the stress–strain relation. The value $\nu = \frac{1}{2}$ thus represents incompressibility. The constitutive equation needs to be modified in this case to

$$\sigma_{ij} = -p\delta_{ij} + 2\mu\varepsilon_{ij}, \tag{8.10}$$

where p, the hydrostatic pressure, is determined as part of the solution of the boundary value problem. As p represents an additional unknown, the kinematic condition of incompressibility must be introduced as an additional equation:

$$\text{div}\mathbf{u} = u_{k,k} = 0 \qquad \text{in } \Omega, \tag{8.11}$$

leading to a ***mixed method***. In the nearly incompressible case, however, the ratio λ/μ is large but not infinite. The compressible theory applies, but some modifications in the discrete case are warranted to help alleviate the propensity for mesh locking.

8.1.1 An intuitive look at mesh locking

To understand mesh locking and why it occurs as incompressibility is approached, it is enough to consider a very simple situation. Figure 8.1a shows a mesh comprised of two linear, triangular elements in which the displacements of three of the four nodes are fixed to be zero due to homogeneous Dirichlet boundary conditions. We assume plane strain conditions, that is there is no strain in the out-of-plane direction (*e.g.*, $\varepsilon_{33} = 0$). Only the fourth node (shown in red) is permitted to move.

Now let us assume that the material is incompressible. This means that (8.11) is satisfied pointwise. By integrating (8.11) over the volume of each element, we conclude that the volume remains unchanged after deformation. By virtue of the plane strain constraint, this in

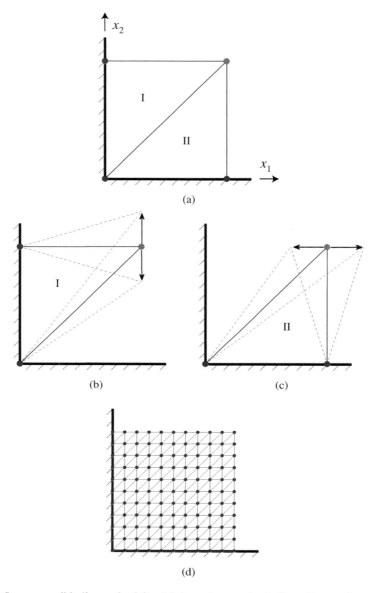

Figure 8.1   Incompressible linear elasticity. (a) A mesh comprised of two linear, triangular elements. The top and left sides are fixed due to homogeneous Dirichlet boundary conditions. (b) The boundary condition on element I constrains the red node to only move in the vertical direction if area is to be conserved. (c) The boundary condition on element II constrains the red node to only move in the horizontal direction if area is to be conserved. Taken together, these constraints dictate that *the position of the node must be fixed,* and both elements are completely "locked." (d) If the two elements in the lower left-hand corner lock, as in (a)–(c), this effectively places a Dirichlet condition on the neighboring elements, which in turn lock as well. The effect propagates, and the entire mesh locks, admitting no displacement at all.

turn means that the area remains unchanged. For element I, this means that the unconstrained node can only move in the vertical direction and must have zero displacement in the horizontal direction, as in Figure 8.1b. For element II, the situation is reversed. The base of the element is fixed and so the height must also be fixed if the element is to preserve its area. Thus, element II admits only horizontal displacements and not vertical displacements, as in Figure 8.1c. Taken together, the two-element mesh admits no displacements at all and so it is completely "locked."

Considering the much larger mesh of Figure 8.1d, the above argument clearly applies to the elements in the lower left-hand corner of the mesh. Once it is determined that they locked, the fixed fourth node acts as a homogeneous Dirichlet condition for the neighboring elements. The same argument extends to them, and they lock as well. The behavior spreads, and no deformation is possible at all. The mesh is fully locked, and any numerical results are meaningless.

If we consider the *nearly* incompressible case, the same logic dictates that the deformations of the linear elements must be very small. If, alternatively, we had a mesh of very high polynomial order elements, then deformations would be possible in which the sides of the elements do not remain straight. In such a case, much larger deformations could take place while the material remains almost incompressible. This is the intuition behind the fact that higher-order elements are less prone to locking.

Heuristically, when the number of constraints placed on an element is large relative to its number of degrees-of-freedom, there is very little for it to do other than lock. Thus, avoidance of locking can be achieved by increasing the ratio of the number of degrees-of-freedom to the number of constraints. The $\bar{B}$ method is a procedure for achieving this objective.

Remark

In the case of plane stress, that is when the out-of-plane stress $\sigma_{33} = 0$, there is no locking problem. The reason for this is that, from (8.8), $0 = \lambda(\varepsilon_{11} + \varepsilon_{22} + \varepsilon_{33}) + 2\mu\varepsilon_{33}$, and, therefore

$$\varepsilon_{33} = \frac{-\lambda}{\lambda + 2\mu}(\varepsilon_{22} + \varepsilon_{33}) \tag{8.12}$$

Using this result in (8.8) yields

$$\sigma_{\alpha\beta} = \tilde{\lambda}\varepsilon_{\gamma\gamma}\delta_{\alpha\beta} + 2\mu\varepsilon_{\alpha\beta}, \tag{8.13}$$

where α, β, and γ range over 1 and 2, and

$$\tilde{\lambda} = \frac{2\mu\lambda}{\lambda + 2\mu}. \tag{8.14}$$

In this case, as $\lambda \to \infty$, $\tilde{\lambda} \to 2\mu$ and there is no singularity in the stress–strain relation.

8.1.2 Strain projection and the $\bar{B}$ method

The strain projection approach, referred to as the $\bar{B}$ method, was introduced by Hughes, 1980. The main idea in the strain projection approach is to additively split the strain tensor into its deviatoric and dilatational (*i.e.*, volumetric) parts

$$\boldsymbol{\varepsilon}(\mathbf{u}) = \boldsymbol{\varepsilon}^{\mathrm{dev}}(\mathbf{u}) + \boldsymbol{\varepsilon}^{\mathrm{dil}}(\mathbf{u}), \tag{8.15}$$

where

$$\boldsymbol{\varepsilon}^{\mathrm{dil}}(\mathbf{u}) = \frac{1}{3}\,(\mathrm{div}\mathbf{u})\,\mathbf{I} \ \text{ or } \ \varepsilon_{ij}^{\mathrm{dil}}(\mathbf{u}) = \frac{1}{3}\frac{\partial u_k}{\partial x_k}\delta_{ij}, \tag{8.16}$$

and $\mathbf{I}$ is the identity tensor.

To achieve an effective formulation in the nearly incompressible case, the dilatational part is replaced by an "improved" dilatational contribution (i.e., a projected one), using a linear projection operator π

$$\bar{\boldsymbol{\varepsilon}}^{\mathrm{dil}}(\mathbf{u}) = \pi\left(\boldsymbol{\varepsilon}^{\mathrm{dil}}(\mathbf{u})\right). \tag{8.17}$$

In terms of $\mathbf{B}$, the strain-displacement matrix introduced in Chapter 4 is replaced by

$$\bar{\mathbf{B}} = \mathbf{B}^{\mathrm{dev}} + \bar{\mathbf{B}}^{\mathrm{dil}}, \tag{8.18}$$

with

$$\mathbf{B}^{\mathrm{dev}} = \mathbf{B} - \mathbf{B}^{\mathrm{dil}}. \tag{8.19}$$

The effect of this new definition is to lower the number of volumetric constraints, mitigating the tendency to lock.

8.1.3 $\bar{B}$, the projection operator, and NURBS

The use of the $\bar{\mathbf{B}}$ method within isogeometric analysis requires further investigation into the choices of the projection operator and the associated space onto which the projection will be performed. Since the technique has been applied mostly to piecewise bilinear and trilinear finite elements, and we want to make intensive use of the properties of high-order k-refined NURBS, these topics need to be studied without any assumption on the order of approximation.

In the discrete case, we have

$$\mathbf{u}^h(\mathbf{x}) = \sum_{A=1}^{n} \mathbf{u}^A N^A(\mathbf{x}), \tag{8.20}$$

likewise

$$\mathbf{w}^h(\mathbf{x}) = \sum_{A=1}^{n} \mathbf{w}^A N^A(\mathbf{x}), \tag{8.21}$$

where N^A are the NURBS basis functions and $\mathbf{u}^A$ and $\mathbf{w}^A$ are the associated control variables. Note that we have temporarily shifted to the convention of making A a superscript for notational clarity in what follows.

In developing the $\bar{\mathbf{B}}$ method for higher-order finite elements and NURBS, we need to define the linear projection operator and the spaces upon which to project the dilatational strain. Throughout, we use the L^2 projection of the strains. For the spaces, we define the following procedure: assume the displacement space is given. We shall refer to it as Q_p, that is quadrilateral, or hexahedral, elements of order p. The continuity of Q_p elements within a patch can be any order k from 0 to $p-1$. We are particularly interested in elements of maximal continuity, namely, C^{p-1}. As we have throughout, we assume an open knot vector construction so that only C^0 continuity is attained across patch interfaces. The basis functions for the projected dilatational strain are taken to be one order lower, and usually one order of continuity lower, namely the space Q_{p-1}, of continuity C^{p-2}. The only exception occurs when $p \geq 2$, but there are lines or surfaces of C^0 continuity *within* a patch. In this situation, the projected space is also taken to be C^0 continuous across those lines or surfaces.

There is nothing fundamental about this choice for the space onto which we project, but it is a particularly convenient one from the point of view of implementation. The goal is to project onto a space that is coarser in some way than the solution space. The current choice of spaces is not even nested, but the lower-order NURBS space onto which we project does have fewer degrees-of-freedom than its k-refined counterpart that is used for the solution space. The decreased degrees-of-freedom and the lower polynomial order of this space renders it sufficiently coarse relative to the solution space. The validity of the choice is argued in Elguedj *et al.*, 2008 and supported by the numerical results obtained.

As an example of what this construction produces, consider the displacement space Q_1. This is the space of bilinear quadrilaterals, or trilinear hexahedra, and is C^0 continuous across element boundaries (which correspond to knots in this case). The space for projected dilatational strain is then Q_0, of continuity class C^{-1}, that is, piecewise constants. This element becomes the classical mean dilatational element (see Hughes and Allik, 1969; Nagtegaal *et al.*, 1974; Hughes, 2000) referred to, herein, as Q_1/Q_0.

In constructing the Q_p/Q_{p-1} spaces, open knot vectors are employed on each patch. Tensor product constructs are utilized so we focus on the situation in each direction separately. We need to specify the order of the space and the knot vector. We begin with the displacement space, assumed to be of order $p \geq 1$. The knot vector, denoted Ξ_p, is assumed to have the following form,

$$\Xi_p = \{\underbrace{0, 0, \ldots, 0}_{p+1 \text{ copies}}, \Xi_{int}, \underbrace{1, 1, \ldots, 1}_{p+1 \text{ copies}}\}, \tag{8.22}$$

where, for simplicity, we have assumed the initial and final knots are located at 0 and 1, respectively. Ξ_{int} denotes the vector of internal knots. The case we are primarily concerned with at present is each internal knot having multiplicity 1 which results in maximal smoothness of continuity class C^{p-1} on each patch. The corresponding knot vector for the projected space, denoted Ξ_{p-1}, is given by

$$\Xi_{p-1} = \{\underbrace{0, 0, \ldots, 0}_{p \text{ copies}}, \Xi_{int}, \underbrace{1, 1, \ldots, 1}_{p \text{ copies}}\}. \tag{8.23}$$

The order of the projected space is taken to be $p - 1 \geq 0$. The span of the projected space is precisely the span of the derivatives of all functions in the displacement space. An example of the spaces Q_p/Q_{p-1}, $p = 1, 2, 3$, in the general case, for a one-dimensional patch of four elements, is given in Figure 8.2. We see that C^{p-1}/C^{p-2} continuity is achieved in all cases.

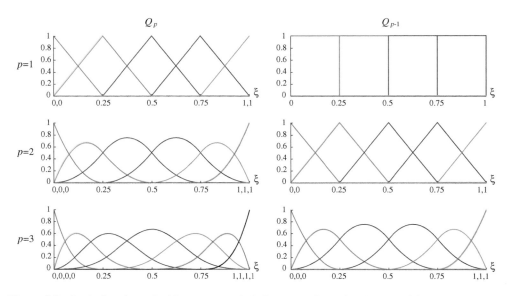

Figure 8.2 Basis functions Q_p/Q_{p-1}, $p = 1, 2, 3$, for a one-dimensional patch of four elements. All cases attain continuity C^{p-1}/C^{p-2}.

The exception to the general case occurs when there are lines or surfaces of only C^0 continuity within a patch, as mentioned previously. Let us assume that Ξ_{int} has one or more knots having multiplicity p, signifying C^0 continuity. Then, in the space Ξ_{p-1}, Ξ_{int} needs to be replaced with $\tilde{\Xi}_{int}$, which is identical to Ξ_{int} except for the knots having multiplicity p; in $\tilde{\Xi}_{int}$ these knots have multiplicity $p - 1$, preserving C^0 continuity of the projected space within each patch. An example of the spaces Q_p/Q_{p-1}, $p = 2, 3$, with a point of C^0 continuity inside the patch, is given in Figure 8.3. Only C^0 continuity is achieved across the repeated knot $\xi = 0.75$, and C^{p-1}/C^{p-2} continuity is achieved elsewhere on the patch interior.

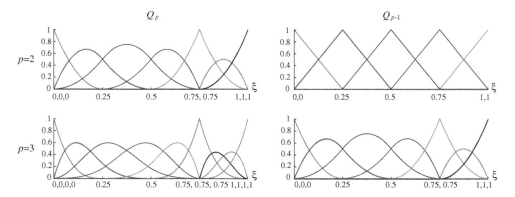

Figure 8.3 Basis functions Q_p/Q_{p-1}, $p = 2, 3$, with a point of C^0 continuity within the patch, for a one-dimensional patch of four elements. All cases attain continuity C^{p-1}/C^{p-2}, except at the repeated knot $\xi = 0.75$ where the continuity is only C^0.

Let us denote the basis functions in the parameter space as $\{\hat{N}^A\}$, and the basis in the physical space as $\{N^A\}$, such that

$$N^A = \hat{N}^A \circ \mathbf{x}^{-1}, \tag{8.24}$$

where $\mathbf{x} : \hat{\Omega} \to \Omega$ is the geometrical mapping, defined via control points $\{\mathbf{x}^A\}$ as

$$\mathbf{x}(\boldsymbol{\xi}) = \sum_A \hat{N}^A(\boldsymbol{\xi})\mathbf{x}^A. \tag{8.25}$$

Using the same geometrical mapping, we construct the "tilde basis" $\{\tilde{N}^A\}$, which corresponds to the projection space, by

$$\tilde{N}_A = \hat{\tilde{N}}_A \circ \mathbf{x}^{-1}, \tag{8.26}$$

where $\{\hat{\tilde{N}}\}$ is the lower order NURBS basis built on the same parametric domain. As described previously, we take the $\hat{\tilde{N}}_A$'s to be one order lower than the $\hat{N}^A$'s to reduce the number of incompressibility constraints.

Note that, even in the case of lowest-order elements (*i.e.*, bilinear and trilinear), we still use the exact geometrical mapping. This means the lowest-order elements are isogeometric and precisely fit curved boundaries. We believe that Barth, 1998 was the first to use this approach, and to demonstrate its effectiveness in compressible fluid calculations.

In the discrete case, (8.17) becomes:

$$\bar{\varepsilon}_{ij}^{dil}(\mathbf{u}^h) = \sum_{A=1}^{\tilde{n}} \tilde{N}_A \tilde{\varepsilon}_{ij}^A, \tag{8.27}$$

where

$$\tilde{\varepsilon}_{ij}^A = \sum_{B=1}^{\tilde{n}} \tilde{M}_{AB}^{-1} \left(\tilde{N}_B, \varepsilon_{ij}^{dil}(\mathbf{u}^h) \right)_\Omega = \sum_{B=1}^{\tilde{n}} \tilde{M}_{AB}^{-1} \int_\Omega \tilde{N}_B \varepsilon_{ij}^{dil}(\mathbf{u}^h) d\Omega, \tag{8.28}$$

that is

$$\bar{\varepsilon}_{ij}^{dil}(\mathbf{u}^h) = \sum_{A,B=1}^{\tilde{n}} \sum_{C=1}^{n} \tilde{N}_A \tilde{M}_{AB}^{-1} \int_\Omega \tilde{N}_B \frac{\partial N^C}{\partial x_k} d\Omega \, u_k^C \delta_{ij}, \tag{8.29}$$

and $\tilde{\mathbf{M}}$ is the "mass" matrix of the tilde basis, namely

$$\tilde{M}_{AB} = \left(\tilde{N}_A, \tilde{N}_B \right)_\Omega = \int_\Omega \tilde{N}_A \tilde{N}_B d\Omega. \tag{8.30}$$

In summary, the procedure corresponds to L^2 projection of ε_{ij}^{dil} onto the $\{\tilde{N}\}$ basis.

8.1.3.1 Implementational aspects

We consider aspects of solving the global matrix system in this section.

For an isotropic homogeneous linear elastic material, with Hooke's law given by (8.8), the discrete version of the bilinear form is given by:

$$\bar{a}(\mathbf{w}^h, \mathbf{u}^h) = \sum_{A,B=1}^{n} \left(w_i^A \int_{\Omega} N_{,j}^A \hat{c}_{ijkl} N_{,l}^B d\Omega \, u_k^B \right.$$

$$\left. + \frac{1}{3}(3\lambda + 2\mu) \sum_{C,D=1}^{\tilde{n}} w_i^A \, (N_{,i}^A, \tilde{N}_C) \tilde{M}_{CD}^{-1} (\tilde{N}_D, N_{,k}^B) \, u_k^B \right), \tag{8.31}$$

where

$$\hat{c}_{ijkl} = \mu \left(\delta_{ik}\delta_{jl} + \delta_{il}\delta_{jk} - \frac{2}{3}\delta_{ij}\delta_{kl} \right). \tag{8.32}$$

Due to the inverse of $\tilde{\mathbf{M}}$, the second term in (8.31) increases the population of the stiffness matrix on each patch for $p \geq 2$. Note, as we always assume use of patches constructed with open knot vectors, the displacement field is continuous across patch interfaces, but no smoother. Consequently, the tilde basis will be discontinuous across patches and $\tilde{\mathbf{M}}^{-1}$ will be uncoupled from patch to patch. Nevertheless, if we use a direct solver to solve the global equation system, we need to account for increased coupling of the equations due to $\tilde{\mathbf{M}}^{-1}$ on each patch. There are at least two ways to circumvent the effect of the increased coupling. One is to use an iterative strategy that does not require the assembly of the stiffness matrix, such as conjugate gradients, to solve the global problem. Within each conjugate gradient iteration, a direct solver can be used to evaluate $\tilde{\mathbf{M}}^{-1}$ patch-wise, retaining its sparse band-profile structure. This procedure can be used to solve very large problems. It has been used extensively in these calculations and found it to be very efficient. A second possibility is to replace $\tilde{\mathbf{M}}$ with a diagonal, or "lumped" approximation. This would only need to be done in the left-hand-side matrix, and so would be interpreted as a "preconditioner," see Saad, 2003. In this case, the band-profile structure of the preconditioner would be only slightly larger than for the system constructed without projection. Using the consistent $\tilde{\mathbf{M}}$ on the right-hand-side would ensure the full accuracy of the projection procedure. Convergence would require one or more iterations, but this involves only a forward reduction and back substitution for each additional iteration with an existing factorized array when employing a direct solver.

8.1.4 Infinite plate with circular hole under in-plane tension

Let us consider a plane-strain infinite plate with a hole under tension. This problem has been studied previously in Chapter 4 assuming an isotropic compressible linear elastic medium. It is interesting from the geometrical point of view because quadratic NURBS can exactly represent the circular hole and the existence of an analytical solution allows us to focus on the convergence rates that the proposed method can attain without geometrical approximation. In the nearly incompressible regime, this problem has not been studied extensively. Some results in the incompressible limit using meshless methods can be found in Huerta and Fernandez-Mendez, 2001 and Dolbow and Belytschko, 1999. The infinite plate is modeled by a quarter plate. The geometry, loading, boundary conditions and parameters are shown in Figure 4.1. The exact solution is given by (4.46)–(4.48). The value for ν was chosen to be 0.49999, very close to the incompressible limit, in order to study the convergence of the $\bar{B}$ isogeometric

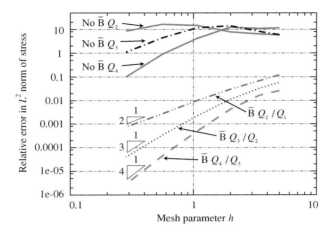

Figure 8.4 Plate with a circular hole. Convergence curves in the relative L^2 norm of stress with and without $\bar{\text{B}}$ for various NURBS orders obtained from k-refinement.

analysis for that case. A rational quadratic basis is the minimum order capable of representing the exact geometry. The sequence of meshes used for the convergence study is the same as that used in the compressible case; see Figure 4.3 in Chapter 4.

We define the relative error as the error normalized by the corresponding value of the exact solution. Convergence results for the relative error in the L^2 norm of stress are shown in Figure 8.4. The cubic and quartic NURBS are obtained from k-refinement of the coarsest quadratic mesh. The mesh parameter h is defined as the maximum distance, in the physical space, between diagonally opposite knot locations. As can be seen, the $\bar{\text{B}}$ method obtains good convergence rates with relatively coarse meshes. Note that the standard displacement based formulation performs relatively poorly and needs comparatively fine meshes to attain convergence rates equivalent to what is obtained with the $\bar{\text{B}}$ formulation. Even when seemingly optimal asymptotic rates of convergence are attained in the case of the standard Q_p elements, the error is four orders of magnitude greater than for the corresponding projected Q_p/Q_{p-1} elements. This result clearly shows that optimal rate of convergence is not the only issue to be considered.

8.2 $\bar{\text{F}}$ formulation for nonlinear elasticity

8.2.1 Constitutive equations

The central idea of this approach is, as done in the geometrically linear case, to split the tensor that measures the deformation into its deviatoric (volume preserving) and volumetric-dilatational parts. In the finite deformation case, the deformation gradient $\mathbf{F}$ is the relevant tensor, and, contrary to the small deformation case, the split is multiplicative rather than additive. This multiplicative decomposition has been exploited previously by Flory, 1961[1]; Hughes *et al.*, 1975; Simo *et al.*, 1985; Simo and Taylor, 1991 (within a three field Hu-Washizu principle); and more recently by de Souza Neto *et al.*, 1996, 2005 in an alternative $\bar{\text{F}}$ approach. The work presented here shares features with these techniques. The intent is for it to be a simple pure displacement formulation, but having greater generality and a more rigorous theoretical base than previous approaches.

The multiplicative split of the deformation gradient is written

$$\mathbf{F} = \mathbf{F}^{\text{dil}}\mathbf{F}^{\text{dev}}, \tag{8.33}$$

where

$$\det\mathbf{F} = J = \det\mathbf{F}^{\text{dil}} \text{ and } \det\mathbf{F}^{\text{dev}} = 1, \tag{8.34}$$

which leads to:

$$\mathbf{F}^{\text{dev}} = J^{-1/3}\mathbf{F} \text{ and } \mathbf{F}^{\text{dil}} = J^{1/3}\mathbf{I}. \tag{8.35}$$

A modified deformation gradient $\bar{\mathbf{F}}$ is defined in terms of the deviatoric part of the deformation gradient $\mathbf{F}^{\text{dev}}$ and a modified dilatational part of the deformation gradient $\bar{\mathbf{F}}^{\text{dil}}$:

$$\bar{\mathbf{F}} = \bar{\mathbf{F}}^{\text{dil}}\mathbf{F}^{\text{dev}}, \tag{8.36}$$

where

$$\bar{\mathbf{F}}^{\text{dil}} = \pi(\mathbf{F}^{\text{dil}}) = \pi(J^{1/3})\mathbf{I} = \overline{J^{1/3}}\mathbf{I}, \tag{8.37}$$

with π a linear projection operator identical to the one proposed previously for the linear case. Combining (8.35)–(8.37) results in the projected deformation gradient, $\bar{\mathbf{F}}$:

$$\bar{\mathbf{F}} = \alpha\mathbf{F}, \tag{8.38}$$

$$\alpha = \left(\overline{J^{1/3}}\right)\Big/\left(J^{1/3}\right). \tag{8.39}$$

The examples will assume hyperelastic homogeneous material behavior for which there exists a free-energy function[2] Ψ that depends on the Cauchy–Green tensor $\mathbf{C} = \mathbf{F}^T\mathbf{F}$ and from which the second Piola–Kirchhoff stress tensor is derived as:

$$\mathbf{S} = 2\frac{\partial\Psi(\mathbf{C})}{\partial\mathbf{C}}. \tag{8.40}$$

The standard additive decomposition of Ψ (see, *e.g.*, Simo and Hughes, 1998) into a volumetric part depending only on J and an isochoric part is used:

$$\Psi(J, \mathbf{C}) = \Psi^{\text{dil}}(J) + \Psi^{\text{iso}}(J, \mathbf{C}). \tag{8.41}$$

A full derivation of the variational formulation incorporating $\bar{\mathbf{F}}$ in this nonlinear setting is beyond the scope of this book. Please see Elguedj *et al.*, 2008 for details.

8.2.2 Pinched torus

This example again exploits the ability of NURBS to exactly represent conic sections. It consists of the pinching of a toroidal solid, and is similar to an example proposed in Chavan *et al.*, 2007.

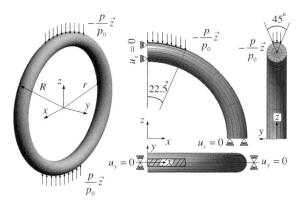

Figure 8.5 Pinched torus: geometry, quarter mesh, loading, and boundary conditions.

The geometry, loading, boundary conditions, and mesh are shown in Figure 8.5, and the material parameters are given in Table 8.1. The material model used is Neo-Hookean following the additive decomposition of the stored energy function given in (8.41). The isochoric and volumetric parts of Ψ are (see, e.g., Simo and Hughes, 1998):

$$\Psi(J, \mathbf{C}) = U(J) + \Psi^{\text{iso}}(J^{-2/3}\mathbf{C}) \tag{8.42}$$

$$U(J) = \frac{1}{2}\kappa \left(\frac{1}{2}(J^2 - 1) - \ln J \right) \tag{8.43}$$

$$\Psi^{\text{iso}}(J^{-2/3}\mathbf{C}) = \frac{1}{2}\mu \left(J^{-2/3}\text{tr}[\mathbf{C}] - 3 \right) \tag{8.44}$$

where κ is the bulk modulus and μ the shear modulus. Due to symmetry conditions, only one quarter of the structure is considered, with the corresponding symmetry boundary conditions applied. The quarter mesh with $4 \times 16 \times 2$ elements (that is 4 elements in the "large" circumferential direction, 16 elements in the "small" circumferential direction and 2 elements in the radial direction) shown in Figure 8.5 with quadratic and cubic NURBS is used.

Table 8.1 Pinched torus: material properties and boundary conditions

Shear modulus μ	5.67 MPa
Bulk modulus κ	$2.8333 \ 10^3$ MPa
Inner radius r	8 m
Outer radius R	10 m
Reference pressure p_0	0.195 MPa
BC in plane $x = 0$	$u_x = 0$
BC in plane $y = 0$	$u_y = 0$
BC in plane $z = 0$	$u_z = 0$

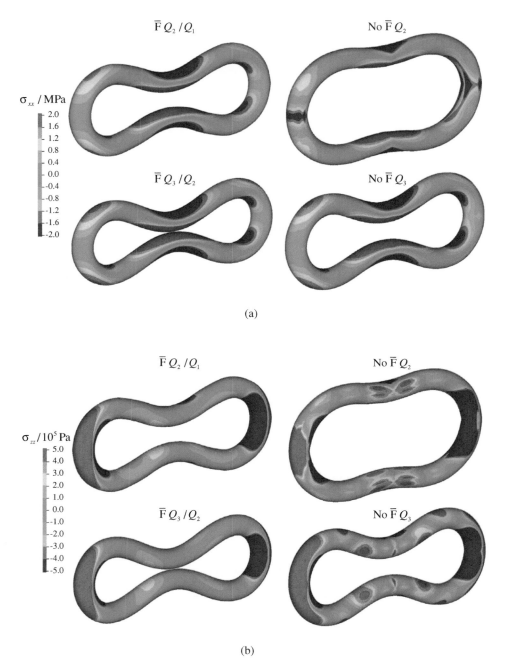

(a)

(b)

Figure 8.6 Pinched torus: Cauchy stress tensor components (a) σ_{xx} and (b) σ_{zz} on the deformed configuration with and without $\bar{F}$ for C^1-quadratic functions and C^2-cubic functions.

The σ_{xx} and σ_{zz} components of the Cauchy stress tensor plotted on the final deformed configuration with and without $\bar{F}$ for C^1-quadratic and C^2-cubic functions are shown in Figure 8.6. The differences on the final deformed configurations show that Q_2 without $\bar{F}$ suffers locking. Although $\bar{F}$ Q_2/Q_1 and Q_3 without $\bar{F}$ look similar, we can see that $\bar{F}$ Q_3/Q_2 improves the result considerably. The stress contours for the component σ_{xx} of the Cauchy stress tensor shows typical oscillations due to locking for Q_2 without $\bar{F}$. These oscillations are not observed in the three other cases for this component. However, we can see on the bottom part of Figure 8.6 that oscillations are present for both quadratic and cubic meshes without $\bar{F}$ for the component σ_{zz} of the Cauchy stress tensor. Note that, for both components and both orders of approximation, the results with $\bar{F}$ do not present such oscillations: the stresses are smooth and very similar for Q_2/Q_1 and Q_3/Q_2.

A number of additional examples have been solved in Elguedj *et al.*, 2008.

Notes

1. Paul Flory (1910–1985) received the Nobel Prize in chemistry in 1974 "for his fundamental work, both theoretical and experimental, in the physical chemistry of macromolecules."
2. The free-energy function is also called the stored energy or strain energy function.

9

Fluids

Many of the biggest challenges in computational mechanics are encountered when trying to model the behavior of fluids. This is in part due to the wide range of scales present in such problems, and also to the fact that these scales frequently interact with each other in complex ways. Failure to properly represent these interactions can result in inaccurate and/or unstable calculations. The keys to success when performing computational fluids analysis are accuracy *and* robustness. These attributes may or may not be possessed by the methods and functions used to approximate solutions. NURBS are functions that satisfy both of these criteria and seem to be an ideal basis for fluid mechanical applications. Methods are another matter. Both will be discussed in this chapter.

9.1 Dispersion analysis

Fluids problems typically feature a combination of advective and diffusive behavior. Let us begin by considering the spectral properties of NURBS applied to the limit cases of pure advection and pure diffusion, in order to assess their accuracy. Compared with standard FEA, NURBS exhibit superior results in both regimes, suggesting that NURBS might deliver better quality results when applied to more general fluid dynamics applications. This will indeed be demonstrated throughout the chapter.

9.1.1 Pure advection: the first-order wave equation

To determine the performance of NURBS applied to flow problems, which by their very definition contain advective phenomena, a natural starting point is the first-order wave equation, or pure advection. Here we compare *analytic* solutions to the discrete equations arrived at by finite element and NURBS treatments of the problem.

A linear dispersive system is one that admits solutions of the form (see Whitham, 1974)

$$\phi = a\cos(kx - \omega t), \tag{9.1}$$

where the frequency ω is a real function of the wavenumber k, with the specific form of $\omega(k)$ being determined by the system. If the phase speed $\omega(k)/k$ depends on k, rather than being a constant, the system is said to be "dispersive." For the first-order wave equation posed on an

Isogeometric Analysis: Toward Integration of CAD and FEA by J. A. Cottrell, T. J. R. Hughes, Y. Bazilevs
© 2009, John Wiley & Sons, Ltd

infinite domain, namely,

$$\frac{\partial \phi}{\partial t} + c\frac{\partial \phi}{\partial x} = 0, \quad \text{for } x \in (-\infty, +\infty), \tag{9.2}$$

the frequency $\omega = kc$, and thus any dispersion in a numerical solution is artificial. Every Fourier mode should travel to the right at speed c (hence the name "pure advection"), and any deviations from this constant velocity are artifacts of the numerics.

The infinite domain is modeled by a finite domain with periodic boundary conditions. The effect of this approximation is moot as the local support of the basis functions localizes the analysis. For both finite elements and NURBS, a solution of the form

$$\phi = \sum_{A=1}^{n_{np}} \phi_A(t) N_A(x), \tag{9.3}$$

is sought. In the case where N_A is a standard finite element basis function, its coefficient ϕ_A is associated with the value of the function at the node x_A, whereas for the non-interpolatory NURBS basis, ϕ_A is a control variable.

A stencil is arrived at in a manner that is somewhat analogous to the vibration analysis of Chapter 5. For either finite elements or NURBS, (9.3) is substituted into (9.2), multiplied by basis function N_A, and integrated to obtain

$$\int_0^L N_A \sum_{B=1}^{n_{np}} (\dot{\phi}_B N_B + c\phi_B N_B')dx = 0, \tag{9.4}$$

where the superposed dot denotes differentiation with respect to t, and the prime superscript denotes differentiation with respect to x. The integration is performed analytically to obtain a single equation, rather than assembling a matrix system numerically as done in previous chapters. Solving this equation provides important analytical information about the numerical method.

Linear finite elements and linear NURBS are identical, so the quadratic case is investigated first. Assuming a uniform mesh with element length h and that the N_A's are C^1 quadratic NURBS functions (actually, B-splines in this simple scenario), the integration in (9.4) is performed yielding

$$\frac{1}{120}(\dot{\phi}_{A-2} + 26\dot{\phi}_{A-1} + 66\dot{\phi}_A + 26\dot{\phi}_{A+1} + \dot{\phi}_{A+2})$$

$$+ \frac{c}{24h}(-\phi_{A-2} - 10\phi_{A-1} + 10\phi_{A+1} + \phi_{A+2}) = 0. \tag{9.5}$$

As in Vichnevetsky and Bowles, 1982, ϕ_A is expressed as

$$\phi_A = e^{i(k^h Ah - \omega t)}, \tag{9.6}$$

where k^h is the discrete wave number, an approximation to $k = \omega/c$, and $i = \sqrt{-1}$. Substituting this into (9.5) and simplifying yields

$$\frac{-i\omega}{120}(e^{-2i\theta} + 26e^{-i\theta} + 66 + 26e^{i\theta} + e^{2i\theta})$$

$$+\frac{c}{24h}(-e^{-2i\theta} - 10e^{-i\theta} + 10e^{i\theta} + e^{2i\theta}) = 0, \tag{9.7}$$

where $\theta = k^h h$. Rearranging and recalling that $(e^{i\alpha} + e^{-i\alpha})/2 = \cos\alpha$ and $(e^{i\alpha} - e^{-i\alpha})/2i = \sin\alpha$ results in

$$\omega(\cos 2\theta + 26\cos\theta + 33) - \frac{5c}{h}(\sin 2\theta + 10\sin\theta) = 0. \tag{9.8}$$

Finally, solving for $k/k^h = \omega^h/\omega$ gives

$$\frac{k}{k^h} = \frac{5(10\sin\theta + \sin 2\theta)}{\theta(33 + 26\cos\theta + \cos 2\theta)}. \tag{9.9}$$

For the classical quadratic finite element, the situation is more complicated as the basis function N_A can take on two forms. If N_A corresponds to an end node (i.e., A odd), then performing the integration in (9.4) results in

$$\frac{1}{10}(-\dot{\phi}_{A-2} + 2\dot{\phi}_{A-1} + 8\dot{\phi}_A + 2\dot{\phi}_{A+1} - \dot{\phi}_{A+2})+$$

$$2c\frac{\phi_{A+1} - \phi_{A-1}}{2h} - u\frac{\phi_{A+2} - \phi_{A-2}}{4h} = 0. \tag{9.10}$$

For the case where N_A is associated with a center node (i.e., A even), performing the same steps yields

$$\frac{1}{10}(\dot{\phi}_{A-1} + 8\dot{\phi}_A + \dot{\phi}_{A+1}) + c\frac{\phi_{A+1} - \phi_{A-1}}{2h} = 0. \tag{9.11}$$

Following Gresho and Sani, 1998, let

$$\phi_A(t) = \left[\frac{1 + (-1)^A}{2} + \beta\frac{1 - (-1)^A}{2}\right]e^{i(k^h A h - \omega t)}. \tag{9.12}$$

Substituting (9.12) into (9.11), solving the latter for β and using that result in (9.10), yields[1]

$$\frac{k}{k^h} = \frac{-2\sin 2\theta \pm \sqrt{(1 - \cos 2\theta)(19 - \cos 2\theta)}}{\theta(3 - \cos 2\theta)}. \tag{9.13}$$

See Gresho and Sani, 1998 for a discussion on selecting "+" or "−" in (9.13).

Plots of the phase error $k/k^h = \omega^h/\omega$ for these two quadratic cases, as well as C^2 cubic NURBS and linears, are shown in Figure 9.1 (see Section 5.7.2 of Chapter 5 for a discussion of the duality between the dispersion analysis of k/k^h and the spectrum analysis of ω^h/ω). We see that the quadratic finite elements actually overshoot the exact solution for part of the

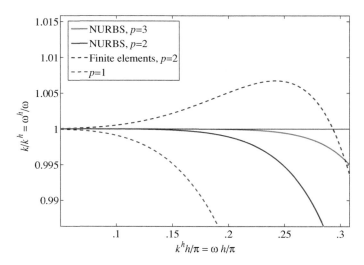

Figure 9.1   The first-order wave equation. Phase errors versus non-dimensional wave numbers. Comparison of linear and quadratic finite elements, C^1 quadratic NURBS, and C^2 cubic NURBS.

domain whereas the NURBS solution is considerably more accurate. The cubic NURBS are better still. For a fixed wavenumber, the error in the phase speed goes as $O(h^4)$ for C^0 quadratic finite elements and as $O(h^6)$ for the C^1 quadratic NURBS. In general, the error is $O(h^{2p})$ for classical C^0 finite elements of order p, $p > 1$, and $O(h^{2p+2})$ for C^{p-1} NURBS of order p, $p \geq 1$ (see Vichnevetsky and Bowles, 1982). Note, this acknowledges the fact that linear finite elements, that is, $p = 1$, are superconvergent, in that they achieve $O(h^4)$ phase error (see Gresho and Sani, 1998). These results illustrate the superiority of NURBS over classical finite elements for advective processes governed by the first-order wave equation.

9.1.2 Pure diffusion: the heat equation

The behavior of NURBS applied to purely diffusive phenomena may be determined by studying the heat equation:

$$\frac{\partial \phi}{\partial t} = \kappa \frac{\partial^2 \phi}{\partial x^2}, \quad \text{for } x \in (-\infty, +\infty), \tag{9.14}$$

and proceeding as in the case of the first-order wave equation. This time ϕ_A is written

$$\phi_A = e^{(ik^h Ah - \omega t)}. \tag{9.15}$$

The dispersion analysis is performed for finite elements and NURBS using basis functions of order $p = 2$ through $p = 4$. For completeness, the solution using linear elements is shown as well, though for linear elements there is no difference between finite elements and NURBS. Results are presented in Figure 9.2.

The superior behavior of NURBS basis functions compared with finite elements is once again evident. In this case, the finite element results depict an accurate acoustical branch and

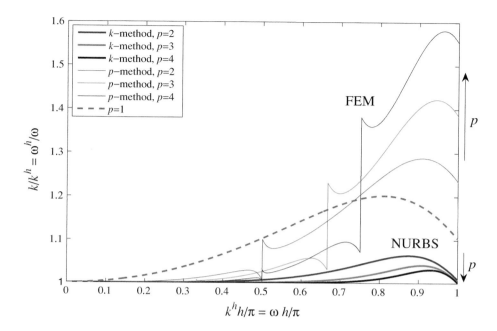

Figure 9.2 The heat equation. Phase errors versus non-dimensional wave numbers. Comparison of classical C^0-continuous finite elements and NURBS for $p = 1$ to 4.

inaccurate optical branches (see Brillouin, 1953, and Chapter 5). It is very important to observe the trends in Figure 9.2. For finite elements, the optical branches *diverge* as p is increased. That is, the errors in the higher wave numbers become greater as p is increased. On the other hand, for NURBS, *the entire spectrum converges* as p is increased. These opposite trends are likely very important in applications in which the *entire* discrete spectrum participates significantly in the solution. These results demonstrate the superiority of NURBS over classical finite elements for diffusive processes governed by the heat equation. The combination of results for advective and diffusive processes suggests that NURBS may be capable of attaining better accuracy than classical finite elements in representing turbulence, as will be demonstrated in Section 9.4.

Figure 9.2 may look familiar. To machine precision, it is *exactly* the same as Figure 5.7, obtained numerically in Chapter 5 for the longitudinal vibrations of an elastic rod. The fact that these two distinct types of phenomena lead to identical results may be understood from the duality of spectral and dispersion analyses described in Chapter 5. In both cases, the behavior of NURBS and FEA functions applied to second-order spatial derivative operators in one dimension were examined.

9.2 The variational multiscale (VMS) method

In the introduction to this chapter, we mentioned the instabilities that can arise due to the failure of a numerical method to represent all of the scales present in a problem. In a finite-dimensional setting, it is simply impossible to capture all of the features of a system, and that which is missing can, for certain classes of problems, have a deleterious effect on the ability to accurately model the scales that are otherwise within reach. The framework in which to understand and address these issues is the ***variational multiscale*** (VMS) method[2]. VMS deserves to be the topic of a book unto itself, and only the surface of this rich and active area of research will be scratched. For further information, see Hughes *et al.*, 2004 and Hughes and Sangalli, 2007. However, an attempt will be made to provide a brief explanation of its motivations so that the formulations employed in the fluids examples throughout the remainder of the chapter may be better understood.

9.2.1 Numerical example: linear advection–diffusion

Though we will discuss the advection–diffusion equation in more detail in Section 9.3, let us present an example to motivate the VMS framework. We model the advection and diffusion of concentration $u : \overline{\Omega} \to [0, 1]$ of a species in an incompressible fluid by

$$\mathbf{a} \cdot \nabla u - \nabla \cdot (\mathbf{k}\nabla u) = f \quad \text{in} \quad \Omega \tag{9.16a}$$

$$u = g \quad \text{on} \quad \Gamma_D \tag{9.16b}$$

$$-\nabla u \cdot \mathbf{n} = h \quad \text{on} \quad \Gamma_N \tag{9.16c}$$

where the (divergence-free) velocity field is $\mathbf{a} = \mathbf{a}(\mathbf{x})$, $\mathbf{k}$ is the diffusivity tensor, f is a source term, and g and h are prescribed Dirichlet and Neumann boundary data, respectively. Specifically, consider the two-dimensional problem setup shown in Figure 9.3, where Ω is the bi-unit square $[0, 1] \times [0, 1]$.

Let $\theta = \tan^{-1}(2)$, and note that $\|\mathbf{a}\| = 1$. Whether the problem is more "advection-dominated," or "diffusion-dominated" depends on the size of κ. When it is advection-dominated, boundary layers near the outflow boundaries are expected, along with an internal

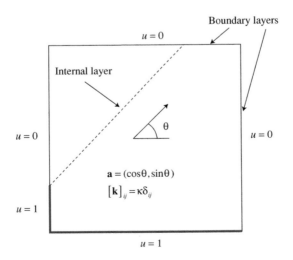

Figure 9.3   Advection skew to mesh. Problem description and data.

layer aligned with the flow direction. In more diffusion-dominated cases, all such features are expected to be "smeared," with no large gradients or sharp layers.

It turns out that the important parameter in the Galerkin method is the **element Péclet number**, α, defined as

$$\alpha = \frac{\|a\|h}{2\kappa}, \tag{9.17}$$

where h is the edge-length of an element. Clearly, the element Péclet number is a non-dimensional measure of the competition between advective and diffusive effects on the length scale of the mesh. When α is greater than 1 (less than 1, respectively) the situation is referred to as advection-dominated (diffusion-dominated, respectively).

Figure 9.4 shows two results for this problem on a 20×20 mesh of linear elements using a classical Galerkin's method approach. In Figure 9.4a, the diffusion ($\kappa = 10^{-1}$, in this case) is not drastically out of proportion with the magnitude of the advective velocity. The element Péclet number is $\alpha = 1(0.05)/(2(0.01)) = 0.25$, and a very reasonable looking solution with good accuracy throughout the domain is obtained. Similar accuracy would be expected if κ were to grow, even if it were to completely dominate. In Figure 9.4b, the diffusivity has been decreased ($\kappa = 10^{-3}$ and so $\alpha = 1(0.05)/(2(0.001)) = 25$) and the advection dominates. Oscillatory behavior polluting the solution throughout the domain is observed. This result is completely inaccurate, and the values of concentration below zero and above one do not even make physical sense. This oscillatory behavior is a manifestation of "instability," referred to previously. This is an example of the fact that when the features of the solution to certain classes of problems fall significantly below the resolution of the mesh that is being used, Galerkin's method becomes unstable and exhibits spurious oscillations. They have been the bane of many finite element researchers and the topic of thousands of papers over the years. One interpretation of the cause of the instabilities is the failure to account for the effect of the scales that are too small to be represented explicitly with the basis employed.

9.2.2 The Green's operator

In an effort to grasp the salient issues, the following problem for an abstract linear operator $\mathcal{L}$ may be considered,

$$\mathcal{L}u = f \quad \text{in} \quad \Omega, \tag{9.18a}$$

$$u = b \quad \text{on} \quad \Gamma, \tag{9.18b}$$

where we have changed the notation for the Dirichlet condition in order to reserve the letter g for further duty below. We can re-express (9.18) in a variational form as: Find $u \in \mathcal{S} = \mathcal{V}$ such that, for all $w \in \mathcal{V}$

$$a(w, u) = L(w), \tag{9.19}$$

where

$$a(w, u) = (w, \mathcal{L}u), \tag{9.20}$$

$$L(w) = (w, f), \tag{9.21}$$

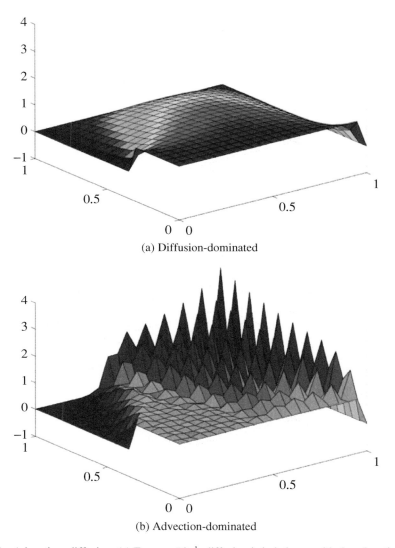

Figure 9.4 Advection–diffusion. (a) For $\kappa = 10^{-1}$, diffusion is in balance with the advection, and the Galerkin solution is reasonable. (b) For $\kappa = 10^{-3}$ advection dominates, spurious oscillations emerge, and the solution is completely unacceptable.

and

$$(w, u) \equiv \int_{\Omega} w(\mathbf{x}) u(\mathbf{x}) \, d\mathbf{x}. \tag{9.22}$$

We have used $\mathcal{S}$ and $\mathcal{V}$ to denote the trial and weighting spaces, as in previous chapters.

Analytically, we can think of solving (9.18) by means of a ***global Green's operator***, $\mathcal{G}$, such that

$$u = \mathcal{G}f. \tag{9.23}$$

Frequently, we will represent this global Green's operator by the classical Green's function, $g : \Omega \times \Omega \to \mathbb{R}$, such that

$$u(\mathbf{y}) = \int_{\Omega} g(\mathbf{x}, \mathbf{y}) f(\mathbf{x}) \, d\mathbf{x}. \tag{9.24}$$

The problem used to define g is

$$\mathcal{L}^* g(\mathbf{x}, \mathbf{y}) = \delta(\mathbf{x} - \mathbf{y}) \quad \mathbf{x}, \mathbf{y} \in \Omega, \tag{9.25a}$$

$$g(\mathbf{x}, \mathbf{y}) = 0 \quad \mathbf{x} \in \Gamma, \tag{9.25b}$$

where $\delta(\mathbf{x} - \mathbf{y})$ is the Dirac delta distribution and $\mathcal{L}^*$ is the adjoint of operator $\mathcal{L}$, as defined by the relationship

$$\left(\mathcal{L}^* w, u \right) = (w, \mathcal{L}u) \quad \forall u, w \in \mathcal{S}. \tag{9.26}$$

If we had an expression for g, we would have no need for a numerical method, but its existence is a tool that we can use to gain understanding of the problem.

9.2.3 A multiscale decomposition

Rather than pursuing an expression for the global Green's function (which might be very difficult to find), let us instead consider a direct sum decomposition of the solution space $\mathcal{S}$ into a finite dimensional subspace $\bar{\mathcal{S}}$ that we will refer to as the "coarse-scale space," and an infinite dimensional subspace $\mathcal{S}'$, called the "fine-scale space," such that

$$\mathcal{S} = \bar{\mathcal{S}} \oplus \mathcal{S}'. \tag{9.27}$$

Thus, for all $u \in \mathcal{S}$, we have a unique decomposition

$$u = \bar{u} + u', \tag{9.28}$$

with $\bar{u} \in \bar{\mathcal{S}}$ and $u' \in \mathcal{S}'$.

In practice, we will identify $\bar{\mathcal{S}}$ with $\mathcal{S}^h$, the space spanned by the NURBS or FEA basis, and we will think of $\mathcal{S}'$ as being comprised of the remaining subgrid scales that the basis is incapable of representing. Schematically, the idea is represented in Figure 9.5. At this point, however, we have not provided enough information to make the decomposition (9.28) unique. We have not stated *how* the coarse scales will fit the exact solution. For example, $\bar{u}$ could interpolate u at a discrete set of points, or be a best fit in the L^2 or H^1 norms. To remove any ambiguity, we introduce a linear projector $\mathcal{P} : \mathcal{S} \to \bar{\mathcal{S}}$ such that

$$\bar{u} = \mathcal{P}u, \tag{9.29}$$

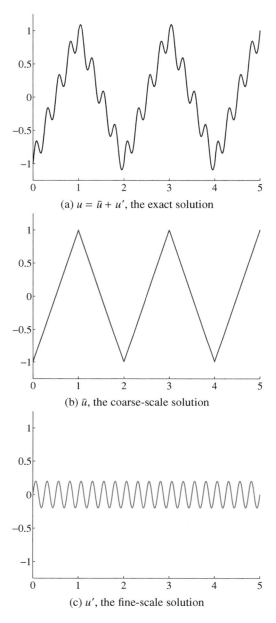

(a) $u = \bar{u} + u'$, the exact solution

(b) $\bar{u}$, the coarse-scale solution

(c) u', the fine-scale solution

Figure 9.5   A multiscale decomposition of a function u into its coarse-scale component $\bar{u}$, given here by piecewise linear interpolation, and its fine-scale component $u' = u - \bar{u}$, which is the component of u that the linear basis is incapable of representing.

and thus

$$u' = u - \bar{u} = (\mathcal{I} - \mathcal{P})u = \mathcal{P}'u, \tag{9.30}$$

where $\mathcal{P}' = \mathcal{I} - \mathcal{P}$.

The importance of this projector cannot be overstated. Saying that $\bar{u}$ is a finite element approximation of u does not tell us anything about the manner in which they are related, even when the basis has been clearly specified. It is the specific choice of projector that closes the loop and determines the type of optimality with which we hope to fit the exact solution. It is also crucial to note that it is not u or u' but $\bar{u}$ that we will seek to approximate with the finite element solution, u^h. The goal is for u^h to be the optimal approximation possible given the choice of basis, where optimality is defined by (9.29). Of course, by its very definition $\bar{u}$ exists in the discrete space, but the challenge is to find it.

9.2.4 The variational multiscale formulation

We insert (9.28) into (9.19), and consider the analogous splitting for the weighting function, $w = \bar{w} + w'$. Rearranging, taking advantage of linearity, and recalling that (9.19) holds for all $w \in \mathcal{V}$ leads us to two separate problems,

$$a(\bar{w}, \bar{u}) + a(\bar{w}, u') = L(\bar{w}) \quad \forall \bar{w} \in \mathcal{V}, \tag{9.31}$$

and

$$a(w', u') = L(w') - a(w', \bar{u}) \quad \forall w' \in \mathcal{V}'. \tag{9.32}$$

If u' was known, the **coarse-scale problem**, (9.31), would be an exact, finite-dimensional problem for the coarse-scale solution, $\bar{u}$. That is, recalling that $\bar{u} \in \bar{\mathcal{S}}$ and therefore can be represented by a linear combination of basis functions, we could set up a matrix equation in the standard way and solve numerically to obtain a $\bar{u}$ that would satisfy (9.29). Unfortunately, we are not given u'. We might hope to solve the **fine-scale problem**, (9.32), in order to obtain it, but this problem is infinite-dimensional.

Formally, we can think of the solution u' to (9.32) as being given by a **fine-scale Green's operator**, $\mathcal{G}'$, analogously to (9.23), such that

$$u' = \mathcal{G}'R'(\bar{u}), \tag{9.33}$$

where R', the residual[3] of the coarse scales projected onto the fine scales, satisfies

$$\left(w', R'(\bar{u})\right) = L(w') - a(w', \bar{u}) \quad \forall w' \in \mathcal{V}'. \tag{9.34}$$

Moreover, as in the global case, we may think of $\mathcal{G}'$ as being represented by a fine-scale Green's function $g' : \Omega \times \Omega \to \mathbb{R}$ such that

$$u'(\mathbf{y}) = \int_\Omega g'(\mathbf{x}, \mathbf{y})R'\left(\bar{u}(\mathbf{x})\right) d\mathbf{x}. \tag{9.35}$$

Clearly, the exact form of $\mathcal{G}'$ must be dependent upon the projector $\mathcal{P}$. It was shown by Hughes and Sangalli, 2007 to depend upon the global Green's function $\mathcal{G}$ as well, and to take the form

$$\mathcal{G}' = \mathcal{G} - \mathcal{G}\mathcal{P}^T \left(\mathcal{P}\mathcal{G}\mathcal{P}^T\right)^{-1} \mathcal{P}\mathcal{G}. \tag{9.36}$$

Though expression (9.36) is quite elegant, for it to be of practical use we need an expression for $\mathcal{G}$. Of course, if we had that, we would have no need for $\mathcal{G}'$, or for a numerical method in the first place, as we could simply solve for u directly as in (9.23). Though intractable, this multiscale formulation has still given us quite a bit of insight into the nature of the fine-scale solution. In particular, (9.32), (9.33), and (9.35) all tell us that the fine scales are driven by the residual of the coarse scales. This corresponds to the intuitive notion that the more accurately $\bar{u}$ approximates u, the smaller u' should become. Additionally, (9.36) informs us as to exactly how the choice of the projector affects the fine scales.

With the projector specified, $\bar{u}$ and u' are well defined entities with an exact mathematical definition. It is the task of the variational multiscale method to obtain a numerical expression $u^h \approx \bar{u}$ such that (9.29) is approximately satisfied. As we have already noted, u^h and $\bar{u}$ both exist in the same space $\bar{S}$; we are simply trying to devise a numerical method that will select the member of this space that satisfies the optimality condition. This is a very active area of research and there are many techniques in the literature. They all hinge upon using what analytical knowledge we do possess of u', or more generally, of (9.32), such that *we may approximate the interaction of the fine and coarse scales that appears in the coarse-scale problem* (9.32). Namely, obtaining a quality approximation $u^h \approx \bar{u}$ depends upon approximating the term $a(\bar{w}, u')$ that appears in the coarse-scale problem. Such modeling of this interaction between the fine and coarse scales will be pointed out where it arises in the applications considered throughout the remainder of this chapter.

9.2.5 Reconciling Galerkin's method with VMS

Examining (9.31), it is obvious that Galerkin's method is precisely equivalent to the assumption that $u' = 0$. Whether we are able to reach a level of resolution that is fine enough to make this approximation reasonable depends on the details of the specific problem under investigation, and it may not be obvious until calculations are performed. Given how completely unusable results such as those in Figure 9.4b can be, it is reasonable to ask why Galerkin's method has enjoyed the success that it has. The answer is that, though it was not conceived with this goal in mind, Galerkin's method is optimal in a very natural sense when applied to the symmetric elliptic problems that dominated the early days of the development of the finite element method.

In the cases where the bilinear form is symmetric (*e.g.*, the Laplace equation, linear elasticity, etc.), this bilinear form is also an inner product on the solution space[4]. As such, it induces a norm (frequently called the "energy norm" in solids applications), given by

$$\|u\|_E^2 = a(u, u). \tag{9.37}$$

As this norm is naturally induced by the weak form of the problem we are attempting to solve, it is an appropriate norm in which to seek optimality of the solution. The optimality condition could be expressed as a projector, as in Section 9.2.3, or we can simply seek to minimize the error between the exact solution u and finite-dimensional solution $\bar{u} = \sum_{B=1}^{n_{eq}} u_B N_B$ in the energy norm. We are trying to find the $\bar{u} \in \bar{S}$ such that $a(u - \bar{u}, u - \bar{u}) = a(u', u')$ is minimized. Setting the derivative with respect to the coefficients equal to zero, we have that for $A = 1, \ldots, n_{eq}$

$$
\begin{aligned}
0 &= \frac{\partial}{\partial u_A}[a(u - \bar{u}, u - \bar{u})] \\
&= \frac{\partial}{\partial u_A}[a(u, u) - 2a(\bar{u}, u) + a(\bar{u}, \bar{u})] \\
&= \left[\frac{\partial}{\partial u_A}a(u, u) - 2\frac{\partial}{\partial u_A}a(\bar{u}, u) + a\frac{\partial}{\partial u_A}(\bar{u}, \bar{u})\right] \\
&= -2a(N_A, u) + 2a(N_A, \bar{u}) \\
&= -2a(N_A, u - \bar{u}).
\end{aligned}
\tag{9.38}
$$

Dividing by -2 and substituting $u' = u - \bar{u}$, we obtain

$$
a(N_A, u') = 0, \quad A = 1, \ldots, n_{eq}, \tag{9.39}
$$

and thus the fine-scale solution u' is orthogonal to the entire coarse-scale space with respect to the inner product $a(\cdot, \cdot)$. This means that, despite the fact that $u' \neq 0$, the interaction of the fine scales with the coarse scales, $a(\bar{w}, u')$, is equal to zero. Thus, the implicit Galerkin assumption that $u' = 0$ is false, but it inadvertently leads to the correct formulation when the bilinear form is symmetric. Failure to appreciate the subtlety of this "two wrongs make a right" coincidence is the reason Galerkin's method has had its vociferous adherents, even for the classes of problems in which this coincidence no longer holds.

When the bilinear form is no longer symmetric, as is the case in the majority of fluids problems, it does not constitute an inner product and therefore does not induce a natural norm for the problem. In these cases, if we seek optimality with respect to any reasonable norm, we find that $a(\bar{w}, u') \neq 0$. If we do not account for this term in one fashion or another, we are inevitably left with an *ad hoc*, inaccurate method, and we must accustom ourselves to results such as Figure 9.4b.

9.3 Advection–diffusion equation

Turning our attention from theoretical developments back towards applications, let us consider the advection–diffusion equation. This equation is frequently the starting point for research in fluids as many of the difficulties encountered in more complicated nonlinear systems, such as the Navier–Stokes equations, also appear in this simple linear setting (recall Figure 9.4). This is a model problem, but a rich one. We will begin by formulating the problem, then take a brief aside to introduce streamline upwind/Petrov–Galerkin (SUPG) stabilization, and finally turn our attention to the behavior of NURBS on the two-dimensional problem setup described previously in Figure 9.3.

9.3.1 Formulating the problem

We begin with the strong form of the advection–diffusion equation as in (9.16). As usual, we multiply by a test function and integrate to obtain the weak form

$$a(w, u) = L(w), \tag{9.40}$$

where

$$a(w, u) = - \int_\Omega \nabla w \cdot (\mathbf{a}u - \kappa \nabla u) \, d\Omega + \int_{\Gamma_N} w u \mathbf{a} \cdot \mathbf{n} \, d\Gamma + \int_{\Gamma_N} w h \, d\Gamma, \tag{9.41}$$

and

$$L(w) = \int_\Omega w f \, d\Omega. \tag{9.42}$$

9.3.2 The streamline upwind/Petrov–Galerkin (SUPG) method

Many years before the theoretical framework of the variational multiscale method explained the origins of the spurious oscillations that plagued fluids calculations, researchers developed techniques that were designed to suppress the oscillations while attempting to minimize any negative impact on the overall accuracy of the solution. Though these approaches lacked a complete theoretical grounding, a lot of practical progress was made. The most successful of these classical stabilized methods is the streamline upwind/Petrov–Galerkin method (SUPG). The motivations behind the development of SUPG are found in the original archival journal paper on the subject, Brooks and Hughes, 1982. In the first paper on VMS (Hughes, 1995) it was seen that SUPG and VMS are equivalent in the very special case of advection–diffusion in one dimension with piecewise linear basis functions and a piecewise constant forcing function. They diverge somewhat in other, more complicated cases, but SUPG remains an effective method that has enjoyed wide popularity in both academic and commercial settings.

Let us denote the set of element interiors $\Omega^{\text{int}} \subset \Omega$. This is simply the domain with all of the element boundaries removed. SUPG augments the Galerkin form of the advection–diffusion equation with an additional term, applied only on Ω^{int}, such that we are tasked with solving

$$a(w^h, u^h) + \left(\mathcal{L}_{\text{adv}} w^h, \tau(\mathcal{L}u^h - f) \right)_{\Omega^{\text{int}}} = L(w^h), \tag{9.43}$$

where $\mathcal{L}_{\text{adv}} w^h = \mathbf{a} \cdot \nabla w^h$ is the advective part of the operator acting on the weighting function, and τ is a **stabilization parameter**.

We can understand some of the effectiveness of this method by comparing it with VMS. First, note that (without realizing it) SUPG is implicitly approximating the fine-scale field u' by the stabilization parameter times the residual of the coarse scales. Consider (9.33) and (9.35), which tell us that u' is given by an *integral* operator acting on the residual of the coarse scales projected into the fine-scale space. We can approximate this by an *algebraic* operator acting directly on the residual of the numerical solution. That is

$$\mathcal{G}' R'(\bar{u}) \approx \tau R(u^h), \tag{9.44}$$

where $R(u^h) = f - \mathcal{L}u^h$, and so

$$u' \approx \tau(f - \mathcal{L}u^h) \tag{9.45}$$

Selection of an appropriate τ in various situations has been the subject of much research, but for advection–diffusion we will use the original formulation from Brooks and Hughes, 1982.

If we temporarily assume homogeneous boundary conditions, then

$$a(w, u) = (w, \mathcal{L}u) = (\mathcal{L}^*w, u), \tag{9.46}$$

where the adjoint operator $\mathcal{L}^*$ is given by

$$\mathcal{L}^* = -\mathcal{L}_{\text{adv}} + \mathcal{L}_{\text{diff}} = -\mathbf{a} \cdot \nabla - \kappa \Delta. \tag{9.47}$$

Recall, however, that in the advection-dominated cases we have $\|\mathbf{a}\|h >> \kappa$. In such cases, $\mathcal{L}_{\text{diff}}$ has negligible effect and $\mathcal{L}^* \approx -\mathcal{L}_{\text{adv}}$. Combining this with (9.45) we see that the SUPG stabilization term is a reasonable approximation of the multiscale interaction term of VMS:

$$\left(\mathcal{L}_{\text{adv}}w^h, \tau(\mathcal{L}u^h - f)\right)_{\Omega^{\text{int}}} \approx a(\bar{w}, u'). \tag{9.48}$$

Thus, despite the fact that such an interpretation was only available more than a decade after SUPG's initial introduction, we can now see why it has been such an effective method. (The relationship between SUPG and VMS has been clarified in Hughes and Sangalli, 2007).

Remark

For comprehensive treatments of stabilized and variational multiscale methods, see Hughes *et al.*, 2004.

9.3.3 Numerical example: advection–diffusion in two dimensions, revisited

Isogeometric analysis is fundamentally a higher-order approach and one might not expect good behavior in situations with unresolved interior and boundary layers. Recalling Figure 2.13 from Chapter 2, we remark that oscillations in polynomial-based finite element methods tend to become more pronounced as polynomial order is increased. This is the reason that most practical fluids formulations employ lower-order, typically constant and linear, interpolation of flow variables. However, the variation diminishing property of the Dirichlet boundary condition specification, plus the notion of k-refinement, leads to some remarkable results in the case of NURBS.

Let us again consider the problem setup we saw in Figure 9.3. The global Péclet number, $Pe = aL/\kappa = 10^6$. When this number is greater than one, advection dominates and diffusion is only important in very small layers. In the present case, diffusion is important in a region of thickness $O(Pe^{-1}\ln Pe)$ in the outflow boundary layers and $O(Pe^{-1/2}\ln Pe)$ in the internal layer (see Wahlbin, 1991, pp. 468). In all calculations the mesh is uniform, consisting of a 20×20 grid of square elements, with element side length $h = 1/20 = 0.05$. Refinement is performed by the k-method, and solutions from $p = 1$ to $p = 12$ are calculated. In all cases, the standard SUPG formulation is used with $\tau = h_a/(2a)$, where h_a is the element length in the direction of the flow velocity which, in the present case, is simply, $h_a = h/\max\{\cos\theta, \sin\theta\}$.

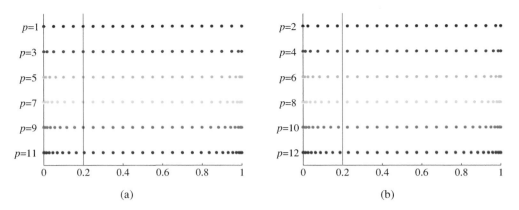

Figure 9.6 The y-coordinate of the control points along the left edge of the domain. (a) Odd polynomial orders. (b) Even polynomial orders.

The boundary condition is set by specifying the control variables. On the top and right edges of the domain, all control variables are set to 0 and the boundary condition is exactly satisfied along these edges. On the bottom, the control variable corresponding to the lower right-hand corner is set to 0 and the remainder are set to 1. The result is that the boundary value is identically 1 up to the last element in which it smoothly decreases to 0 at the corner. The left-hand-side boundary is more interesting. If we think of the control variables as control points in $\mathbb{R}^3$ defining the surface plot of the solution, where the x and y coordinates have been fixed by the two-dimensional geometrical mapping and are no longer to be chosen by the user, then what we have done along the left side of the domain is to set the z-component (our actual control variable) equal to 1 for each control point that falls in the interval $[0, 0.2]$, and equal to 0 if it falls in $[0.2, 1]$. The locations of the control points are shown in Figure 9.6. Note the clustering of points near the edges of the domain. This is necessary to maintain a linear parameterization of the domain despite the use of open knot vectors. The resulting boundary conditions are shown in Figure 9.7. For $p = 1$, the boundary condition is interpolated, whereas

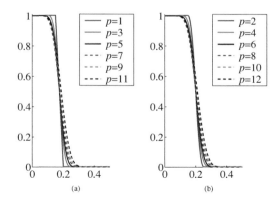

Figure 9.7 Dirichlet boundary conditions along the left edge of the domain. (a) Odd polynomial orders. (b) Even polynomial orders.

for $p > 1$ it is fit to the control variables in monotone fashion as the variation diminishing property of B-splines prevents the curve from over- and undershooting. We wish to emphasize that k-refinement produces non-nested solution spaces, which prevents us from having exactly the same boundary condition at each stage of the refinement process. As a result of this technique, the discontinuity "smears" about the location 0.2. For odd polynomial orders with $p \leq 9$, a control point falls directly on 0.2, whereas for even polynomial orders one does not (see Figure 9.6). When p is larger than 9, the aforementioned clustering of control points seen in Figure 9.6 spreads sufficiently to disrupt this pattern.

We wish to assess the ability of NURBS to deal with unresolved boundary and interior layers. We present the results for $\theta = 45°$ for $p = 1$, $p = 5$, $p = 8$, and $p = 12$ in Figure 9.8. (See Hughes *et al.*, 2005 for a more complete discussion and additional results.) Two views are presented for each p, one in which the plotting routine sampled the solution with a 100×100 grid of uniformly distributed points and one in which it is sampled with a 21×21 uniform grid. In the former case the plot is Phong shaded, and in the latter it is represented by bilinear interpolation on each element and the element edges are drawn. The philosophy behind the dual views is that the 100×100 grid plots are a more faithful rendering of the higher-order cases, whereas the 21×21 point piecewise bilinear interpolates are the type of plots that have appeared in numerous research articles over the years and these may be more easily visually compared with results in the literature.

For $p = 1$, there are noticeable oscillations. This demonstrates that classical stabilized methods alone are unable to achieve accurate solutions in advection–dominated cases with unresolved layers. As one examines the results, it is clear that they improve as p increases and are converging toward monotone results with quite sharp layers. One might have expected that oscillations would increase with increasing p but this is not the case. This is certainly due in part to monotone treatment of the boundary condition, but it is apparent that the high continuity of the basis obtained through k-refinement plays an essential role. The conclusion to be drawn is that higher-order NURBS functions are both accurate and robust, even in the presence of unresolved features.

9.4 Turbulence

The difficulty of stabilizing a numerical method in the presence of under-resolved features is even greater for nonlinear problems. Not only can the types of spurious oscillations that we saw with advection–diffusion affect the accuracy of the solution, but we must worry about convergence of the nonlinear solver as well. In this section, we examine incompressible turbulence – a highly nonlinear application that is characterized by rich behavior through an exceptionally wide range of scales. Success in capturing the character of the solution relies on two key components: a basis capable of accurately representing both large and small scales and a formulation that encapsulates the effect of the scales that are simply beyond reach. NURBS-based isogeometric analysis, paired with the variational multiscale method, provides both.

Much of the traditional research in turbulence has focused on understanding fundamental *physical* behavior of the system through numerical simulations that have typically made use of very simple geometries and high-order spectral or compact finite difference methods (see, e.g., Lele, 1992; Moin, 2001). The underlying function spaces utilized in spectral methods are of high continuity (C^∞ in the cases of Fourier series and global polynomials). While such approaches are capable of accuracy across many scales, they are exceedingly restrictive in the

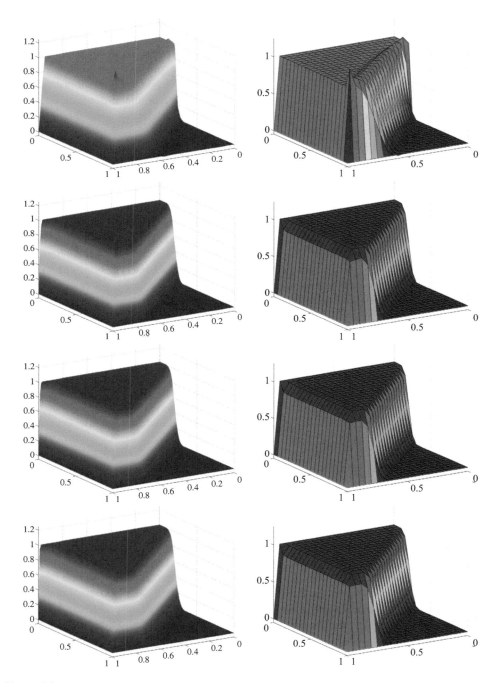

Figure 9.8   Advection skew to the mesh, $\theta = 45°$. The mesh is 20×20 in all cases. Top to bottom: results for $p = 1$, $p = 5$, $p = 8$, and $p = 12$. Left: plot with 100×100 sampling points, Phong shaded. Right: plot with 21×21 sampling points.

types of geometries that can be considered. Turbulent flows, however, are also of great interest in general geometry industrial applications. These are typically computed using finite volume and finite element methods, which employ low-order approximation functions that are at most C^0-continuous. Clearly, this as an opportunity for isogeometric analysis to bridge these worlds by combining geometrical flexibility with the ability to use functions of higher-order and higher-continuity. In this section, however, we will restrict ourselves to simple geometries in an effort to isolate the effects of continuity and compare how smooth C^1-continuous quadratic NURBS functions perform as compared with their C^0-continuous quadratic counterparts.

9.4.1 Incompressible Navier–Stokes equations

The incompressible Navier–Stokes equations can be expressed in terms of the linear momentum equations and incompressibility constraint given by

$$\frac{\partial \mathbf{u}}{\partial t} + \nabla \cdot (\mathbf{u} \otimes \mathbf{u}) + \nabla p - \nabla \cdot (2\nu \nabla^s \mathbf{u}) - \mathbf{f} = \mathbf{0} \qquad \text{in } \Omega, \tag{9.49}$$

$$\nabla \cdot \mathbf{u} = 0 \qquad \text{in } \Omega, \tag{9.50}$$

where

$$\nabla^s \mathbf{u} = \frac{1}{2}(\nabla \mathbf{u} + \nabla \mathbf{u}^T), \tag{9.51}$$

$\mathbf{f}$ is the force (per unit mass), ν is the kinematic viscosity, $\mathbf{u}$ is the velocity vector, and p is the pressure divided by the density.

Note that one may use the incompressibility constraint to simplify the momentum equation as

$$\frac{\partial \mathbf{u}}{\partial t} + \mathbf{u} \cdot \nabla \mathbf{u} + \nabla p - \nu \Delta \mathbf{u} - \mathbf{f} = \mathbf{0} \qquad \text{in } \Omega. \tag{9.52}$$

We assume for simplicity of presentation that $\mathbf{u} = \mathbf{0}$ on Γ and $\int_\Omega p(t) \, d\Omega = 0$ for all $t \in (0, T)$. Following the standard approach, we seek to recast (9.49) as a variational formulation. Let $\mathcal{V}$ denote both the trial solution and weighting function spaces, which are assumed to be the same. Multiplication by a test function and integration lead to a weak form of the incompressible Navier–Stokes equations: Find $\mathbf{U} = \{\mathbf{u}, p\} \in \mathcal{V}$ such that $\forall \mathbf{W} = \{\mathbf{w}, q\} \in \mathcal{V}$,

$$a(\mathbf{W}; \mathbf{U}) = L(\mathbf{W}) \tag{9.53}$$

where

$$a(\mathbf{W}; \mathbf{U}) = \left(\mathbf{w}, \frac{\partial \mathbf{u}}{\partial t}\right)_\Omega - (\nabla \mathbf{w}, \mathbf{u} \otimes \mathbf{u})_\Omega + (q, \nabla \cdot \mathbf{u})_\Omega$$

$$- (\nabla \cdot \mathbf{w}, p)_\Omega + \left(\nabla^s \mathbf{w}, 2\nu \nabla^s \mathbf{u}\right)_\Omega \tag{9.54}$$

and

$$L(\mathbf{W}) = (\mathbf{w}, \mathbf{f})_\Omega. \tag{9.55}$$

Note that $a(\cdot\,;\,\cdot)$ is no longer a bilinear form. The semicolon is used to denote the fact that it is linear with respect to the weighting function, to the left of the semicolon, but nonlinear with respect to the solution.

9.4.2 Multiscale residual-based formulation of the incompressible Navier–Stokes equations employing the advective form

As in Section 9.2, we consider a multiscale direct-sum decomposition of $\mathcal{V}$ into coarse-scale and fine-scale subspaces, $\mathcal{V}^h = \bar{\mathcal{V}}$ and $\mathcal{V}'$, respectively,

$$\mathcal{V} = \mathcal{V}^h \oplus \mathcal{V}', \tag{9.56}$$

where we have assumed from the outset that the coarse-scale space is given by the span of the basis functions to be used in the calculations. Again, to obtain a unique decomposition in (9.56), we require the aid of a linear projection operator $\mathcal{P}$, that gives $\mathbf{U}^h = \mathcal{P}\mathbf{U} \in \mathcal{V}^h$ and $\mathbf{U}' = (\mathcal{I} - \mathcal{P})\mathbf{U} \in \mathcal{V}'$ from a given $\mathbf{U} \in \mathcal{V}$.

Following the VMS methodology, we restrict the weighting space to $\mathcal{V}^h$ in (9.53) in order to obtain a finite-dimensional problem for the coarse scales. Employing the direct-sum decomposition (9.56) for the solution space yields the coarse-scale equation: Find $\mathbf{U}^h = \{\mathbf{u}^h, p^h\} \in \mathcal{V}^h$ such that $\forall \mathbf{W}^h = \{\mathbf{w}^h, q^h\} \in \mathcal{V}^h$,

$$a(\mathbf{W}^h; \mathbf{U}^h + \mathbf{U}') = L(\mathbf{W}^h). \tag{9.57}$$

We now see the manner in which the large scales depend on $\mathbf{U}' = \{\mathbf{u}', p'\}$, but the coupling is more complex than for advection–diffusion due to the nonlinearity of $a(\cdot\,;\,\cdot)$ with respect to its second argument.

Expanding and rearranging (9.57), the problem has become: Find $\mathbf{U}^h \in \mathcal{V}^h$, such that $\forall \mathbf{W}^h \in \mathcal{V}^h$,

$$\left(\mathbf{w}^h, \frac{\partial \mathbf{u}^h}{\partial t}\right)_\Omega - \left(\nabla \mathbf{w}^h, \mathbf{u}^h \otimes \mathbf{u}^h\right)_\Omega + \left(q^h, \nabla \cdot \mathbf{u}^h\right)_\Omega$$

$$- \left(\nabla \cdot \mathbf{w}^h, p^h\right)_\Omega + \left(\nabla^s \mathbf{w}^h, 2\nu\nabla^s \mathbf{u}^h\right)_\Omega - \left(\mathbf{w}^h, \mathbf{f}\right)_\Omega$$

$$+ \left(\mathbf{w}^h, \frac{\partial \mathbf{u}'}{\partial t}\right)_\Omega - \left(\nabla \mathbf{w}^h, \mathbf{u}^h \otimes \mathbf{u}'\right)_\Omega - \left(\nabla \mathbf{w}^h, \mathbf{u}' \otimes \mathbf{u}^h\right)_\Omega$$

$$- \left(\nabla \mathbf{w}^h, \mathbf{u}' \otimes \mathbf{u}'\right)_\Omega + \left(q^h, \nabla \cdot \mathbf{u}'\right)_\Omega$$

$$- \left(\nabla \cdot \mathbf{w}^h, p'\right)_\Omega + \left(\nabla^s \mathbf{w}^h, 2\nu\nabla^s \mathbf{u}'\right)_\Omega = 0 \tag{9.58}$$

We make the simplifying assumption that $\left(\mathbf{w}^h, \frac{\partial \mathbf{u}'}{\partial t}\right)_\Omega = 0$. Note, however, that Codina *et al.*, 2007 have demonstrated that it is beneficial to incorporate this effect in modeling the fine scales. The term $\left(\nabla^s \mathbf{w}^h, 2\nu\nabla^s \mathbf{u}'\right)_\Omega$ may be omitted by selecting a projector that enforces the orthogonality of the coarse and fine scales in the semi-norm induced by this term (see Hughes and Sangalli, 2007).

Let us elaborate on the convective terms in (9.58). Assuming incompressibility of the velocity field, namely, $\nabla \cdot (\mathbf{u}^h + \mathbf{u}') = 0$, we compute:

$$-\left(\nabla \mathbf{w}^h, \mathbf{u}^h \otimes \mathbf{u}^h\right)_{\Omega} - \left(\nabla \mathbf{w}^h, \mathbf{u}^h \otimes \mathbf{u}'\right)_{\Omega}$$

$$-\left(\nabla \mathbf{w}^h, \mathbf{u}' \otimes \mathbf{u}^h\right)_{\Omega} - \left(\nabla \mathbf{w}^h, \mathbf{u}' \otimes \mathbf{u}'\right)_{\Omega}$$

$$= -\left(\nabla \mathbf{w}^h, \mathbf{u}^h \otimes (\mathbf{u}^h + \mathbf{u}')\right)_{\Omega}$$

$$-\left(\nabla \mathbf{w}^h, \mathbf{u}' \otimes \mathbf{u}^h\right)_{\Omega} - \left(\nabla \mathbf{w}^h, \mathbf{u}' \otimes \mathbf{u}'\right)_{\Omega}$$

$$= \left(\mathbf{w}^h, (\mathbf{u}^h + \mathbf{u}') \cdot \nabla \mathbf{u}^h\right)_{\Omega}$$

$$-\left(\nabla \mathbf{w}^h, \mathbf{u}' \otimes \mathbf{u}^h\right)_{\Omega} - \left(\nabla \mathbf{w}^h, \mathbf{u}' \otimes \mathbf{u}'\right)_{\Omega} \tag{9.59}$$

As previously, we model the fine scales by a scaling parameter multiplying the residual of the coarse scales, see Bazilevs *et al.*, 2007a:

$$\mathbf{U}' \approx \boldsymbol{\tau} \mathbf{R}(\mathbf{U}^h), \tag{9.60}$$

where $\boldsymbol{\tau}$ is a 4×4 matrix (in three spatial dimensions) and $\mathbf{R}(\mathbf{U}^h)$ is a 4×1 vector that collects momentum and continuity residuals of the Navier–Stokes equations,

$$\mathbf{R}(\mathbf{U}^h) = \{\mathbf{r}_M^T(\mathbf{u}^h, p^h), r_C(\mathbf{u}^h)\}^T, \tag{9.61}$$

in which

$$\mathbf{r}_M(\mathbf{u}^h, p^h) = \mathbf{f} - \frac{\partial \mathbf{u}^h}{\partial t} - \mathbf{u}^h \cdot \nabla \mathbf{u}^h - \nabla p^h + \nu \Delta \mathbf{u}^h, \tag{9.62}$$

$$r_C(\mathbf{u}^h) = -\nabla \cdot \mathbf{u}^h \tag{9.63}$$

With $\mathbf{x} = \{x_i\}_{i=1}^d$ denoting the coordinates of element Ω^e in physical space, $\boldsymbol{\xi} = \{\xi_i\}_{i=1}^d$ denoting the coordinates of element $\hat{\Omega}^e$ in parametric space, and $\tilde{\boldsymbol{\xi}} = \{\tilde{\xi}_i\}_{i=1}^d$ denoting the coordinates of the parent element, $\tilde{\Omega}^e$ (recall Figure 3.4 in Chapter 3), we assume $\mathbf{x} = \mathbf{x}(\tilde{\boldsymbol{\xi}})$: $\tilde{\Omega}^e \rightarrow \Omega^e$ to be a continuously differentiable map with a continuously differentiable inverse. We define $\boldsymbol{\tau}$ as follows:

$$\boldsymbol{\tau} = diag(\tau_M, \tau_M, \tau_M, \tau_C), \tag{9.64}$$

where

$$\tau_M = (\frac{4}{\Delta t^2} + \mathbf{u}^h \cdot \mathbf{G}\mathbf{u}^h + C_I \nu^2 \mathbf{G} : \mathbf{G})^{-1/2}, \tag{9.65}$$

$$\tau_C = (\mathbf{g} \cdot \tau_M \mathbf{g})^{-1}, \tag{9.66}$$

$\mathbf{G}$ is a second rank metric tensor given by

$$\mathbf{G} = \frac{\partial \tilde{\boldsymbol{\xi}}}{\partial \mathbf{x}}^T \frac{\partial \tilde{\boldsymbol{\xi}}}{\partial \mathbf{x}}, \tag{9.67}$$

and $\mathbf{g}$ is a vector obtained from the column sums of $\frac{\partial \tilde{\boldsymbol{\xi}}}{\partial \mathbf{x}}$,

$$\mathbf{g} = \{g_i\}$$

$$g_i = \sum_{j=1}^{d} \left(\frac{\partial \tilde{\boldsymbol{\xi}}}{\partial \mathbf{x}}\right)_{ji}. \tag{9.68}$$

As an example of $\partial \tilde{\boldsymbol{\xi}}/\partial \mathbf{x}$, consider the case when the element under consideration is a cube with edge length h. The parent element is scaled such that $\partial \tilde{\boldsymbol{\xi}}/\partial \mathbf{x} = 2h^{-1}\mathbf{I}$, where $\mathbf{I}$ is the identity matrix. See Bazilevs et $al.$, 2007a for further details.

The definition of τ_M in (9.65) is inspired by the theory of stabilized methods for advection–diffusion–reaction systems (see, $e.g.$, Hughes and Mallet, 1986; Shakib et $al.$, 1991). The definition of τ_C comes from the small-scale Shur complement operator for the pressure. In the definition of τ_M (9.65), C_I is a positive constant, independent of the mesh size, that derives from an element-wise inverse estimate (see, $e.g.$, Johnson, 1987; Ern and Guermond, 2004).

Combining equations (9.58)–(9.60), we obtain the discrete formulation: Find $\mathbf{U}^h \in \mathcal{V}^h$, such that $\forall \mathbf{W}^h \in \mathcal{V}^h$,

$$\left(\mathbf{w}^h, \frac{\partial \mathbf{u}^h}{\partial t}\right)_{\Omega} + \left(\mathbf{w}^h, (\mathbf{u}^h - \tau_M \mathbf{r}_M) \cdot \nabla \mathbf{u}^h\right)_{\Omega} + \left(q^h, \nabla \cdot \mathbf{u}^h\right)_{\Omega}$$

$$- \left(\nabla \cdot \mathbf{w}^h, p^h\right)_{\Omega} + \left(\nabla^s \mathbf{w}^h, 2\nu \nabla^s \mathbf{u}^h\right)_{\Omega} - \left(\mathbf{w}^h, \mathbf{f}\right)_{\Omega}$$

$$+ \left(\mathbf{u}^h \cdot \nabla \mathbf{w}^h + \nabla q^h, \tau_M \mathbf{r}_M\right)_{\Omega} + \left(\nabla \cdot \mathbf{w}^h, \tau_C \nabla \cdot \mathbf{u}^h\right)_{\Omega}$$

$$- \left(\nabla \mathbf{w}^h, \tau_M \mathbf{r}_M \otimes \tau_M \mathbf{r}_M\right)_{\Omega} = 0. \tag{9.69}$$

9.4.3 Turbulent channel flow

To examine the effects of continuity, we examine a turbulent channel flow at Reynolds number $Re_\tau = 590$ based on the friction velocity and the channel half-width (for results at $Re_\tau = 180$ and $Re_\tau = 590$ on a coarser mesh, see Akkerman et $al.$, 2008). To assess the accuracy of the calculations, comparison is made with the direct numerical simulation (DNS) of Moser et $al.$, 1999.

The computational domain for this problem is a rectangular box, and the flow is driven by a constant pressure gradient in the stream-wise direction. Periodic boundary conditions are imposed in the stream-wise and span-wise directions, commonly referred to as homogeneous directions. A no-slip boundary condition is applied at the walls. This no-slip boundary condition is enforced strongly, that is, the discrete velocity is set to zero at the walls. An alternative approach is to enforce Dirichlet boundary conditions weakly. As discussed in Chapter 3, this is accomplished by appropriately augmenting the semi-discrete equations (9.69) by terms that enforce the no-slip condition weakly (see Bazilevs and Hughes, 2007; Bazilevs et $al.$, 2007b, 2008b for additional details). Though not employed in the computations presented herein, weak enforcement of Dirichlet boundary conditions is an extremely powerful technique that should be considered in many application areas.

The domain size is 2π, 2, and $4/3\pi$ in the stream-wise, wall-normal, and span-wise directions, respectively. The corresponding DNS computation was carried out on a domain of

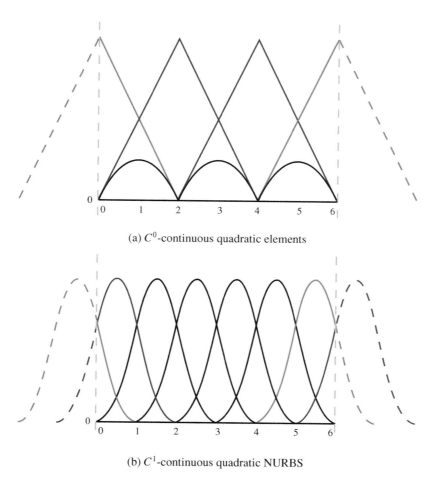

(a) C^0-continuous quadratic elements

(b) C^1-continuous quadratic NURBS

Figure 9.9 Basis functions employed in homogeneous directions. Periodic boundary conditions are imposed by associating each function that is non-zero on one of the boundaries of the computational domain (denoted by the gray vertical lines) with a function that is non-zero on the opposite boundary. For example, a function drawn in red that is non-zero at the left boundary will be given exactly the same control variables and the red function at the right boundary, thus treating them as one unique function within the code.

the same size with $128 \times 129 \times 128$ spectral functions in the stream-wise, wall-normal and span-wise direction, respectively.

Computations were carried out employing C^0- and C^1-continuous quadratic discretizations, keeping the number of degrees of freedom nearly the same in both cases. For the C^0 case a mesh of 32^3 elements was used, which gave $64 \times 65 \times 64$ basis functions in the discrete space, whereas for the C^1 case a mesh of 64^3 elements was employed, which led to a discrete space comprised of $64 \times 66 \times 64$ basis functions (the open knot vector construction is responsible for the extra basis function in the wall-normal direction). Figures 9.9a and 9.9b show the types of functions used in the stream-wise and span-wise direction for the C^0 and C^1 calculations, respectively (for the sake of visual clarity, fewer elements are shown than were used in the computation). Periodic boundary conditions are enforced by ensuring that the coefficient (control variable) multiplying any function with support at one edge of the computational domain is identical to the coefficient of an appropriate function on the opposite edge of the computational domain. This ensures that the analysis code treats them as one single function. Note that in the C^1 case, the functions are chosen such that continuity at the periodic boundary is not degraded.[5] The mesh is non-uniform in the wall-normal direction (not shown), with smaller elements placed near the boundary for increased resolution. The C^0 basis is analogous to that of Figure 9.9a, but stretched toward the wall. For the C^1 case, open knot vectors are employed and the same stretching is employed.

The semi-discrete equations (9.69) are advanced in time using the generalized-α method (see Chung and Hulbert, 1993; Jansen *et al.*, 1999). We use meshes that are stretched in the wall-normal direction according to a hyperbolic function to cluster points near the wall. Moreover, in the definition of τ_M (9.65) we set C_I to 36.

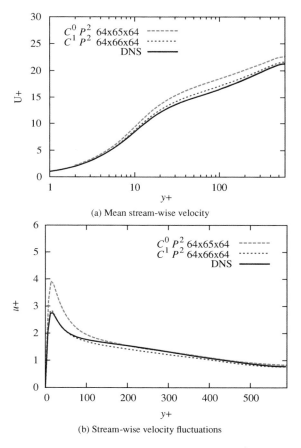

(a) Mean stream-wise velocity

(b) Stream-wise velocity fluctuations

Figure 9.10 Turbulent channel flow at $Re_\tau = 590$ computed on a 64^3 element mesh. Comparison of C^0- versus C^1-continuous discretizations.

Numerical results are reported in the form of statistics of the mean velocity and root-mean-square velocity fluctuations. The statistics were computed by sampling the velocity field at the mesh knots and averaging the solution in time as well as in the homogeneous directions. The meshes were chosen such that the number of degrees-of-freedom for both quadratic discretizations are approximately the same. All computational results are presented in non-dimensional wall units.

Figure 9.10a illustrates that the mean flow obtained with the C^1-continuous discretization is quite a bit more accurate. The stream-wise velocity fluctuations in Figure 9.10b are also better in the case of C^1 quadratics. This confirms the intuition gleaned in Section 9.1: the smooth NURBS functions appear more accurate per degree-of-freedom than their C^0 counterparts when applied to systems with advective and diffusive phenomena taking place across a wide range of scales.

Notes

1. Note that if we had considered C^0 quadratic NURBS instead of C^0 quadratic finite elements, the stencil would have been different, but the results for k/k^h would be exactly the same. This is because C^0 NURBS basis functions are different from the classical finite element basis functions, but the *space* they span is exactly the same.
2. The letters "VMS" stand for Variational MultiScale. Some researchers also refer to the Variational Multiscale Method by the abbreviation "VMM."
3. The sign associated with the residual is a matter of preference. Some take it to be $f - \mathcal{L}u$, while others use $\mathcal{L}u - f$. Here, we use the former, but use care when consulting the literature.
4. Recall that after we define a lift that satisfies the Dirichlet conditions, the terms involving that lift are moved to the right hand side of the equation and treated as data. From that point on, both the weighting functions and the solutions that we seek are zero on the Dirichlet boundary. Thus it makes sense to speak of the solution and weighting spaces as being identical.
5. In practice, this can be accomplished using open knot vectors by restricting the basis to act only in linear combinations that reproduce C^1-periodic behavior. This technique is completely analogous to the approach to local refinement discussed in Chapter 3, Section 3.5.1. The appropriate restriction is given by treating the C^{-1} basis at the boundary as the result of repeatedly replicating a knot in a C^1 basis, and making the degrees-of-freedom of the finer C^{-1} basis slaves to those of the coarser C^1 basis.

10

Fluid–Structure Interaction and Fluids on Moving Domains

In fluid–structure interaction (FSI), both the fluid and the structural equations emanate from conservation laws posed on moving domains (note that the very movement of the domains is a result of the structure and the fluid interacting with each other). There are many ways to write these laws, for example, on the material, referential, and current/physical domain. As a result, various discretization techniques exist for FSI, including but not limited to arbitrary Lagrangian–Eulerian (ALE), space–time, and particle FEM (PFEM) approaches. Each of these approaches has advantages and disadvantages depending on the application. In this section we will focus on the ALE formulation. See Bazilevs *et al.*, 2008a and references therein for additional details.

There are two main classes of solution algorithms for ALE formulations of fluid–structure interaction, staggered and monolithic. In the staggered approaches, the solid and fluid equations are solved in uncoupled fashion. Typically, the motion of the solid defines the geometry of the fluid domain. This can create problems in certain situations (see Bazilevs *et al.*, 2008a for elaboration). The advantage of the staggered approach is that one can combine existing, independently developed solid and fluid computer programs. The drawback is that the passing of information from one code to another can lead to time-stepping instabilities or, in the case when iteration is utilized, lack of convergence of iterates. Many examples of these phenomena have been reported, and various special "fixes" have been proposed. Monolithic procedures endeavor to solve the fluid–structure interaction system in fully-coupled fashion. This results in a large system of equations compared with the staggered approach. The benefit is improved numerical stability and more rapid convergence of iterates. This is the approach described herein. For further information, see Bazilevs *et al.*, 2008a.

10.1 The arbitrary Lagrangian–Eulerian (ALE) formulation

The structure is treated as a nonlinear elastic solid in the ***Lagrangian*** description governed by the equations of elastodynamics. By "Lagrangian description" we mean that the geometrical mapping of the solid domain moves with the material such that each parametric coordinate

Isogeometric Analysis: Toward Integration of CAD and FEA by J. A. Cottrell, T. J. R. Hughes, Y. Bazilevs
© 2009, John Wiley & Sons, Ltd

refers to the same solid particle throughout its deformation. This is the most utilized approach in structural analysis, and is the one adopted previously in Chapter 8, Section 8.2.

We shall assume the fluid to be viscous and incompressible, governed by the incompressible Navier–Stokes equations. Heretofore we have treated fluids problems in an ***Eulerian*** setting. That is, the mesh remained fixed in the region of interest while the particles of the fluid flowed through it. We would like to again use an Eulerian description of the fluid domain in the FSI calculations, but we also seek to maintain a compatible discretization with the solid domain, in which the mesh is moving with the material. This is accomplished by allowing the mesh to move in the fluid domain in such a way as to avoid becoming excessively distorted as the solid domain moves. This is referred to as the arbitrary Lagrangian–Eulerian (ALE) formulation.

ALE equations mandate the specification of the motion of the mesh in the region of the fluid. This motion is found by considering the fluid domain to be a fictitious elastic solid and solving an auxiliary static linear elasticity boundary value problem for which the fluid–solid interface displacement acts as a Dirichlet boundary condition (see, e.g., Johnson and Tezduyar, 1994). We know from the movement of the solid how the boundary of the fluid mesh must move (parts of the boundary of the fluid region may also have motion prescribed independent of the motion of the solid; naturally, this is handled by Dirichlet boundary conditions as well), and we use the fictitious elasticity problem posed on the fluid domain to move the mesh in a way that is compatible with the prescribed motion at the boundary while preserving the topology on the interior. In practice, this problem is solved monolithically with the FSI problem, but the coupling is one-way from the FSI problem to the elasticity problem for the mesh. At each step, the solution to the FSI problem dictates how the boundary for the elasticity problem must move.

It is interesting to consider the appropriate elastic coefficients for the moving mesh problem. They should be selected such that the fluid mesh quality is preserved for as long as possible. In particular, mesh quality can be preserved by dividing the elastic coefficients by the Jacobian of the element mapping, effectively increasing the stiffness of the smaller elements, which are typically placed near the fluid–solid interface (see Tezduyar *et al.*, 1992).

The kinematic compatibility (*i.e.*, "no-slip") condition between the fluid and the solid must be enforced. That is, the fluid velocity must be equal to the velocity of the solid at the interface. The coupled FSI problem is written in a variational form such that the stress compatibility condition at the fluid–solid interface is enforced weakly. Note also that the formulation in the fluid domain must be carefully written to incorporate the relative motion of the mesh. For a general discussion of ALE, the reader is referred to Hughes *et al.*, 1981; Donea *et al.*, 1982; LeTallec and Mouro, 2001; Farhat *et al.*, 2001; Farhat and Geuzaine, 2004; Bazilevs *et al.*, 2008a and references therein.

10.2 Inflation of a balloon

Bazilevs *et al.*, 2008a consider a three-dimensional benchmark example, originally proposed by Tezduyar and Sathe, 2007, belonging to a class of problems known as "flows in enclosed domains." For such problems, the boundary of the fluid subdomain is composed of two parts, an inflow and a fluid–solid interface. For incompressible fluids, conservation of mass results in the condition that the inflow flow-rate must equal the rate of change of the volume of the fluid domain (see, e.g., Kuttler *et al.*, 2006). In loosely coupled approaches, where the solutions of the fluid and the solid subproblems are obtained in a staggered fashion, this condition is lost during each subiteration, often leading to divergence of the calculations. In this section

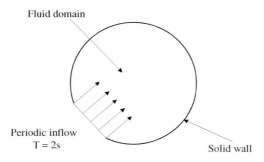

Figure 10.1   Inflation of a balloon. Problem setup.

it is shown that the strongly-coupled, NURBS-based procedures advocated in Bazilevs *et al.*, 2008a have no difficulty dealing with this situation.

The problem setup is illustrated in Figure 10.1. An initially spherical balloon is inflated, with the inflow velocity being given by a cosine function with a period of 2 s and an amplitude varying from 0 m/s to 2 m/s. The problem geometry, boundary conditions and material parameters are as in Tezduyar and Sathe, 2007 and Bazilevs *et al.*, 2008a. The mesh of the initial configuration, comprised of 10,336 quadratic NURBS elements, is shown in Figure 10.2. Note that, because NURBS are used to define the analysis-suitable geometry, the spherical balloon geometry is represented exactly.

For the motion of the mesh, one can take advantage of the parametric definition of the geometry. In this case E^m, the fictitious elastic modulus for the mesh motion problem, is set to be an exponentially increasing function of the parameter defining the radial direction, thus effectively "stiffening" the fluid elements near the fluid–solid boundary, thereby preserving the shape of the fluid mesh in this region.

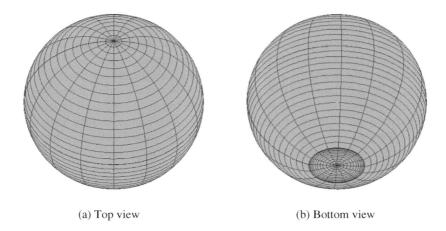

(a) Top view   (b) Bottom view

Figure 10.2   Inflation of a balloon. NURBS mesh of the balloon in both top and bottom views.

The computation is advanced for 14 inflow cycles, during which the volume of the balloon grows by a factor of approximately five with respect to its initial value. Figures 10.3–10.5 show snapshots of fluid velocity vectors superposed on the pressure contours at a planar cut through the diameter of the sphere. The flow is initially axially symmetric, although it is apparent that the symmetry of the solution breaks down towards the end of the computation. This is not surprising as the Reynolds number of the flow, based on the initial diameter of the balloon and the maximum inflow speed, is 4×10^5.

Figure 10.6(a) shows the inflow flow-rate versus the rate of change of the fluid domain volume, which are expected to be the same. On the scale of the plot they are indistinguishable. A closer examination of the error between the inflow flow-rate and the rate of change of the fluid domain volume reveals that the relative error in the quantities is on the order of $10^{-4} - 10^{-3}$, which is attributable to the fact that the nonlinear equations are solved up to a tolerance of this order (see Figure 10.6(b)). Note that the results are, on average, slightly less accurate during the last few periods of the simulation, which is attributable to the loss of radial symmetry in the solution. Also note that the error has the same sign, that is, the rate of change of the fluid domain volume is always greater than the inflow flow-rate. This suggests that in the discrete setting there is a tendency of the balloon to expand slightly faster (i.e., overcompensate) than dictated by the inflow flow-rate.

10.3 Flow in a patient-specific abdominal aorta with aneurysm

One of the interesting application areas of FSI technology has been in the modeling of arterial blood flow. In particular, patient specific modeling allows the opportunity to address complex geometrical issues, intricate flow patterns, and interesting biophysics all within the same analysis.

10.3.1 Construction of the arterial cross-section

Blood vessels are tubular objects and so we employ a sweeping method to construct meshes for isogeometric analysis. A solid NURBS description of a single arterial branch is obtained by extrusion of a circular curve along the vessel path, projection onto the true surface, and filling the volume radially inward. Arterial systems engender various branchings and inter-sections, which are handled with a template-based approach described in detail in Zhang *et al.*, 2007. Application of these procedures generates multi-patch, trivariate descriptions of patient-specific arterial geometries that are also analysis suitable.

A central feature of the approach is a construction of an arterial cross-section template that is based on the NURBS definition of the circular surface. Here we focus on the construction of the cross-section template as it relates to fluid–structure interaction analysis of arterial blood flow. We identify the area occupied by the blood, or the fluid region, and the arterial wall, or the solid region. Fluid and solid regions are separated by the luminal surface, or the fluid–solid interface. Figure 10.7 shows an example of a NURBS mesh for a circular cross-section with both fluid and solid regions present. Recall that NURBS elements are defined as areas enclosed between isoparametric lines (*i.e.*, knot spans). Note that the isoparametric lines correspond to radial and circumferential directions. For purposes of analysis we separate the fluid and the solid region by a C^0 surface as the solution is not expected to have regularity beyond C^0 at the interface. Knot vectors and control points for the cylindrical template for this arterial mesh can be found in Appendix 10.A at the end of this chapter.

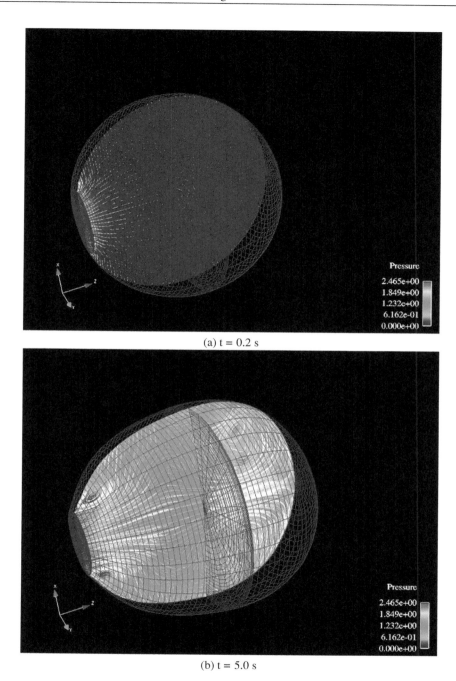

(a) t = 0.2 s

(b) t = 5.0 s

Figure 10.3 Inflation of a balloon. Mesh deformation and fluid velocity vectors superposed on the pressure plotted on a planar cut through the diameter of the balloon.

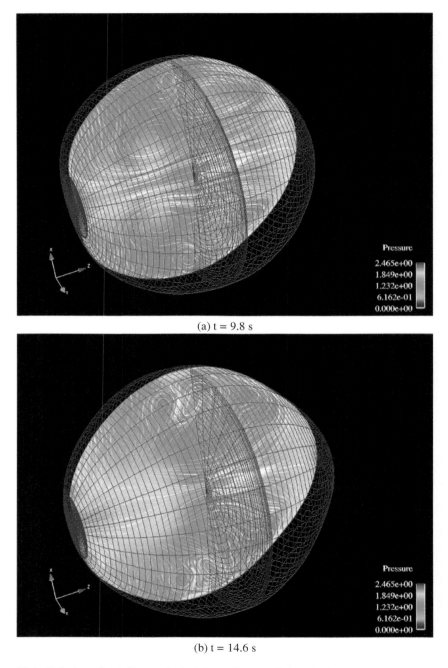

(a) t = 9.8 s

(b) t = 14.6 s

Figure 10.4   Inflation of a balloon. Mesh deformation and fluid velocity vectors superposed on the pressure plotted on a planar cut through the diameter of the balloon.

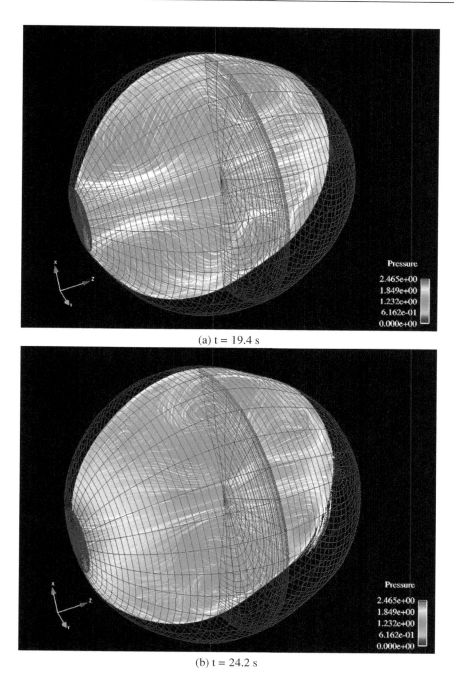

(a) t = 19.4 s

(b) t = 24.2 s

Figure 10.5 Inflation of a balloon. Mesh deformation and fluid velocity vectors superposed on the pressure plotted on a planar cut through the diameter of the balloon.

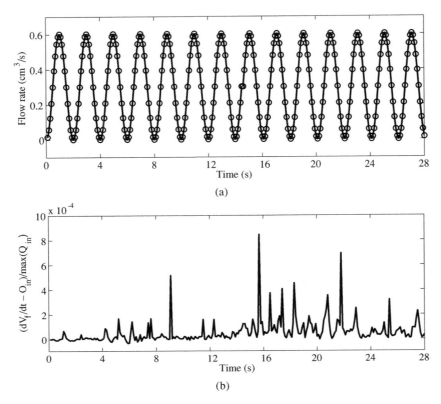

Figure 10.6 Inflation of a balloon. (a) Plot of the volumetric inflow rate versus the rate of change of the fluid domain volume. (b) Plot of the relative error in the flow rates that is attributable to convergence tolerances employed in the calculations.

Human arteries are not exactly circular, hence projection of the template onto the true surface is necessary. Only control points that govern the cross-section geometry are involved in the projection process, while the underlying parametric description of the cross-section stays unchanged. The end result of this construction is shown in Figure 10.7, which illustrates the mapping of the template cross-section onto the patient-specific geometry. Here the isoparametric lines are somewhat distorted so as to conform to the true geometry, while the topology of the fluid and solid subdomains is preserved along with their interface. It is worth noting that cross-sections of healthy arteries are nearly circular, so little distortion of the template is required to accurately capture the true geometry in this case.

Compared to the standard finite element method, the current method has significant benefits for analysis of blood flow in arteries, both in terms of accuracy and implementational convenience. It is well known in fluid mechanics that steady, laminar, incompressible flow in a straight circular pipe that is driven by a constant pressure gradient develops a parabolic profile in the radial direction and has no dependence on the circumferential or axial directions. NURBS discretization is capable of exactly representing this solution profile pointwise, in contrast to standard finite element discretizations.

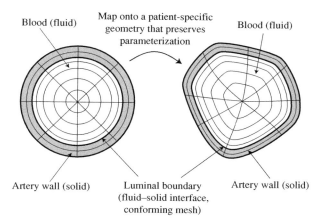

Figure 10.7   Arterial cross-section template based on a NURBS mesh of a circle that is subsequently mapped onto a patient-specific geometry. Fluid and solid regions are identified and separated by an interface. For analysis purposes, basis functions are made C^0-continuous at the fluid–solid interface. Note that the topology of the fluid and the solid subdomains remains unchanged.

Parametric definition of the geometry is not only attractive from the mesh refinement point of view, it is also beneficial in arterial blood flow applications for the following reasons:

- In the fluid region it allows one to build high quality structured boundary layer meshes near arterial walls. This is crucial for overall accuracy of the fluid–structural simulation as well as for obtaining accurate wall quantities, which play an important role in predicting the onset and development of vascular disease.
- In the solid region it allows for a natural representation of material anisotropy of the arterial wall because the parametric coordinates are aligned with the axial, circumferential and wall-normal directions. See Holzapfel, 2004 for arterial wall material modeling which accounts for anisotropic behavior.
- Parametric mesh definition in the fluid region allows for a straightforward specification of the elastic mesh parameters used for the mesh-movement problem. For example, we "stiffen" the mesh near the fluid–structure interface so as to preserve boundary-layer elements during mesh motion.

10.3.2 Numerical results

A patient-specific geometry was obtained using 64-slice CT angiography and was provided to us by T. Kvamsdal and J.H. Kaspersen of SINTEF, Norway. The geometrical model, which contains some of the major branches of a typical abdominal aorta, is shown in Figure 10.8(a). Note that one of the renal arteries is missing in the model because the patient had only one kidney. The fluid properties are: $\rho^f = 1.06$ g/cm^3, $\mu^f = 0.04$ g/cm s. The solid has the density $\rho^s = 1$ g/cm^3, Young's modulus, $E = 4.144 \times 10^6$ dyn/cm^2, and Poisson's ratio, $\nu = 0.45$. The computational mesh, consisting of 44,892 quadratic NURBS elements, is shown in Figure 10.8(c). Two quadratic NURBS elements and four C^1-continuous basis functions are used for through-thickness resolution of the arterial wall (see Figure 4.9).

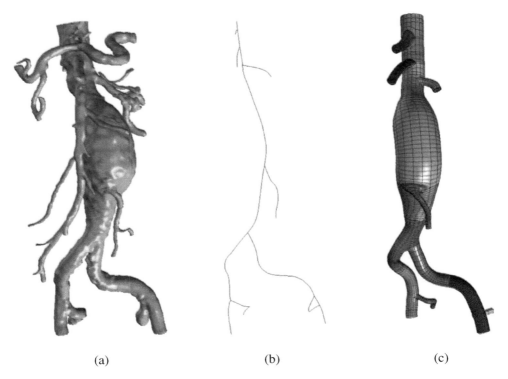

(a)   (b)   (c)

Figure 10.8   Flow in a patient-specific abdominal aorta with aneurysm. (a) Patient-specific imaging data; (b) Skeleton of the NURBS mesh; (c) Smoothed and truncated NURBS model and mesh. In (c), every NURBS patch is assigned a different color. For more details of geometrical modeling for isogeometric analysis of blood flow the reader is referred to Zhang *et al.*, 2007.

A periodic flow waveform, with period $T = 1.05$ s, is applied at the inlet of the aorta, while resistance boundary conditions are applied at all outlets. The solid is fixed at the inlet and at all outlets. Material and flow rate data, as well as resistance values are taken from Figueroa *et al.*, 2006. Wall thickness for this model is taken to be 15% of the nominal radius of each cross-section of the fluid domain model.

Figure 10.9 shows snapshots of the velocity field plotted on the moving domain at two different times during the heart cycle. The flow field is quite complex and fully three-dimensional, especially in diastole. The velocity magnitude is largest near the inflow and is significantly lower in the aneurysmal region. This occurs in part due to the fact that a significant percentage of the flow goes to the upper branches of the abdominal aorta and the increase in the cross-sectional area of the vessel associated with the aneurysm.

Figure 10.10 shows the so-called oscillatory shear index (OSI) distribution at the luminal surface. OSI is defined as (see, e.g., Taylor *et al.*, 1998, 1999),

$$\text{OSI} = \frac{1}{2} \left(1 - \frac{\tau_{\text{mean}}}{\tau_{\text{abs}}} \right),$$
(10.1)

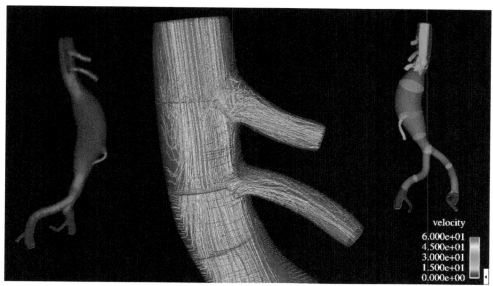

(a) Flow as the heart contracts during systole

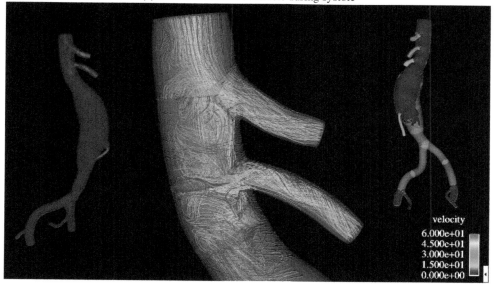

(b) Flow as the heart relaxes during diastole

Figure 10.9   Flow in a patient-specific abdominal aorta with aneurysm. Large frame: fluid velocity vectors colored by their magnitude, zoom on the top portion of the artery. Left small frame: volume rendering of the velocity magnitude. Right small frame: fluid velocity isosurfaces.

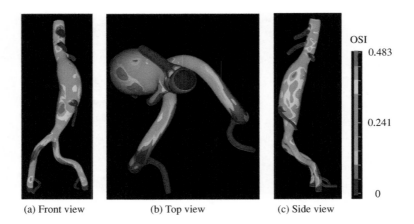

(a) Front view   (b) Top view   (c) Side view

Figure 10.10   Flow in a patient-specific abdominal aorta with aneurysm. Oscillatory shear index (OSI) plotted in three different views.

where, denoting by $\boldsymbol{\tau}_s$ the wall shear stress vector,

$$\tau_{\text{mean}} = \left| \frac{1}{T} \int_0^T \boldsymbol{\tau}_s dt \right|, \tag{10.2}$$

and

$$\tau_{\text{abs}} = \frac{1}{T} \int_0^T |\boldsymbol{\tau}_s| dt. \tag{10.3}$$

Note that OSI is largest in the aneurysm region, especially along the posterior wall, indicating that wall shear stress is highly oscillatory there. Low time-averaged wall shear stress, in combination with high shear stress temporal oscillations, as measured by the OSI, are identified with regions of high probability of occurrence of atherosclerotic disease.

10.4 Rotating components

Applications involving flows with rotating components, for example, ship propellers, cooling fans, heat exchangers, etc., have a great practical significance in various branches of engineering. As a result, robust and accurate simulation techniques are necessary to predict and analyze the behavior and physical characteristics of these systems. Computation of flows around rotating objects engenders two difficulties compared with computation of flows on stationary domains: 1) obtaining discretizations that are compatible with the relative motion of rotating and fixed components and 2) deriving the discrete formulation and solution spaces for the flow fields in question. The unique combination of the geometrical flexibility and accuracy possessed by NURBS on problems of computational fluid dynamics facilitates addressing these problems in an elegant and effective manner.

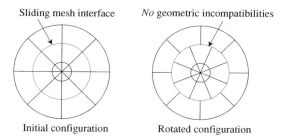

Figure 10.11   Embedding of a rotating component in a stationary flow domain using NURBS discretizations.

As we have seen many times by now, NURBS are capable of exactly representing all conic sections, including circular and cylindrical surfaces. This method for problems of rotating components consists of embedding a rotating body in a circular (in 2D) or a cylindrical (in 3D) domain, which, in turn, is placed inside the surrounding flow domain. While the surrounding flow domain is stationary, the subdomain that contains a rotating body spins with it. A key observation for the developments is that the interface between the rotating and stationary subdomains is unique, it remains circular or cylindrical at all times, and it can be *exactly* represented in the space of quadratic or higher-order NURBS functions. In this approach, unlike in standard finite elements, geometric compatibility between the rotating and stationary subdomains is exact. This situation is illustrated in Figures 10.11 and 10.12.

Although geometric compatibility is naturally attained by using a NURBS representation, imposing solution compatibility directly in the solution space is too restrictive with respect to the kinds of meshes and time step sizes we wish to employ in the computations. Hence, we abandon compatible discretizations at the stationary and rotating subdomain interface and devise a numerical technique that imposes continuity of the discrete solution weakly. For this purpose, we borrow ideas from the discontinuous Galerkin (DG) methodology, as we did for the weak enforcement of boundary conditions in Chapter 3. It should be noted that using ideas from DG methods to impose solution compatibility in the presence of incompatible meshes and multi-physics phenomena has been exploited, for example, by Wriggers and Zavarise, 2007 for solid mechanics and contact, and by Hansbo and Hermansson, 2003 and Hansbo *et al.*, 2004 for fluid–structure interaction.

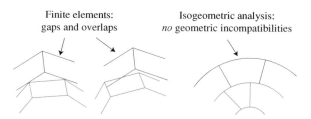

Figure 10.12   Embedding of a rotating component in a stationary flow domain: comparison between NURBS and finite element discretizations.

10.4.1 Coupling of the rotating and stationary domains

In this section we give a semi-discrete formulation of the problem that couples stationary and rotating parts of the domain. We first consider the individual subproblems of the incompressible fluid on the stationary and rotating domains. We then state the coupled formulation at the continuous level. We close the section with a statement of the coupled problem at the semi-discrete level and discuss implementation details.

10.4.1.1 Incompressible Navier–Stokes equations on the stationary and rotating domains

We begin by considering a weak formulation of the incompressible Navier–Stokes equations posed on a stationary domain Ω_s. We follow the approach of Bazilevs and Hughes, 2008, which may be consulted for additional details. We note that there are some differences between this formulation and the one described in Chapter 9, Section 9.4. In particular, the present formulation is written in the advective form whereas the one in Chapter 9, Section 9.4 is written in conservation form. Let $V(\Omega_s)$ denote the trial solution and weighting function spaces, which are assumed to be the same. The variational formulation on a stationary domain is stated as follows: Find a velocity–pressure pair, $\mathbf{U} = \{\boldsymbol{u}, p\} \in V(\Omega_s)$, such that for all weighting functions $\boldsymbol{W} = \{\boldsymbol{w}, q\} \in V(\Omega_s)$

$$B_s(\boldsymbol{W}, \mathbf{U}) = F_s(\boldsymbol{W}),\tag{10.4}$$

where

$$B_s(\boldsymbol{W}, \mathbf{U}) = \left(\boldsymbol{w}, \rho\frac{\partial \boldsymbol{u}}{\partial t} + \rho\boldsymbol{u}\cdot\nabla\boldsymbol{u}\right)_{\Omega_s} + (q, \nabla\cdot\boldsymbol{u})_{\Omega_s} - (\nabla\cdot\boldsymbol{w}, p)_{\Omega_s}\tag{10.5}$$
$$+ \left(\nabla^s\boldsymbol{w}, 2\mu\nabla^s\boldsymbol{u}\right)_{\Omega_s},$$

and

$$F_s(\boldsymbol{W}) = (\boldsymbol{w}, \rho\boldsymbol{f})_{\Omega_s}.\tag{10.6}$$

In (10.5), μ is the dynamic viscosity, ρ is the density, $\boldsymbol{f}$ is the body force per unit mass, and $\nabla^s = \frac{1}{2}(\nabla + \nabla^T)$.

Variational equations (10.4)–(10.6) imply satisfaction of the linear momentum equations and the incompressibility constraint, namely,

$$\mathcal{L}_s(\boldsymbol{u}, p) - \rho\boldsymbol{f} = \mathbf{0}\qquad \text{in } \Omega_s,\tag{10.7}$$

and

$$\nabla\cdot\boldsymbol{u} = 0\qquad \text{in } \Omega_s,\tag{10.8}$$

where

$$\mathcal{L}_s(\boldsymbol{u}, p) = \rho\frac{\partial \boldsymbol{u}}{\partial t} + \rho\boldsymbol{u}\cdot\nabla\boldsymbol{u} + \nabla p - \nabla\cdot(2\mu\nabla^s\boldsymbol{u}).\tag{10.9}$$

In the case of the rotating domain, denoted by $\Omega_r(t)$, the weak formulation becomes: Find $\mathbf{U} = \{\boldsymbol{u}, p\} \in \mathcal{V}(\Omega_r(t))$, such that for all weighting functions $\mathbf{W} = \{\boldsymbol{w}, q\} \in \mathcal{V}(\Omega_r(t))$

$$B_r(\mathbf{W}, \mathbf{U}; \boldsymbol{v}) = F_r(\mathbf{W}), \tag{10.10}$$

where

$$B_r(\mathbf{W}, \mathbf{U}; \boldsymbol{v}) = \left(\boldsymbol{w}, \rho \frac{\partial \boldsymbol{u}}{\partial t}|_y + \rho(\boldsymbol{u} - \boldsymbol{v}) \cdot \nabla \boldsymbol{u} \right)_{\Omega_r(t)} \tag{10.11}$$

$$+ (q, \nabla \cdot \boldsymbol{u})_{\Omega_r(t)} - (\nabla \cdot \boldsymbol{w}, p)_{\Omega_r(t)} + \left(\nabla^s \boldsymbol{w}, 2\mu \nabla^s \boldsymbol{u} \right)_{\Omega_r(t)},$$

and

$$F_r(\mathbf{W}) = (\boldsymbol{w}, \rho \boldsymbol{f})_{\Omega_r(t)}. \tag{10.12}$$

In the above equations $\Omega_r(t)$ is a configuration at time t that is an image of some referential configuration $\hat{\Omega}_r$ under a time-dependent mapping $\boldsymbol{\phi} : \mathbb{R}^3 \times \mathbb{R} \to \mathbb{R}^3$ called the motion. In (10.11), $\boldsymbol{v}$ is the velocity of $\Omega_r(t)$ in the spatial description defined as

$$\boldsymbol{v} = \hat{\boldsymbol{v}} \circ \boldsymbol{\phi}^{-1}, \tag{10.13}$$

where $\hat{\boldsymbol{v}}$ is the velocity of $\Omega_r(t)$ in the referential description given as

$$\hat{\boldsymbol{v}} = \frac{\partial \boldsymbol{\phi}(\boldsymbol{y}, t)}{\partial t}|_y, \tag{10.14}$$

$\circ$ denotes composition, the $\boldsymbol{y}$'s are the referential coordinates, and $\boldsymbol{\phi}^{-1}$ is understood as the inverse of the mapping $\boldsymbol{\phi}$ at a fixed time.

Variational equation (10.10) implies satisfaction of the linear momentum equations and of the incompressibility constraint, namely

$$\mathcal{L}_r(\boldsymbol{u}, p; \boldsymbol{v}) - \rho \boldsymbol{f} = \mathbf{0} \qquad \text{in } \Omega_r(t), \tag{10.15}$$

and

$$\nabla \cdot \boldsymbol{u} = 0 \qquad \text{in } \Omega_r(t), \tag{10.16}$$

where

$$\mathcal{L}_r(\boldsymbol{u}, p) = \rho \frac{\partial \boldsymbol{u}}{\partial t} + \rho(\boldsymbol{u} - \boldsymbol{v}) \cdot \nabla \boldsymbol{u} + \nabla p - \nabla \cdot (2\mu \nabla^s \boldsymbol{u}). \tag{10.17}$$

For a rotation, the mapping $\boldsymbol{\phi}$ takes on a particularly simple form:

$$\boldsymbol{\phi} = \boldsymbol{R}(t)(\boldsymbol{y} - \boldsymbol{y}_0) + \boldsymbol{y}_0, \tag{10.18}$$

where $\boldsymbol{y}_0$ is a fixed point, and $\boldsymbol{R}(t)$ is a rotation matrix satisfying

$$\boldsymbol{R}(t)^T \boldsymbol{R}(t) = \boldsymbol{I} \tag{10.19}$$

and

$$\det \mathbf{R}(t) = 1. \tag{10.20}$$

Taking the referential time derivative of ϕ defined by (10.18), the velocity of $\Omega_r(t)$ in the referential description becomes

$$\hat{v} = \frac{\partial \mathbf{R}(t)}{\partial t}(\mathbf{y} - \mathbf{y}_0), \tag{10.21}$$

and the acceleration of $\Omega_r(t)$ is

$$\hat{a} = \frac{\partial^2 \mathbf{R}(t)}{\partial t^2}(\mathbf{y} - \mathbf{y}_0). \tag{10.22}$$

Remark
Variational equation (10.10) pertains to an arbitrary Lagrangian–Eulerian (ALE) description of the incompressible fluid flow on a moving domain. We use ALE in this work to handle rotating domain motion rather than expressing the equations in a co-rotational frame of reference. This enables the formulation to be used for more general motions in addition to rotations.

10.4.1.2 Continuous problem

Consider a domain $\Omega(t) = \Omega_s \cup \Omega_r(t)$, where Ω_s is the stationary subdomain, and $\Omega_r(t)$ is a subdomain that contains a rotating component. We assume that $\Omega_r(t)$ rotates inside $\Omega(t)$ with the speed of rotation of the rotating component that is embedded in it. We refer to $\Gamma_{sr} = \Omega_s \cap \Omega_r(t)$, the interface between the stationary and rotating domains, as the "sliding interface." Γ_{sr} is a circular surface in two spatial dimensions, and a cylindrical surface in three spatial dimensions. Note that, although $\Omega_r(t)$ undergoes a rotating motion, Γ_{sr} does not change with time.

We state the continuous problem as follows: Find $\mathbf{U}_s = \{\mathbf{u}_s, p_s\} \in \mathcal{V}(\Omega_s)$ and $\mathbf{U}_r = \{\mathbf{u}_r, p_r\} \in \mathcal{V}(\Omega_r(t))$, such that for all $\mathbf{W}_s = \{\mathbf{w}_s, q_s\} \in \mathcal{V}(\Omega_s)$ and $\mathbf{W}_r = \{\mathbf{w}_r, q_r\} \in \mathcal{V}(\Omega_r(t))$

$$B_s(\mathbf{W}_s, \mathbf{U}_s) + B_r(\mathbf{W}_r, \mathbf{U}_r; \mathbf{v}) - F_s(\mathbf{W}_s) - F_r(\mathbf{W}_r) = 0, \tag{10.23}$$

subject to

$$\mathbf{u}_s = \mathbf{u}_r \qquad \text{on } \Gamma_{sr} \tag{10.24}$$

and

$$\mathbf{w}_s = \mathbf{w}_r \qquad \text{on } \Gamma_{sr}. \tag{10.25}$$

Variational equations (10.23), together with (10.24) and (10.25), imply satisfaction of linear momentum and incompressibility in both subdomains, as well as compatibility of tractions at the interface between the subdomains, namely

$$\mathcal{L}_s(\boldsymbol{u}_s, p_s) - \rho \boldsymbol{f} = \boldsymbol{0} \qquad \text{in } \Omega_s, \tag{10.26}$$

$$\nabla \cdot \boldsymbol{u}_s = 0 \qquad \text{in } \Omega_s, \tag{10.27}$$

$$\mathcal{L}_r(\boldsymbol{u}_r, p_r; \boldsymbol{v}) - \rho \boldsymbol{f} = \boldsymbol{0} \qquad \text{in } \Omega_r(t), \tag{10.28}$$

$$\nabla \cdot \boldsymbol{u}_r = 0 \qquad \text{in } \Omega_r(t), \tag{10.29}$$

$$-p_s \boldsymbol{n}_s + 2\mu \nabla^s \boldsymbol{u}_s \cdot \boldsymbol{n}_s - p_r \boldsymbol{n}_r + 2\mu \nabla^s \boldsymbol{u}_r \cdot \boldsymbol{n}_r = \boldsymbol{0} \qquad \text{on } \Gamma_{sr}, \tag{10.30}$$

where $\boldsymbol{n}_s$ and $\boldsymbol{n}_r$ are the unit outward normal vectors to the stationary and rotating subdomains, respectively. Also note, $\boldsymbol{n}_s = -\boldsymbol{n}_r$.

10.4.1.3 Discrete formulation

Let Ω_s and $\Omega_r(t)$ be decomposed into NURBS elements. The discretization of $\Omega_r(t)$ is obtained by simply applying a rotation to $\Omega_r(0)$, a configuration of the rotating subdomain at initial time. Note that, due to the affine covariance property of NURBS (see Chapter 2), the rotation needs to be applied only to the control mesh of $\Omega_r(0)$, which is a very simple operation. Discretization of Ω_s and $\Omega_r(t)$ induces two separate discretizations of Γ_{sr}, one coming from the stationary side and one from the rotating side. These two discretizations may be combined by means of h-refinement, which is done by inserting knots from both Ω_s and $\Omega_r(t)$ into Γ_{sr}. Note that the knots that need to be inserted from the rotating patch change their parametric locations in the stationary patch due to the relative motion of two subdomains. See Figure 10.13.

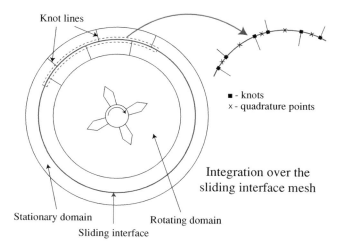

Figure 10.13 Embedding of a rotating component in a stationary flow domain.

Combining the two interface meshes into a finer mesh is necessary in order to accurately compute interface integrals that are present in the formulation described later in this section. Gauss quadrature is performed over the elements of the combined mesh. On the interiors of these interface integral elements, basis functions coming from the rotating and stationary sides of the domain are smooth and, as a result, Gaussian integration gives accurate results. See Figure 10.13.

Let the stationary and rotating subdomains be decomposed into n_{es} and n_{er} elements, respectively. Let the knot insertion procedure at a given time t generate n_{eb} boundary faces on Γ_{sr}. We discretize (10.23) together with (10.24) and (10.25) over the finite-dimensional NURBS spaces as follows: Find $\mathbf{U}_s = \{\boldsymbol{u}_s, p_s\} \in \mathcal{V}^h(\Omega_s)$ and $\mathbf{U}_r = \{\boldsymbol{u}_r, p_r\} \in \mathcal{V}^h(\Omega_r(t))$, such that for all $\boldsymbol{W}_s = \{\boldsymbol{w}_s, q_s\} \in \mathcal{V}^h(\Omega_s)$ and $\boldsymbol{W}_r = \{\boldsymbol{w}_r, q_r\} \in \mathcal{V}^h(\Omega_r(t))$,

$$B_s^{MS}(\boldsymbol{W}_s, \mathbf{U}_s) + B_r^{MS}(\boldsymbol{W}_r, \mathbf{U}_r; v) - F_s(\boldsymbol{W}_s) - F_r(\boldsymbol{W}_r)$$

$$- \sum_{eb=1}^{n_{eb}} ((\boldsymbol{w}_s - \boldsymbol{w}_r), \boldsymbol{t}(\boldsymbol{u}_s, p_s, \boldsymbol{u}_r, p_r))_{\Gamma_{eb}}$$

$$- \sum_{eb=1}^{n_{eb}} (\tilde{\boldsymbol{t}}_u(\boldsymbol{w}_s, q_s, \boldsymbol{w}_r, q_r), (\boldsymbol{u}_s - \boldsymbol{u}_r))_{\Gamma_{eb}}$$

$$+ \sum_{eb=1}^{n_{eb}} ((\boldsymbol{w}_s - \boldsymbol{w}_r)\tau_B, (\boldsymbol{u}_s - \boldsymbol{u}_r))_{\Gamma_{eb}} = 0, \qquad (10.31)$$

where

$$B_s^{MS}(\boldsymbol{W}_s, \mathbf{U}_s) = B_s(\boldsymbol{W}_s, \mathbf{U}_s)$$

$$+ \sum_{e=1}^{n_{es}} ((\boldsymbol{u}_s \cdot \nabla \boldsymbol{w}_s + \nabla q_s/\rho)\tau_{MS}, \mathcal{L}_s(\boldsymbol{u}_s, p_s) - \rho \boldsymbol{f})_{\Omega_e}$$

$$- \sum_{e=1}^{n_{es}} (\boldsymbol{w}_s \tau_{MS}, (\mathcal{L}_s(\boldsymbol{u}_s, p_s) - \rho \boldsymbol{f}) \cdot \nabla \boldsymbol{u}^s)_{\Omega_e}$$

$$- \sum_{e=1}^{n_{es}} (\nabla \boldsymbol{w}_s, \tau_{MS}(\mathcal{L}_s(\boldsymbol{u}_s, p_s) - \rho \boldsymbol{f}) \otimes \tau_{MS}(\mathcal{L}_s(\boldsymbol{u}_s, p_s) - \rho \boldsymbol{f}))_{\Omega_e}$$

$$+ \sum_{e=1}^{n_{es}} (\nabla \cdot \boldsymbol{w}_s, \tau_{CS} \nabla \cdot \boldsymbol{u}_s)_{\Omega_e}, \qquad (10.32)$$

and

$$B_r^{MS}(\boldsymbol{W}_r, \mathbf{U}_r; \boldsymbol{v}) = B_r(\boldsymbol{W}_r, \mathbf{U}_r; \boldsymbol{v})$$

$$+ \sum_{e=1}^{n_{er}} (((\boldsymbol{u}_r - \boldsymbol{v}) \cdot \nabla \boldsymbol{w}_r + \nabla q_r/\rho)\tau_{Mr}, \mathcal{L}_r(\boldsymbol{u}_r, p_r; \boldsymbol{v}) - \rho \boldsymbol{f})_{\Omega_e}$$

$$- \sum_{e=1}^{n_{er}} (\boldsymbol{w}_r \tau_{Mr}, (\mathcal{L}_r(\boldsymbol{u}_r, p_r; \boldsymbol{v}) - \rho \boldsymbol{f}) \cdot \nabla \boldsymbol{u}^r)_{\Omega_e}$$

$$- \sum_{e=1}^{n_{er}} (\nabla \boldsymbol{w}_r, \tau_{Mr}(\mathcal{L}_r(\boldsymbol{u}_r, p_r; \boldsymbol{v}) - \rho \boldsymbol{f}) \otimes \tau_{Mr}(\mathcal{L}_r(\boldsymbol{u}_r, p_r; \boldsymbol{v}) - \rho \boldsymbol{f}))_{\Omega_e}$$

$$+ \sum_{e=1}^{n_{er}} (\nabla \cdot \boldsymbol{w}_r, \tau_{Cr} \nabla \cdot \boldsymbol{u}_r)_{\Omega_e}. \tag{10.33}$$

(10.32) and (10.33) are the discrete semilinear forms corresponding to the variational multiscale residual-based formulation of the incompressible Navier–Stokes equations in the stationary and rotating subdomains, respectively. The finite-dimensional NURBS spaces denoted by $\mathcal{V}^h(\Omega_s) \subset \mathcal{V}(\Omega_s)$ and $\mathcal{V}^h(\Omega_r(t)) \subset \mathcal{V}^h(\Omega_r(t))$. For precise definitions of τ_{Ms}, τ_{Mr}, τ_{Cs}, and τ_{Cr} the reader is referred to Bazilevs *et al.*, 2007a, although it should be noted that in the ALE setting $\boldsymbol{u}_r - \boldsymbol{v}$ is used as the advective velocity in the definition of τ_{Mr}, and, as a result, in τ_{Cr}. The τ's are designed by asymptotic scaling arguments (Barenblatt, 1979), developed within the theory of stabilized methods (see, e.g., Brooks and Hughes, 1982; Hughes and Mallet, 1986; Shakib *et al.*, 1991; Tezduyar, 2003). They may also be viewed as approximations to the small-scale Green's operator within the theory of multiscale methods introduced by Hughes *et al.*, 1998 and studied in detail by Hughes and Sangalli, 2007.

The last three terms of (10.31) are associated with the weak imposition of solution continuity. Operators $\boldsymbol{t}(\boldsymbol{u}_s, p_s, \boldsymbol{u}_r, p_r)$ and $\tilde{\boldsymbol{t}}_u(\boldsymbol{w}_s, q_s, \boldsymbol{w}_r, q_r)$, acting on the solution and weighting functions, are the tractions defined as

$$\boldsymbol{t}(\boldsymbol{u}_s, p_s, \boldsymbol{u}_r, p_r) = (- p_s \boldsymbol{n}_s + \mu(\nabla \boldsymbol{u}_s + \nabla \boldsymbol{u}_s^T)\boldsymbol{n}_s$$

$$+ p_r \boldsymbol{n}_r - \mu(\nabla \boldsymbol{u}_r + \nabla \boldsymbol{u}_r^T)\boldsymbol{n}_r)/2, \tag{10.34}$$

and

$$\tilde{\boldsymbol{t}}_u(\boldsymbol{w}_s, q_s, \boldsymbol{w}_r, q_r) = (q_s \boldsymbol{n}_s + \mu(\nabla \boldsymbol{w}_s + \nabla \boldsymbol{w}_s^T)\boldsymbol{n}_s$$

$$- q_r \boldsymbol{n}_r - \mu(\nabla \boldsymbol{v}_r + \nabla \boldsymbol{v}_r^T)\boldsymbol{n}_r)/2$$

$$+ (\boldsymbol{w}_s - \boldsymbol{w}_r)(\{\boldsymbol{u}_s \cdot \boldsymbol{n}_s\}_- + \{\boldsymbol{u}_r \cdot \boldsymbol{n}_r\}_-). \tag{10.35}$$

In (10.31), $\{A\}_-$ denotes the negative part of A, that is, $\{A\}_- = A$ if $A < 0$ and $\{A\}_- = 0$ if $A \geq 0$. τ_B is defined as

$$\tau_B = \frac{1}{2}\left(\frac{C_s \mu}{h_s} + \frac{C_r \mu}{h_r}\right), \tag{10.36}$$

where h_s and h_r are the element lengths in the normal direction to the sliding interface in the stationary and rotating domains, respectively, and are explicitly given as

$$h_s = 2\left(n_s^T \frac{\partial \xi}{\partial x}^T \frac{\partial \xi}{\partial x} n_s\right)^{-1/2} \quad \text{and} \quad h_r = 2\left(n_r^T \frac{\partial \xi}{\partial x}^T \frac{\partial \xi}{\partial x} n_r\right)^{-1/2}, \tag{10.37}$$

where $\frac{\partial \xi}{\partial x}$ is the inverse Jacobian of the element mapping between the parent and physical domains, and C_s and C_r are positive constants arising in the element-wise inverse estimates (see, e.g., Ern and Guermond, 2004). The parametric mapping, $x(\xi)$, in this case is defined by the local element, that is, the region between consecutive knots. In the simplest geometries, h_s and h_r become the radial distances between knots in the physical space.

Remarks
1. The last three terms on the left-hand side of (10.31) are associated with the weak enforcement of continuity of the velocities and tractions at the interface between the stationary and rotating subdomains. The form of these terms is inspired by the selective interior penalty Galerkin (SIPG) discontinuous method of Wheeler, 1978. The third-to-last term in (10.31) is the so-called consistency term: When deriving the Euler–Lagrange equations corresponding to (10.31), integration-by-parts yields a term that is canceled by the consistency term. The second-to-last term in (10.31) is the so-called adjoint-consistency term: If the exact solution of the adjoint problem is inserted into equation (10.31) in place of the test function, (10.31) is satisfied identically; see Arnold et al., 2002 for details on adjoint consistency. The last term of (10.31) penalizes the discrete version of (10.24).
2. Note that the terms involving the pressure trial solution and weighting function appear in a "skew-symmetric" form in (10.31). This form renders these terms stability-neutral, without upsetting adjoint-consistency of the formulation. Reversal of the sign of the pressure weighting function terms leads to numerical instability.

10.4.2 Numerical example: two propellers spinning in opposite directions

The problem description is given in Figure 10.14. Two four-blade propellers, with blades pitched at $5°$ angles, are rotating at a constant angular velocity in opposite directions, as shown in the figure. No-slip boundary conditions are applied at the propeller surfaces as well as the outer edges of the box. The flow is characterized by a Taylor number $Ta \approx 150,000$, defined as

$$Ta = 4\omega^2 R^4 / v^2, \tag{10.38}$$

where $\omega = 2\pi f$ is the angular velocity, f is the cyclic frequency, R is a characteristic dimension of the propeller blades, and v is the kinematic viscosity of the fluid. Here f is 0.05, R is 2.5, and v is 0.01. This Taylor number is sufficiently high for convective instabilities to set in and create complex flow structures known as Taylor vortices. The Reynolds number of the flow, based on the velocity of the tip of the propeller blades and R, is 196.

Figure 10.15 shows the computational mesh in the reference configuration. Note that even in the reference configuration the mesh between the rotating and stationary subdomains is non-matching. 4360 quadratic NURBS elements were employed in the computation. The

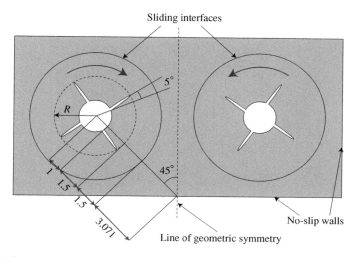

Figure 10.14 Two propellers spinning in opposite directions. Problem setup and dimensions.

problem was solved using a three-dimensional code with boundary conditions prescribed to ensure a two-dimensional response. The semi-discrete equations were advanced in time using the generalized-α method (see Chapter 7, Section 7.3, and the original references, Chung and Hulbert, 1993; Jansen *et al.*, 1999). To complete the specification of the problem, we set the values of the inverse constants to $C_s = C_r = 4$ (see (10.36)).

The flow was impulsively started and it took several propeller revolutions before the vortical structures appeared. Figure 10.16 shows several snapshots of the flow field at various times after the flow symmetry was broken by the onset of Taylor vortices. It should be noted that

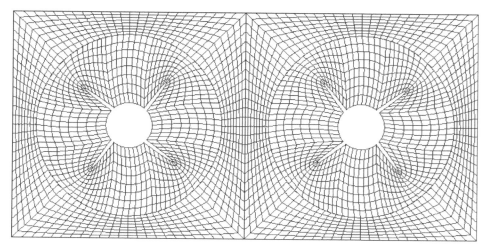

Figure 10.15 Two propellers spinning in opposite directions. Computational mesh in the reference configuration.

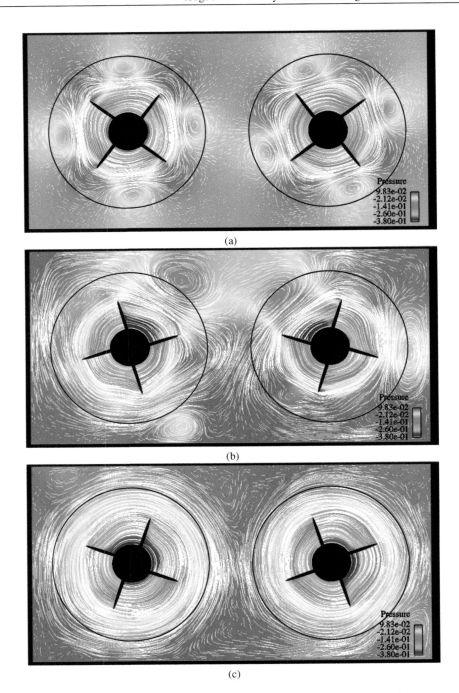

(a)

(b)

(c)

Figure 10.16   Two propellers spinning in opposite directions. Snapshots of the velocity vectors super-posed on the contours of fluid pressure. In the early stages of the simulation, the flow loses symmetry and becomes complex. The later stages are characterized by the appearance of smaller vortices.

although a discontinuous discretization of the fields is employed at the interface between the rotating and stationary domains, the velocity field is virtually continuous as is evidenced by the continuity of the flow vectors in Figure 10.16. Furthermore, the method builds what appears to be a nearly continuous pressure field at the interface, although this condition is not explicitly built into the formulation. We conjecture that this behavior is in part due to the geometric compatibility at the interface engendered by the NURBS-based discretization.

Appendix 10.A A geometrical template for arterial blood flow modeling

Patient-specific blood flow modeling begins with geometry construction. This approach is often template based, beginning with a simple cylindrical model incorporating both the solid arterial wall and the fluid domain. As described in detail in Zhang *et al.*, 2007, this template – matched with the appropriate templates for the various types of arterial bifurcations – is deformed to match patient-specific data obtained from medical imaging procedures.

Here we present the basic arterial template. We begin with the simplest possible description. Beginning at the coarsest level of discretization allows the user the most flexibility in being able to refine in whatever manner suits the need of the specific application.

For the simple cylindrical case shown in Figure 10.A.1, we choose the ξ-direction in the parameter space to correspond to the circumferential direction, the η-direction to correspond with the radial direction, and the ζ-direction to correspond to the axial direction. Thus, we require the associated polynomial orders to be $p = 2, q = 1$, and $r = 1$, respectively. Recalling the circular template from Chapter 2, we take the knot vector in the circumferential direction to be

$$\Xi = \{0, 0, 0, 1, 1, 2, 2, 3, 3, 4, 4, 4\}. \tag{10.A.1}$$

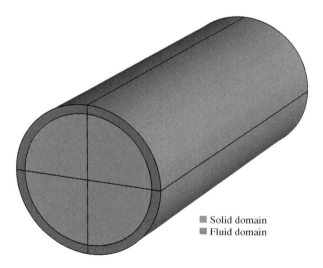

Solid domain
Fluid domain

Figure 10.A.1 Artery template mesh. Both the solid artery wall and the enclosed fluid domain are modeled.

We must consider that we have both a fluid domain and a solid domain, necessitating two elements in the radial direction. The typical arterial wall has a thickness that is between about 7% and 17% of the exterior radius (see, *e.g.*, Humphrey, 2002). Splitting the difference, we take the knot vector in the radial direction to be

$$\mathcal{H} = \{0, 0, 0.88, 1, 1\}, \tag{10.A.2}$$

where the first, larger element will be the fluid domain, and the smaller element will be the arterial wall. We need only a single element in the axial direction, thus

$$\mathcal{Z} = \{0, 0, 1, 1\}. \tag{10.A.3}$$

For illustrative purposes, let us assume a unit radius for the outer arterial wall, and length of 5. Clearly, these numbers are not of particular importance in and of themselves, and the application will dictate the most reasonable values. The control points for the mesh are given in Table 10.A.1.

Table 10.A.1 Control points for the artery template

i	j	$\mathbf{B}_{i,j,1}$	$\mathbf{B}_{i,j,2}$	$w_{i,j,1}$	$w_{i,j,2}$
1	1	$(0, 0, 0)$	$(0, 0, 5)$	1	1
1	2	$(0.88, 0, 0)$	$(0.88, 0, 5)$	1	1
1	3	$(1, 0, 0)$	$(1, 0, 5)$	1	1
2	1	$(0, 0, 0)$	$(0, 0, 5)$	$1/\sqrt{2}$	$1/\sqrt{2}$
2	2	$(0.88, 0.88, 0)$	$(0.88, 0.88, 5)$	$1/\sqrt{2}$	$1/\sqrt{2}$
2	3	$(1, 1, 0)$	$(1, 1, 5)$	$1/\sqrt{2}$	$1/\sqrt{2}$
3	1	$(0, 0, 0)$	$(0, 0, 5)$	1	1
3	2	$(0, 0.88, 0)$	$(0, 0.88, 5)$	1	1
3	3	$(0, 1, 0)$	$(0, 1, 5)$	1	1
4	1	$(0, 0, 0)$	$(0, 0, 5)$	$1/\sqrt{2}$	$1/\sqrt{2}$
4	2	$(-0.88, 0.88, 0)$	$(-0.88, 0.88, 5)$	$1/\sqrt{2}$	$1/\sqrt{2}$
4	3	$(-1, 1, 0)$	$(-1, 1, 5)$	$1/\sqrt{2}$	$1/\sqrt{2}$
5	1	$(0, 0, 0)$	$(0, 0, 5)$	1	1
5	2	$(-0.88, 0, 0)$	$(-0.88, 0, 5)$	1	1
5	3	$(-1, 0, 0)$	$(-1, 0, 5)$	1	1
6	1	$(0, 0, 0)$	$(0, 0, 5)$	$1/\sqrt{2}$	$1/\sqrt{2}$
6	2	$(-0.88, -0.88, 0)$	$(-0.88, -0.88, 5)$	$1/\sqrt{2}$	$1/\sqrt{2}$
6	3	$(-1, -1, 0)$	$(-1, -1, 5)$	$1/\sqrt{2}$	$1/\sqrt{2}$
7	1	$(0, 0, 0)$	$(0, 0, 5)$	1	1
7	2	$(0, -0.88, 0)$	$(0, -0.88, 5)$	1	1
7	3	$(0, -1, 0)$	$(0, -1, 5)$	1	1
8	1	$(0, 0, 0)$	$(0, 0, 5)$	$1/\sqrt{2}$	$1/\sqrt{2}$
8	2	$(0.88, -0.88, 0)$	$(0.88, -0.88, 5)$	$1/\sqrt{2}$	$1/\sqrt{2}$
8	3	$(1, -1, 0)$	$(1, -1, 5)$	$1/\sqrt{2}$	$1/\sqrt{2}$
9	1	$(0, 0, 0)$	$(0, 0, 5)$	1	1
9	2	$(0.88, 0, 0)$	$(0.88, 0, 5)$	1	1
9	3	$(1, 0, 0)$	$(1, 0, 5)$	1	1

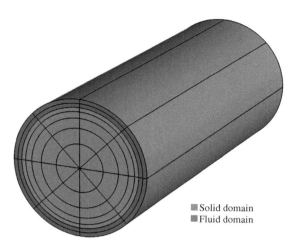

Figure 10.A.2 The artery template should be refined somewhat before deformation to fit the patient-specific geometry. Smaller elements in the fluid domain near the fluid–solid boundary allow for better resolution of the resulting boundary layer.

Before deforming the template to fit patient-specific data, refinement will be required. Elements will need to be added in the axial direction in order to have any flexibility in fitting a prescribed geometry. Similarly, the order must be elevated as physical arteries are usually curved. The approach to take here is k-refinement, where we order elevate first and then insert the new elements. This will allow the smoothest fit of the data, and give the nicest domain upon which to perform out analysis. Artificial corners in the geometry would drastically alter the nature of the blood flow within the artery, degrading accuracy. We will also want to refine in the radial direction in order to ensure maximal accuracy in the calculations. It is important that we accurately resolve the solution near the fluid–structure boundary, and so we might consider biasing the refinements in such a way that the smallest elements in the fluid domain are near that boundary. Again, k-refinement will provide the best results. See an example of a refined mesh in Figure 10.A.2.

To refine the mesh for analysis purposes, we can insert knots in each of the knot vectors and determine new control points according to (2.1) and (2.2). We can also elevate the order of the functions, but we need to repeat knots at the solid–fluid interface to maintain C^0-continuity there.

One final point of note regarding this template is the fact that the mapping has a degeneracy along the axis of the cylinder. This is due to the fact that many different control points are located at the same point in physical space. When solving a system of equations, we map the control variables associated with these repeated control points to a single control variable, thus assuring *a priori* that the solution will not be multi-valued along the axis.

The current construction is very intuitive to work with as the parametric directions correspond perfectly to the circumferential, radial, and axial directions. More importantly, it facilitates construction of a very nice boundary layer mesh next to the fluid–structure boundary. This is vital to obtaining accurate, rapidly converging solutions. Meshes of this kind have been used successfully in a number of analyses.

11

Higher-order Partial Differential Equations

Applications involving differential operators of order greater than two have not historically lent themselves well to finite element analysis. The variational statements of such problems involve second derivatives, necessitating the use of a globally C^1-continuous basis. The difficulty in constructing bases in a general setting has relegated the study of such equations to the realm of finite-differences and spectral methods, both of which are viable methods, but far more limited than FEA in their scope and flexibility.

As we have seen throughout this book, with NURBS based isogeometric analysis, we have a higher-order accurate, robust method with tremendous geometric flexibility and compactly supported basis functions, all while maintaining the possibility of higher-order continuity. Thus, it is an ideal technology for the study of equations involving higher-order differential operators (see, *e.g.*, Auricchio *et al.*, 2007 for a stream function approach to isogeometric analysis of incompressible elastic solids). In this chapter, we focus on the example of the Cahn–Hilliard phase-field model, which has been used most frequently to simulate the segregation of a binary alloy system, but also has found use in applications as diverse as image processing, planet formation, and cancer growth.

11.1 The Cahn–Hilliard equation

Two different approaches have been used to describe phase transition phenomena: *sharp-interface* models and *phase-field* (diffuse-interface) models. Traditionally, the evolution of interfaces, such as the liquid–solid interface, has been modeled using sharp-interface models, as we saw in the previous chapter on fluid–structure interaction. Such an approach requires the resolution of a moving boundary problem, separate differential equations hold in each phase, and certain quantities may suffer jump discontinuities across the interface.

Phase-field models provide an alternative description for phase-transition phenomena by approximating the interface as being diffuse such that it does not need to be tracked explicitly. Such models can be derived from classical irreversible thermodynamics. Utilizing asymptotic expansions for vanishing interface thickness, it can be shown that classical sharp-interface models, including physical laws at interfaces and multiple junctions, are recovered in the limit. In order to capture the physics of the problem, the transition regions (diffuse interfaces) have to be extremely thin. The phase-field model we examine has its origins in the work of Cahn and Hilliard, 1958.

11.1.1 The strong form

Let $\Omega \subset \mathbb{R}^d$ be an open set, where $d = 2$ or 3. The boundary is composed of two complementary parts $\Gamma = \overline{\Gamma_g \cup \Gamma_s}$. A binary mixture is contained in Ω and c denotes the concentration of one of its components. The object is to find $c : \bar{\Omega} \times (0, T) \to \mathbb{R}$ such that

$$\frac{\partial c}{\partial t} = \nabla \cdot (M_c \nabla (\mu_c - \lambda \Delta c)) \qquad \text{in } \Omega \times (0, T), \qquad (11.1a)$$

$$c = g \qquad \text{on } \Gamma_g \times (0, T), \qquad (11.1b)$$

$$M_c \nabla (\mu_c - \lambda \Delta c) \cdot \mathbf{n} = s \qquad \text{on } \Gamma_s \times (0, T), \qquad (11.1c)$$

$$M_c \lambda \nabla c \cdot \mathbf{n} = 0 \qquad \text{on } \Gamma \times (0, T), \qquad (11.1d)$$

$$c(\mathbf{x}, 0) = c_0(\mathbf{x}) \qquad \text{in } \Omega, \qquad (11.1e)$$

where M_c is the mobility, μ_c represents the chemical potential of a regular solution in the absence of phase interfaces, and λ is a positive constant such that $\sqrt{\lambda}$ represents a length scale of the problem. This length scale is related to the thickness of the interfaces that represent the transition between the two phases.

For the mobility, we adopt the relationship

$$M_c = Dc(1 - c), \qquad (11.2)$$

where D is a positive constant which has dimensions of diffusivity (length^2/time). This is commonly called "degenerate mobility" as pure phases (*i.e.*, $c = 0$ and $c = 1$) have vanishing mobility. This expression appeared in the original derivation of the Cahn–Hilliard equation by Cahn, 1961.

The chemical potential of a uniform solution, μ_c, is a highly nonlinear function of the concentration. Though it is frequently approximated by a polynomial of degree three, we prefer the full, thermodynamically consistent form, *viz.*,

$$\mu_c = \frac{1}{2\theta} \log \frac{c}{1 - c} + 1 - 2c, \qquad (11.3)$$

where $\theta = T_c / T$ is a dimensionless number representing the ratio between the critical temperature, T_c, at which the two phases attain the same composition, and the absolute temperature, T.

It can be shown that for $\theta > 1$ the chemical free energy is non-convex, with two wells, which drives phase segregation into the **binodal points**, the two values of c that constitute local minima. This is the case in which we are interested. Note, however, that these binodal points do not correspond to pure phases (*i.e.*, c is neither one nor zero), but they do represent two distinct states for the system. For $\theta \leq 1$ the free energy has a single well, and only a single state of the system is admitted, with a constant concentration.

11.1.2 The dimensionless strong form

It is convenient to write the Cahn–Hilliard equation in a dimensionless form. To do so, we introduce non-dimensional space and time coordinates

$$x^\star = x/L_0, \quad t^\star = t/T_0, \tag{11.4}$$

where L_0 is a representative length scale and $T_0 = L_0^4/(D\lambda)$. Thus, in dimensionless coordinates, we can rewrite (11.1a) as

$$\frac{\partial c}{\partial t^\star} = \nabla^\star \cdot \left(M_c^\star \nabla^\star \left(\mu_c^\star - \Delta^\star c \right) \right), \tag{11.5}$$

where $M_c^\star = c(1 - c)$ and $\mu_c^\star = \mu_c L_0^2/\lambda$.

Let us identify one further dimensionless number of importance. Defining

$$\alpha = \frac{L_0^2}{3\lambda} \tag{11.6}$$

we have that the thickness of the interface layers will be directly proportional to $\alpha^{-1/2}$. Fixing the value $\theta = 3/2$, which corresponds to a physically relevant case, we have that the value of α completely characterizes the solutions, playing a role somewhat analogous to the Reynolds number in fluid dynamics.

Henceforth we will use (11.5) the dimensionless form of the Cahn–Hilliard equation. Thus, let us drop the superscript $\star$ for the sake of notational convenience.

11.1.3 The weak form

We construct a weak form in the standard way: multiply by a weighting function and integrate by parts. Letting $\mathcal{V}$ denote the trial solution and weighting spaces, which we assume to be identical, we seek $c \in \mathcal{V}$ such that $\forall w \in \mathcal{V}$,

$$B(w, c) = 0, \tag{11.7}$$

where

$$B(w, c) = \left(w, \frac{\partial c}{\partial t} \right)_\Omega + (\nabla w, M_c \nabla \mu_c + \nabla M_c \Delta c)_\Omega + (\Delta w, M_c \Delta c)_\Omega. \tag{11.8}$$

As expected, we discretize (11.8) by a straight-forward application of Galerkin's method, using the NURBS basis with at least C^1 continuity. The second derivatives appearing in (11.8) are precisely the reason for this added continuity. Any approach utilizing functions that are not at least C^1 necessitates a more complex formulation of the problem. An example of a shape function routine capable of generating C^1-continuous NURBS function is contained in Appendix 3.A at the end of Chapter 3. The expression for higher-order derivatives of NURBS basis functions is given by (2.35).

We will use a generalized-α approach to integrate the equation in time (see Chapter 7), with time-step size adaptivity. The adaptivity is needed to reach steady solutions in a reasonable amount of time, as the time step typically varies over many orders of magnitude. See Gomez et al., 2008 for details.

11.2 Numerical results

Studies were performed on the periodic box $\Omega = [0, 1]^d$ for both $d = 2$ and $d = 3$. The domains were intentionally kept geometrically simple in an effort to assess the physical and numerical aspects of the problem. C^1-quadratic uniform B-spline meshes were used. Initial conditions were of the form

$$c_0(\mathbf{x}) = \bar{c} + r, \tag{11.9}$$

where $\bar{c}$ is the constant volume fraction and r is a random variable with uniform distribution in $(-0.05, 0.05)$.

11.2.1 A two-dimensional example

The two-dimensional cases considered tracked the evolution of the system from its complex transient behavior until a steady state was reached. Stationary solutions to the Cahn–Hilliard equation are closely related to the so-called "periodic isoperimetric[1] problem," which is one of the major open problems in geometry, see Hauswirth *et al.*, 2004. In particular, we expect stationary solutions of the Cahn–Hilliard equation to converge (under the appropriate rescaling) to solutions of the isoperimetric problem when $\alpha \to \infty$ and $\theta \to \infty$. In a two-dimensional periodic square, the solution of the periodic isoperimetric problem is well known. For $0 < \bar{c} < 1/\pi$ and $1 - 1/\pi < \bar{c} < 1$ the solutions are circles, while for $1/\pi \leq \bar{c} \leq 1 - 1/\pi$ the solution is a strip.

Numerically, we take the domain to be the two-dimensional periodic square, and we seek solutions to the Cahn–Hilliard equation first with $\alpha = 3000$. Recall from above that we are interested in the case of $\theta = 3/2$. Figure 11.1 shows several snapshots of the time-history of the numerical solution for the case of $\bar{c} = 0.5$ obtained on a mesh of 128^2 quadratic elements. The behavior evolves from the highly random initial condition, becoming more structured in time as the two phases emerge, eventually reaching the steady state of a strip, as we anticipated. It should be noted that the time-step size necessary to resolve the dynamics immediately after the evolution begins is about seven orders of magnitude smaller than the amount of time it takes the system to reach its steady state. Adaptive time stepping is the key to making the problem tractable. The situation in three dimensions is even more dramatic.

11.2.2 A three-dimensional example

The difficulties involved in obtaining numerical solutions to the Cahn–Hilliard equation in three dimensions is much greater than it was in the two-dimensional setting. The topology of the solution is much more complex, and it experiences significant changes as time evolves. Additionally, little is known about the steady state solutions on three-dimensional domains. The isoperimetric problem remains open in three dimensions, though it has been conjectured that solutions take the form of a sphere, a cylinder, or two parallel planes. Even the numerical simulations in the literature have been limited to the early transient behavior in two dimensions. This barrier was broken through in Gomez *et al.*, 2008, whose results for stationary solutions in two and three dimensions appear to be the first of their kind.

As an example, consider the case of $\alpha = 600$ and $\bar{c} = 0.75$ on the three-dimensional periodic cube. Figure 11.2 shows isosurfaces of the solution on a 128^3 mesh at several times during its

(a) $t = 1.971 \cdot 10^{-6}$

(b) $t = 1.609 \cdot 10^{-5}$

(c) $t = 1.390 \cdot 10^{-4}$

(d) Steady state

Figure 11.1 Evolution of the concentration from a randomly perturbed initial condition for $\alpha = 3000$, $\bar{c} = 0.50$. The mesh is comprised of 128^2 quadratic elements.

evolution from the randomly perturbed initial condition to the final steady state. The steady state is a cylinder, corresponding to one of the proposed solutions to the isoperimetric problem.

11.3 The continuous/discontinuous Galerkin (CDG) method

The CDG method was originally proposed by Engel *et al.*, 2002. It was further developed by Hughes and Garikipati, 2004; Wells *et al.*, 2006; Wells and Dung, 2007; Dung and Wells, 2008. The basic idea is to use C^0-continuous finite element basis functions for partial differential equations involving derivatives of higher than second order. In the usual continuous Galerkin finite element method this would not work because basis functions are required to have C^1-continuity. In the CDG method, weak C^1-continuity is accomplished through discontinuous Galerkin treatment of first derivatives. This is convenient

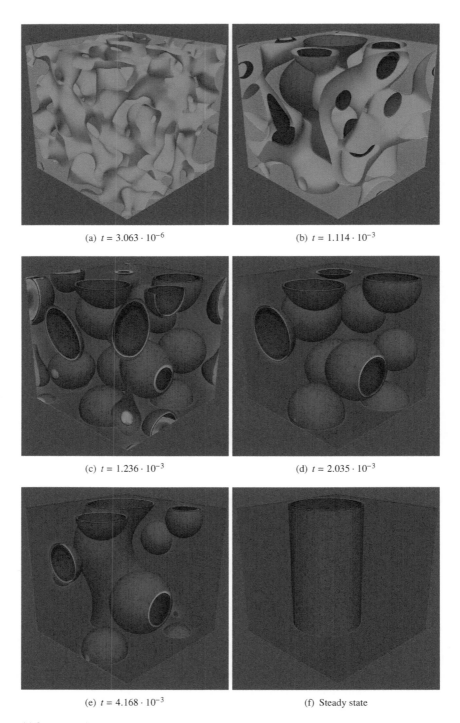

(a) $t = 3.063 \cdot 10^{-6}$

(b) $t = 1.114 \cdot 10^{-3}$

(c) $t = 1.236 \cdot 10^{-3}$

(d) $t = 2.035 \cdot 10^{-3}$

(e) $t = 4.168 \cdot 10^{-3}$

(f) Steady state

Figure 11.2 Evolution of the concentration from a randomly perturbed initial condition for $\alpha = 600$, $\bar{c} = 0.75$. The mesh is comprised of 128^3 quadratic elements.

because C^0-continuous finite element basis functions are ubiquitous and quite simple, whereas C^1-continuous finite element basis functions are few and far between and those that exist are very complex.

The CDG methodology fills an important gap in finite element technology. It also may play a role in isogeometric analysis. There will be times when it is difficult or impossible to create globally smooth parameterizations. A typical case would involve an assemblage of multiple patches joined in only C^0-continuous fashion. In this case, the CDG method could be used to provide weak continuity across patch interfaces.

For an example of the use of discontinuous Galerkin methodology to enforce C^0-continuity across a sliding interface in isogeometric analysis, see Chapter 10, Section 10.4.

Note

1. This has nothing in common with the similarly named "isoparametric concept" discussed in Chapter 3.

12

Some Additional Geometry

This chapter addresses an alternative way to think of B-splines based on polar forms. For many in the analysis community, this approach is less intuitive than the presentation of Chapter 2, and an understanding of this material is not necessary to develop isogeometric analysis applications. Still, many of the most common algorithms are based on the polar form of splines, and the proofs of many theorems rely on it as well. Once the initial hurdle of understanding polar forms has been crossed, many concepts actually become simpler. In particular, one can frequently replace the operation of evaluating polynomial basis functions with the simpler process of linear interpolation between control points, thus leading to a subdivision approach to spline manipulation. Our notation will stray slightly from that which has been used throughout the book in an effort to emphasize what a different viewpoint the use of polar forms represents. All of the developments in this chapter pertain to B-splines in $\mathbb{R}^d$. The relevant extension to NURBS is made by applying all of these concepts to the projective control points of the corresponding B-spline in $\mathbb{R}^{d+1}$.

12.1 The polar form of polynomials

Consider the univariate quadratic function $f(t) = t^2$ and the bilinear function $F(u, v) = uv$. It is obvious that $F(t, t) = f(t)$, and so f is just the restriction of F to the diagonal $u = v$ of the uv-plane. This simple example encapsulates the general principle that every polynomial $g(t)$ of degree p is isomorphic with a symmetric, multiaffine function $G(u_1, \ldots, u_p)$ that satisfies $G(t, \ldots, t) = g(t)$. Such a G is called the ***polar form***[1] of g. Its uniqueness follows from the requirement of symmetry, which demands that $G(u_1, \ldots, u_p)$ remains unchanged under any permutation of its arguments. For example, consider a cubic function of the form

$$g(t) = C_1 t^3 + C_2 t^2 + C_3 t + C_4. \tag{12.1}$$

While there are infinitely many trilinear functions $F(u, v, w)$ such that $F(t, t, t) = g(t)$, the additional requirement of symmetry leaves only one, namely

$$G(u, v, w) = C_1 uvw + \frac{C_2}{3}(uv + uw + vw) + \frac{C_3}{3}(u + v + w) + C_4. \tag{12.2}$$

As desired, $G(t, t, t) = g(t)$ and $G(u, v, w) = G(u, w, v) = G(v, u, w) = G(v, w, u) = G(w, u, v) = G(w, v, u)$.

Isogeometric Analysis: Toward Integration of CAD and FEA   by J. A. Cottrell, T. J. R. Hughes, Y. Bazilevs
© 2009, John Wiley & Sons, Ltd

12.1.1 Bézier curves and the de Casteljau algorithm

Bézier curves (Bézier, 1966, 1967) may be thought of as single element B-spline curves. It follows that they are polynomial curves, rather than *piecewise* polynomials. Bézier curves predate B-splines, and they stand as the technology that revolutionized geometric design by offering intuitive control of the curve through the manipulation of control points.

Bézier curves or order p are built from the famous **Bernstein basis** (Bernstein, 1912). If we define a Bézier curve as a B-spline curve with the knot vector $\Xi = \{0, \ldots, 0, 1, \ldots, 1\}$, where both 0 and 1 appear $p + 1$ times, and apply (2.1) and (2.2), we recover exactly the B-spline basis and so the intuitive notion of a Bézier curve as a one element B-spline is perfectly accurate. The curve is defined by taking a linear combination of the $p + 1$ basis functions, N_i, and the corresponding $p + 1$ control points, $\mathbf{B}_i$, to obtain a polynomial of the form

$$g(t) = \sum_{i=1}^{p+1} \mathbf{B}_i N_i(t). \tag{12.3}$$

Performing the necessary algebra, one may obtain the polar form, $G(u_1, \ldots, u_p)$, corresponding to (12.3). Doing so yields a remarkable result: the control points of the Bézier curve correspond to G evaluated at the corners of the unit hypercube in $\mathbb{R}^p$. Specifically, let $\mathbf{e}_i$ be unit vectors in $\mathbb{R}^p$ such that $(\mathbf{e}_i)_j = \delta_{ij}$, where δ_{ij} is the Kronecker delta. The control points are given by

$$\mathbf{B}_1 = G(\mathbf{0}) = G(0, \ldots, 0) \tag{12.4}$$

and

$$\mathbf{B}_i = G\left(\sum_{j=1}^{i-1} \mathbf{e}_j^T\right). \tag{12.5}$$

As an example, let us consider the case of a cubic Bézier curve (*i.e.*, $p = 3$). We have unit vectors

$$\mathbf{e}_1^T = (1, 0, 0), \tag{12.6}$$

$$\mathbf{e}_2^T = (0, 1, 0), \tag{12.7}$$

$$\mathbf{e}_3^T = (0, 0, 1). \tag{12.8}$$

From (12.4) we have that

$$\mathbf{B}_1 = G(0, 0, 0). \tag{12.9}$$

Inserting (12.6)–(12.8) into (12.5) yields

$$\mathbf{B}_2 = G(\mathbf{e}_1^T) = G(1, 0, 0) \tag{12.10}$$

$$\mathbf{B}_3 = G(\mathbf{e}_1^T + \mathbf{e}_2^T) = G(1, 1, 0), \tag{12.11}$$

$$\mathbf{B}_4 = G(\mathbf{e}_1^T + \mathbf{e}_2^T + \mathbf{e}_3^T) = G(1, 1, 1). \tag{12.12}$$

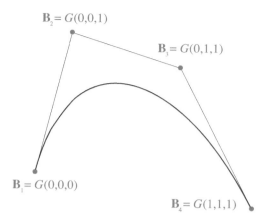

$\mathbf{B}_2 = G(0,0,1)$

$\mathbf{B}_3 = G(0,1,1)$

$\mathbf{B}_1 = G(0,0,0)$

$\mathbf{B}_4 = G(1,1,1)$

Figure 12.1 The control points for a Bézier curve, $g(t)$ with $t \in [0, 1]$, such as the cubic curve shown here, are given by the polar form, G, evaluated at the corners of the hypercube in $\mathbb{R}^p$.

The symmetry of G implies that it is invariant under interchange of any two of its arguments, and thus we also have

$$\mathbf{B}_2 = G(1, 0, 0) = G(0, 1, 0) = G(0, 0, 1), \tag{12.13}$$

$$\mathbf{B}_3 = G(1, 1, 0) = G(1, 0, 1) = G(0, 1, 1). \tag{12.14}$$

An example of such a cubic Bézier curve is shown in Figure 12.1.

A practical application of this alternative viewpoint is found when we seek to evaluate $g(t)$ at a point $t \in (0, 1)$. Recall that G is linear with respect to each of its p arguments, and the control points are evaluations of G at a linearly independent set of points. One can evaluate G at any other point in $\mathbb{R}^p$ by taking linear combinations of these $p + 1$ points. In particular, if we are interested in a point $G(t, \ldots, t) = g(t)$ for $t \in (0, 1)$ that lies on the Bézier curve, we can find it through a process of linear interpolation without ever evaluating any basis functions. This approach results in the famous *de Casteljau algorithm* (de Casteljau, 1959), also known as the "corner cutting algorithm."

Consider the de Casteljau algorithm applied to the cubic Bézier curve of Figure 12.1. As a first step, one can exploit the linearity of G with respect to each of its arguments to calculate

$$G(0, 0, t) = (1 - t) G(0, 0, 0) + t G(0, 0, 1), \tag{12.15}$$

$$G(0, t, 1) = (1 - t) G(0, 0, 1) + t G(0, 1, 1), \tag{12.16}$$

and

$$G(t, 1, 1) = (1 - t) G(0, 1, 1) + t G(1, 1, 1), \tag{12.17}$$

as shown in green in Figure 12.2 for the case of $t = 0.6$. The second step employs not just the linearity of G, but the symmetry as well. Noting that $G(0, t, 1) = G(0, 1, t)$ one can

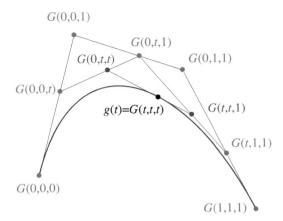

Figure 12.2   The de Casteljau algorithm, also known as the "corner-cutting algorithm," for evaluating points on a Bézier curve. In this example, we seek $g(t) = G(t, t, t)$ for $t = 0.6$. Use of the polar form allows the curve to be evaluated without ever referring to the underlying basis.

calculate

$$G(0, t, t) = (1 - t)\,G(0, 0, t) + t\,G(0, 1, t)$$
$$= (1 - t)\,G(0, 0, t) + t\,G(0, t, 1). \qquad (12.18)$$

Similarly, $G(t, 1, 1) = G(1, t, 1)$ allows one to calculate

$$G(t, t, 1) = (1 - t)\,G(0, t, 1) + t\,G(1, t, 1)$$
$$= (1 - t)\,G(0, t, 1) + t\,G(t, 1, 1). \qquad (12.19)$$

Both of these points are shown in purple in Figure 12.2. Lastly, one obtains the point of interest as

$$G(t, t, t) = (1 - t)\,G(0, t, t) + t\,G(1, t, t)$$
$$= (1 - t)\,G(0, t, t) + t\,G(t, t, 1). \qquad (12.20)$$

The entire process consisted of taking linear combinations of control points, but never required evaluating the recursively defined basis functions. It is particularly efficient and numerically stable, making it a common algorithm in software implementations.

Note that the de Casteljau algorithm would have worked equally well had the starting point been *any* $p + 1$ values of G so long as they were taken at a linearly independent set of points in $\mathbb{R}^p$. Similarly, one could equally easily evaluate the curve for $t \notin [0, 1]$. The particular choice of points and parameter values that we have used, however, is the set that corresponds to the interpretation of control points of a Bézier curve as the coefficients to the Bernstein basis.

12.1.2 Continuity of piecewise curves

Let us consider a Bézier curve constructed on the parametric interval $[a, b]$ instead of $[0, 1]$. This is just an affine mapping, $\phi : \mathbb{R} \to \mathbb{R}$, comprised of a scaling and a translation. As discussed for B-splines in Chapter 2, translating and scaling the knot vector has no effect on the geometry if the control points are unchanged. The parameterization changes by the same affine mapping as has been applied to the knot vector. As the Bézier curve is a special case of a B-spline, the result must still hold, but the polar form inevitably changes. Fortunately, the change is trivial. Simply compose G with the inverse of the mapping, ϕ^{-1}, applied to each of its $p + 1$ dimensions. As expected, the control points (which have not changed) are equal to this new polar form evaluated at the corners of hypercube $\bigotimes_1^{p+1}[a, b]$. Thus, if we reparameterized the example of Figures 12.1 and 12.2 such that $t \in [a, b]$ (this is equivalent to use of the knot vector $\Xi = \{a, a, a, a, b, b, b, b\}$), we have

$$\mathbf{B}_1 = G(a, a, a),$$

$$\mathbf{B}_2 = G(b, a, a) = G(a, b, a) = G(a, a, b),$$

$$\mathbf{B}_3 = G(b, b, a) = G(b, a, b) = G(a, b, b),$$

$$\mathbf{B}_4 = G(b, b, b).$$

Piecewise polynomials can be formed from composite Bézier curves by considering two separate curves where, for example, the first curve is defined for $t \in [a, b)$ and the second curve is defined for $t \in [b, c]$. Figure 12.3 shows two such quadratic curves for the case of $a = 0, b = 1, c = 2$. The composite curve is clearly discontinuous in this case. Experience with B-splines should indicate that C^0 continuity can be obtained, as in Figure 12.4, by simply ensuring that the last control point of the first curve lies at the same position as the first control point of the second curve. As one might expect, this can be translated into a corresponding restriction on the polar form of the two curves.

The general result, presented in Ramshaw, 1989, for a composite Bézier curve comprised of two polynomials of degree p being C^k-continuous at a parameter value r is that

$$G^1(u_1, \ldots, u_k, \underbrace{r, \ldots, r}_{p-k}) = G^2(u_1, \ldots, u_k, \underbrace{r, \ldots, r}_{p-k}) \qquad (12.21)$$

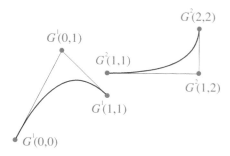

Figure 12.3   A piecewise quadratic curve $g(t)$ for $t \in [0, 2]$ is defined by two separate Bézier curves, the first for $t \in [0, 1)$ and the second for $t \in [1, 2]$. Obviously, the composite curve is discontinuous in this case.

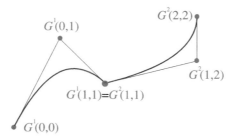

Figure 12.4 C^0-continuity for two quadratic curves meeting at $r = 1$ demands that $G^1(1, 1) = G^2(1, 1)$. This is equivalent to the B-spline condition that the last control point of the first curve and the first control point of the second curve lie at the same position.

That is, if G^1 is the polar form of the first curve, and G^2 is the polar form of the second curve, then the two curves agree to k^{th} order at point r if and only if G^1 and G^2 agree on a set of polar arguments that contain at most k values different from r. One way to think about (12.21) for $k \geq 0$ is that C^k-continuity demands that G^1 and G^2 are equal on a nonempty set of hyperplanes of dimension k that intersect at $(r, \ldots, r)$ (the symmetry of the polar form makes the set on which they are equal richer than just one single hyperplane).

Consider the two trivial cases. If G^1 and G^2 are C^0-continuous, then $G^1(r, \ldots, r) = G^2(r, \ldots, r)$ (as in Figure 12.4) and the polar forms are equal on a zero-dimensional sub-space of $\mathbb{R}^p$, namely the point $(r, \ldots, r)$. If $k = p$, then $G^1(u_1, \ldots, u_p) = G^2(u_1, \ldots, u_p)$, and the polar forms are identical everywhere in $\mathbb{R}^p$. Of course, in this case the resulting curves are identical as well.

The case of $0 < k < p$ is slightly more subtle. For example, consider two curves with $p = 2$ that meet with C^1 continuity at parameter value r, as in Figure 12.5. From (12.21) we see that $G^1(u, r) = G^2(u, r)$ for all values of u. Thus, there is a subspace of dimension 1 on which the polar forms are equal, namely the line in the uv-plane corresponding to $v = r$. Symmetry dictates that the polar forms are identical along the line $u = r$ as well.

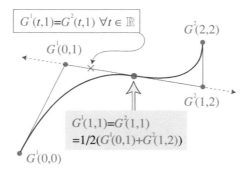

Figure 12.5 C^1-continuity for two quadratic curves meeting at $r = 1$. The control points obey $G^1(1, 1) = G^2(1, 1)$, but the higher order continuity requires that $G^1(1, t) = G^2(1, t)$ for all $t \in \mathbb{R}$.

12.2 The polar form of B-splines

The previous section discussed the constraints that must exist on a set of polynomial curves of order p for them to form a piecewise polynomial curve of prescribed continuity. Each polynomial segment could be interpreted as a Bézier curve with the location of its $p+1$ control points defined in terms of its own polar form. Heuristically speaking, B-splines remove the clutter by eliminating the unnecessary control points from such a construction. The fact that continuity renders some control points unnecessary follows from the constraint imposed by (12.21), which implies that continuity between the segments dictates that some of the Bézier control points will be linearly dependent on the others. This statement will be made precise in a moment, but first consider that we could remove one control point from the curve in Figure 12.4 because $G^1(1, 1)$ and $G^2(1, 1)$ are identical, and thus we have no need to keep track of *both* of them. Taking things one step further, we can remove two control points from Figure 12.5 because $G^1(1, 1)$ and $G^2(1, 1)$ are not only identical, but they are also a linear combination of $G^1(0, 1)$ and $G^2(1, 2)$. As such, we have no need to keep track of *either* of them.

To facilitate the following discussion, let a **multiset** be a collection of values, in any order, where repeated values are allowed. Thus, for example, $\alpha = \{1, 4.3, 2, 4.3\}$ and $\beta = \{4.3, 4.3, 1, 2\}$ both represent the same multiset. We use the term "multiset" instead of "sequence" as the collection is not ordered, and in place of the term "set", as repeated values are allowed. Additionally, let a **super-multiset** of a multiset β be any multiset α such that each item of β appears in α with at least as high of a multiplicity as in β. That is, with $\beta = \{4.3, 4.3, 1, 2\}$, $\alpha_1 = \{4.3, 7, 4.3, 2, 1\}$ is a super-multiset of β, but $\alpha_1 = \{4.3, 7, 8, 2, 1\}$ is not. Using this new terminology, one can restate the continuity condition (12.21) as: Two curves agree to k^{th} order at a point r if their polar forms agree when the multiset of polar arguments is *any* super-multiset of

$$\beta = \{\underbrace{r, \ldots, r}_{p-k}\}. \tag{12.22}$$

12.2.1 Knot vectors and control points

Following Ramshaw, 1989, let $\{t_i\}_{i=1}^{n+p+1}$ be a non-decreasing sequence in $\mathbb{R}$, where we insist that the multiplicity of any value in the sequence be no greater than $p+1$. For each open interval of nonzero measure (t_i, t_{i+1}), let $g_i(t)$ be a polynomial curve of degree p. That is, if $t_i < t_{i+1}$, the piecewise polynomial spline curve $g(t)$ (which we are in the process of building) follows a single polynomial $g_i(t)$ over the interval (t_i, t_{i+1}). The curve g_i will join, on its right end, the curve g_{i+m}, where m is the largest integer such that $t_{i+1} = \cdots = t_{i+m}$. From (12.21), it follows that to form a spline with C^{p-m}-continuity, the polar forms G^i and G^{i+m} of polynomials g_i and g_{i+m}, respectively, must agree on all super-multisets of $\{t_{i+1}, \ldots, t_{i+m}\}$.

Noting that $t_{i+1} = \cdots = t_{i+m}$, this is no more than a restatement of (12.21), which was developed for Bézier curves in the previous section. Applying the same logic to each interval in the **knot vector** $\{t_i\}_{i=1}^{n+p+1}$, however, establishes the connection between the polar forms of the present chapter and the rules regarding continuity and the multiplicity of the knots that have been familiar since Chapter 2: increasing the multiplicity of a knot value decreases the continuity of the spline at that point. To complete the polar viewpoint of B-splines, one must identify the control points with evaluations of the polar forms, as was done for Bézier curves.

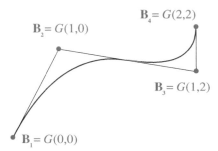

Figure 12.6   The B-spline curve corresponding to the C^1 piecewise quadratic curve in Figure 12.5. The control points follow from the polar forms via (12.23).

Without proof, we present the following result from Ramshaw, 1989: Given a knot vector $\mathbf{T} = \{t_i\}_{i=1}^{n+p+1}$, whose knot values have multiplicity of at most $p + 1$, and given a sequence of n points $\{\mathbf{B}_k\}_{k=1}^{n}$ in $\mathbb{R}^d$, there exists a *unique* spline of degree p and knot vector $\mathbf{T}$ such that

$$\mathbf{B}_k = G^i(t_{k+1}, \ldots, t_{k+p}) \quad \text{for} \quad k \leq i \leq k + p. \tag{12.23}$$

Moreover, this is exactly the spline that would be generated if we were to interpret $\{\mathbf{B}_k\}_{k=1}^{n}$ as control points and use knot vector $\mathbf{T}$ with (2.1) and (2.2) to define a B-spline.

Figure 12.6 shows exactly the same C^1 piecewise quadratic curve as is formed from two separate Bézier curves in Figure 12.5. The knot vector for this B-spline representation is $\mathbf{T} = \{t_1, t_2, t_3, t_4, t_5, t_6, t_7\} = \{0, 0, 0, 1, 2, 2, 2\}$. Using (12.23), one obtains

$$\mathbf{B}_1 = G^1(t_2, t_3) = G^1(0, 0), \tag{12.24}$$

$$\mathbf{B}_2 = G^1(t_3, t_4) = G^1(0, 1), \tag{12.25}$$

$$\mathbf{B}_3 = G^2(t_4, t_5) = G^2(1, 2), \tag{12.26}$$

$$\mathbf{B}_4 = G^2(t_5, t_6) = G^2(2, 2). \tag{12.27}$$

As foreshadowed in the discussion at the beginning of the section, the control points that are required are a subset of those needed to describe the two Bézier curves separately. In the next section, it will be shown that there is no longer a need to distinguish between the different polar forms G_i. Instead one may refer to a single, piecewise-defined polar form G, as in Figure 12.6. Because of the unambiguous mapping between control point $\mathbf{B}_k$ and the polar form evaluated at knots, $G(t_{k+1}, \ldots, t_{k+p})$, we refer to $(t_{k+1}, \ldots, t_{k+p})$ as the *polar label* of control point $\mathbf{B}_k$.

For the quadratic B-spline in Figure 12.6, we saw that the first control point is $\mathbf{B}_1 = G(t_2, t_3)$, while the second control point is $\mathbf{B}_2 = G(t_3, t_4)$. Noting both the symmetry and linearity of

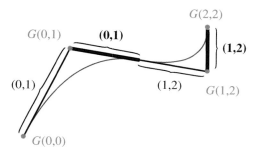

Figure 12.7 Each leg of the control polygon has a parametric interval mapped to it. The interval is easily deduced by looking at the polar argument that differs (recalling symmetry) between consecutive polar labels of the control points.

the polar form, it follows that

$$G(t, t_3) = \frac{t_4 - t}{t_4 - t_2} G(t_2, t_3) + \frac{t - t_2}{t_4 - t_2} G(t_4, t_3)$$

$$= \frac{t_4 - t}{t_4 - t_2} \mathbf{B}_1 + \frac{t - t_2}{t_4 - t_2} \mathbf{B}_2 \tag{12.28}$$

$$\forall t \in (t_2, t_4).$$

In effect, we have created a mapping from the interval (t_2, t_4) to the leg of the control polygon between $\mathbf{B}_1$ and $\mathbf{B}_2$. In this case, $t_2 = t_3 = 0$ and $t_4 = 1$, and so this interval is $(0, 1)$, as depicted in Figure 12.7. Repeating the same logic again results in a mapping from $(0, 2)$ to the second leg of control polygon, and from $(1, 2)$ for the third leg.

In the general case, $\mathbf{B}_k = G(t_{k+1}, \ldots, t_{k+p})$ and $\mathbf{B}_{k+1} = G(t_{k+2}, \ldots, t_{k+p+1})$. Observe that both control points have the polar arguments $t_{k+2}, \ldots, t_{k+p}$, while only $\mathbf{B}_k$ has t_{k+1} and only $\mathbf{B}_{k+1}$ has t_{k+p+1}. Symmetry and linearity lead to the result that

$$G(t, t_{k+2}, \ldots, t_{k+p}) = \frac{t_{k+p+1} - t}{t_{k+p+1} - t_{k+1}} G(t_{k+1}, \ldots, t_{k+p})$$

$$+ \frac{t - t_{k+1}}{t_{k+p+1} - t_{k+1}} G(t_{k+2}, \ldots, t_{k+p+1})$$

$$= \frac{t_{k+p+1} - t}{t_{k+p+1} - t_{k+1}} \mathbf{B}_k + \frac{t - t_{k+1}}{t_{k+p+1} - t_{k+1}} \mathbf{B}_{k+1} \tag{12.29}$$

$$\forall t \in (t_{k+1}, t_{k+p+1}),$$

and thus there is a mapping from the interval (t_{k+1}, t_{k+p+1}) to the leg of the control polygon between $\mathbf{B}_k$ and $\mathbf{B}_{k+1}$.

12.2.2 Knot insertion and the de Boor algorithm

Let us consider what happens when a knot is inserted into the knot vector of an existing B-spline curve. For example, let $t \in (t_i, t_{i+1})$ be the knot to be inserted. Before knot insertion, the curve in this parametric interval is defined by polar form G^i. After insertion, this same segment will be described by two separate polar forms, say, $\bar{G}^i$ and $\bar{G}^{i+1}$. As the curve is completely unchanged, we must have that $G^i = \bar{G}^i = \bar{G}^{i+1}$. More generally, it follows that the piecewise polar form of the entire B-spline curve, G, remains completely unchanged under knot insertion. However, from (12.23) it is clear that the control points depend on the specific knot values contained in the knot vector. Under knot insertion, we obtain the new control points by simply evaluating the old, unchanged polar form at the various multisets of knots from the refined knot vector as dictated by (12.23). This may seem like an impediment as the polar form has not been explicitly constructed. Knowledge of it is limited to the original control points and their polar labels. Fortunately, this is sufficient. As new knot values must necessarily be inserted into existing knot intervals (of course, existing knot values may be repeated as well), each of the new control points can be determined by linearly interpolating between the existing control points using (12.29).

As an example, let us return to the case of the B-spline curve in Figure 12.6. This piecewise quadratic curve has an initial knot vector of $\mathbf{T} = \{0, 0, 0, 1, 2, 2, 2\}$, into which we wish to insert the point $t = \frac{3}{4}$. The new knot vector will be $\bar{\mathbf{T}} = \{0, 0, 0, \frac{3}{4}, 1, 2, 2, 2\}$. The original control points are contained in (12.24)–(12.27). From (12.23) and $\bar{\mathbf{T}}$ we know that the new control points will be

$$\bar{\mathbf{B}}_1 = G(0, 0), \tag{12.30}$$

$$\bar{\mathbf{B}}_2 = G(0, \tfrac{3}{4}), \tag{12.31}$$

$$\bar{\mathbf{B}}_3 = G(\tfrac{3}{4}, 1), \tag{12.32}$$

$$\bar{\mathbf{B}}_4 = G(1, 2), \tag{12.33}$$

$$\bar{\mathbf{B}}_5 = G(2, 2). \tag{12.34}$$

Clearly, $\bar{\mathbf{B}}_1 = \mathbf{B}_1$, $\bar{\mathbf{B}}_4 = \mathbf{B}_3$, and $\bar{\mathbf{B}}_5 = \mathbf{B}_4$. The two remaining points are calculated using (12.29):

$$\bar{\mathbf{B}}_2 = \frac{1 - \frac{3}{4}}{1}\mathbf{B}_1 + \frac{\frac{3}{4} - 0}{1}\mathbf{B}_2, \tag{12.35}$$

$$\bar{\mathbf{B}}_3 = \frac{2 - \frac{3}{4}}{2}\mathbf{B}_2 + \frac{\frac{3}{4} - 0}{2}\mathbf{B}_3. \tag{12.36}$$

The situation is depicted in Figure 12.8. The original control points are in red, while the new control points after insertion of this single knot are shown in green. Note that $G(0, \frac{3}{4})$ is 75% of the way between $G(0, 0)$ and $G(0, 1)$, but $G(1, \frac{3}{4})$ is less than half of the way between $G(0, 1)$ and $G(1, 2)$. Recall that, as in Figure 12.7, the parametric interval mapped onto this second leg of the control polygon is $(0, 2)$, and thus the new control point is only 37.5% of the distance between the old ones.

Consider inserting the knot $t = \frac{3}{4}$ a second time. The new control point added in this case is exactly $G(\frac{3}{4}, \frac{3}{4})$, as shown in black in Figure 12.8. Of course, by the very definition of the polar form, we have $G(\frac{3}{4}, \frac{3}{4}) = g(\frac{3}{4})$, and the control point lies directly on the curve itself. Thus, by inserting the knot until its multiplicity is equal to the polynomial order of the curve and calculating the corresponding control points, we have actually evaluated the B-spline at

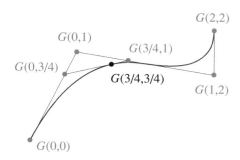

Figure 12.8 The de Boor algorithm for B-spline evaluation is the generalization of the de Casteljau algorithm. It is actually more illuminating as each step of the process corresponds to knot insertion. When the multiplicity of the knot is equal to the polynomial order, the control point lies on the curve. Thus, calculating the control points for such a repeated knot insertion is an alternative, and efficient, method for evaluating points on a B-spline curve. Like the de Casteljau algorithm, it requires no evaluations of basis functions.

this point. This approach is called the ***de Boor algorithm*** (de Boor, 1978). This should be recognized as a generalization of the de Casteljau algorithm. In fact, it sheds light on the de Casteljau algorithm in the regard that, if we interpret a Bézier curve as a single element B-spline, each step of the de Casteljau algorithm amounts to calculating the new control points corresponding to the insertion of a single knot. Of course, Bézier curves in their original form have no notion of knot insertion and so the de Casteljau algorithm predates this interpretation.

12.2.3 Bézier decomposition and function subdivision

As has just been shown, the process of B-spline evaluation by the de Boor algorithm and knot insertion are intimately related. All of the control points needed to actually insert a knot p times are calculated along the way to evaluating the point on the curve. If we continue the process one step farther to raise the multiplicity of the new knot, t, to $p + 1$, then the last control point calculated, $G(t, \ldots, t)$, is simply repeated. This corresponds to splitting the curve into two separate B-splines, each with their own sets of control points (one copy of $G(t, \ldots, t)$ being associated with the curve to the left and the other copy belonging to the curve to the right). If this is done for every knot in the original knot vector, then each of the original polynomial segments of the B-spline (*i.e.*, every element in an isogeometric analysis setting) has been represented by a separate Bézier curve. This process is called ***Bézier decomposition***, and it is utilized in several procedures, including order elevation of a B-spline curve. For the quadratic curve of Figure 12.6, the Bézier control points are exactly those depicted previously in Figure 12.5.

From the polar viewpoint, this process calculates the control points for the Bézier representation of each polynomial segment of the original piecewise polynomial curve from the original B-spline control points. If we leave the polar viewpoint and return to the parametric representation utilized throughout this book, we can extend this concept to relate the polynomial basis functions on each Bézier segment to the original set of piecewise polynomial B-spline basis functions.

To see the connection, note that the control points of the decomposed curve, $\bar{\mathcal{B}} = \{\bar{\mathbf{B}}_i\}$ have been calculated as a linear combination of the original control points, $\mathcal{B} = \{\mathbf{B}_i\}$, which we may represent by the matrix operation

$$\bar{\mathcal{B}} = \mathbf{M}\mathcal{B}. \tag{12.37}$$

Let us similarly collect the basis functions for the Bézier and B-spline representations into vectors $\bar{\mathcal{N}} = \{\bar{\mathbf{N}}_i(t)\}$ and $\mathcal{N} = \{\mathbf{N}_i(t)\}$, respectively, such that we may write the curve as

$$g(t) = \mathcal{B} \cdot \mathcal{N} = \bar{\mathcal{B}} \cdot \bar{\mathcal{N}}. \tag{12.38}$$

Inserting (12.37) into (12.38) yields

$$\begin{aligned}
g(t) &= \mathcal{B} \cdot \mathcal{N} = \bar{\mathcal{B}} \cdot \bar{\mathcal{N}} \\
&= (\mathbf{M}\mathcal{B}) \cdot \bar{\mathcal{N}} \\
&= \mathcal{B}^T \mathbf{M}^T \bar{\mathcal{N}} \\
&= \mathcal{B} \cdot \left(\mathbf{M}^T \bar{\mathcal{N}}\right). \tag{12.39}
\end{aligned}$$

It follows that

$$\mathbf{M}^T \bar{\mathcal{N}} = \mathcal{N}, \tag{12.40}$$

and thus the relationship between the B-spline basis and the basis of the Bézier decomposition is defined by the same set of coefficients that relate the control points for the two cases.

As an example, consider a two element, C^2-continuous, cubic B-spline with the knot vector $\mathbf{T} = \{0, 0, 0, 0, \frac{1}{2}, 1, 1, 1, 1\}$. The basis for such a curve is shown in Figure 12.9. Partitioning the spline into two separate Bézier curves is equivalent to increasing the multiplicity of each knot to $p + 1 = 4$, resulting in knot vector $\bar{\mathbf{T}} = \{0, 0, 0, 0, \frac{1}{2}, \frac{1}{2}, \frac{1}{2}, \frac{1}{2}, 1, 1, 1, 1\}$ and the basis shown in Figure 12.10. Each function of the original basis, $\mathcal{N} = \{N_i\}_{i=1}^5$, may be expressed

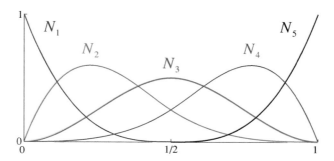

Figure 12.9 C^2-continuous B-spline basis corresponding to knot vector $\mathbf{T} = \{0, 0, 0, 0, \frac{1}{2}, 1, 1, 1, 1\}$. Each of these basis functions can be represented as a linear combination of the basis functions of the Bézier decomposition, shown in Figure 12.10.

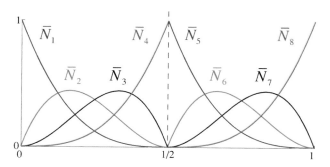

Figure 12.10   The basis functions corresponding to the Bézier decomposition of the B-spline curve built from the basis in Figure 12.9. Note that this is two separate sets of function: one for the polynomial on $[0, \frac{1}{2}]$ and one for the polynomial on $[\frac{1}{2}, 1]$. Equivalently, this may be viewed as a C^{-1}-continuous (*i.e.*, discontinuous) B-spline basis corresponding to knot vector $\bar{\mathbf{T}} = \{0, 0, 0, 0, \frac{1}{2}, \frac{1}{2}, \frac{1}{2}, \frac{1}{2}, 1, 1, 1, 1\}$.

as a linear combination of the new basis functions, $\bar{\mathcal{N}} = \{\bar{N}_i\}_{i=1}^{8}$. To extract the coefficients of this linear relationship, one must first examine the corresponding relationship between the control points, and then use (12.40).

From (12.23), it is clear that the control points for the original curve are given by

$$\mathbf{B}_1 = G(0, 0, 0), \tag{12.41}$$

$$\mathbf{B}_2 = G(0, 0, \tfrac{1}{2}), \tag{12.42}$$

$$\mathbf{B}_3 = G(0, \tfrac{1}{2}, 1), \tag{12.43}$$

$$\mathbf{B}_4 = G(\tfrac{1}{2}, 1, 1), \tag{12.44}$$

$$\mathbf{B}_5 = G(1, 1, 1), \tag{12.45}$$

where G is the polar form for the curve. In practice, the only information available about the specific polar form comes from the specification of these control points. This is important, because as knots are inserted to partition the curve into two separate Bézier curves, the polar labels of the new control points follow trivially by applying (12.23). The actual position of these control points, however, must be calculated by linearly interpolating between the existing control points, just as is in the de Boor algorithm. The first two and the last two of the new control points correspond to points of the original curve:

$$\bar{\mathbf{B}}_1 = G(0, 0, 0) = \mathbf{B}_1, \tag{12.46}$$

$$\bar{\mathbf{B}}_2 = G(0, 0, \tfrac{1}{2}) = \mathbf{B}_2, \tag{12.47}$$

$$\bar{\mathbf{B}}_7 = G(\tfrac{1}{2}, 1, 1) = \mathbf{B}_4, \tag{12.48}$$

$$\bar{\mathbf{B}}_8 = G(1, 1, 1) = \mathbf{B}_5. \tag{12.49}$$

Two more points follow simply from these by virtue of the linearity and symmetry of the polar form:

$$\bar{\mathbf{B}}_3 = G(0, \tfrac{1}{2}, \tfrac{1}{2})$$
$$= \tfrac{1}{2}G(0, 0, \tfrac{1}{2}) + \tfrac{1}{2}G(0, 1, \tfrac{1}{2})$$
$$= \tfrac{1}{2}\mathbf{B}_2 + \tfrac{1}{2}\mathbf{B}_3, \tag{12.50}$$

$$\bar{\mathbf{B}}_6 = G(\tfrac{1}{2}, \tfrac{1}{2}, 1)$$
$$= \tfrac{1}{2}G(\tfrac{1}{2}, 0, 1) + \tfrac{1}{2}G(\tfrac{1}{2}, 1, 1)$$
$$= \tfrac{1}{2}\mathbf{B}_3 + \tfrac{1}{2}\mathbf{B}_4. \tag{12.51}$$

The last two control points for the Bézier decomposition are obtained from the results already computed:

$$\bar{\mathbf{B}}_4 = G(\tfrac{1}{2}, \tfrac{1}{2}, \tfrac{1}{2})$$
$$= \tfrac{1}{2}G(\tfrac{1}{2}, \tfrac{1}{2}, 0) + \tfrac{1}{2}G(\tfrac{1}{2}, \tfrac{1}{2}, 1)$$
$$= \tfrac{1}{2}\bar{\mathbf{B}}_3 + \tfrac{1}{2}\bar{\mathbf{B}}_6$$
$$= \tfrac{1}{4}\mathbf{B}_2 + \tfrac{1}{2}\mathbf{B}_3 + \tfrac{1}{4}\mathbf{B}_4, \tag{12.52}$$
$$\bar{\mathbf{B}}_5 = \bar{\mathbf{B}}_4. \tag{12.53}$$

Concisely, (12.46)–(12.53) can be expressed in the form of (12.37), as

$$\bar{\mathcal{B}} = \mathbf{M}\mathcal{B} \Leftrightarrow \begin{pmatrix} \bar{\mathbf{B}}_1 \\ \bar{\mathbf{B}}_2 \\ \bar{\mathbf{B}}_3 \\ \bar{\mathbf{B}}_4 \\ \bar{\mathbf{B}}_5 \\ \bar{\mathbf{B}}_6 \\ \bar{\mathbf{B}}_7 \\ \bar{\mathbf{B}}_8 \end{pmatrix} = \begin{bmatrix} 1 & 0 & 0 & 0 & 0 \\ 0 & 1 & 0 & 0 & 0 \\ 0 & 1/2 & 1/2 & 0 & 0 \\ 0 & 1/4 & 1/2 & 1/4 & 0 \\ 0 & 1/4 & 1/2 & 1/4 & 0 \\ 0 & 0 & 1/2 & 1/2 & 0 \\ 0 & 0 & 0 & 1 & 0 \\ 0 & 0 & 0 & 0 & 1 \end{bmatrix} \begin{pmatrix} \mathbf{B}_1 \\ \mathbf{B}_2 \\ \mathbf{B}_3 \\ \mathbf{B}_4 \\ \mathbf{B}_5 \end{pmatrix}. \tag{12.54}$$

As seen in (12.38) and (12.40), the coefficient matrix relating the new and old control points in (12.55) is the transpose of the matrix relating the new and old basis functions. Specifically,

$$\mathbf{M}^T\bar{\mathcal{N}} = \mathcal{N} \Leftrightarrow \begin{bmatrix} 1 & 0 & 0 & 0 & 0 & 0 & 0 & 0 \\ 0 & 1 & 1/2 & 1/4 & 1/4 & 0 & 0 & 0 \\ 0 & 0 & 1/2 & 1/2 & 1/2 & 1/2 & 0 & 0 \\ 0 & 0 & 0 & 1/4 & 1/4 & 1/2 & 1 & 0 \\ 0 & 0 & 0 & 0 & 0 & 0 & 0 & 1 \end{bmatrix} \begin{pmatrix} \bar{\mathbf{N}}_1 \\ \bar{\mathbf{N}}_2 \\ \bar{\mathbf{N}}_3 \\ \bar{\mathbf{N}}_4 \\ \bar{\mathbf{N}}_5 \\ \bar{\mathbf{N}}_6 \\ \bar{\mathbf{N}}_7 \\ \bar{\mathbf{N}}_8 \end{pmatrix} = \begin{pmatrix} \mathbf{N}_1 \\ \mathbf{N}_2 \\ \mathbf{N}_3 \\ \mathbf{N}_4 \\ \mathbf{N}_5 \end{pmatrix}. \tag{12.55}$$

This symmetry of the relationship between the refined and unrefined control points and the refined and unrefined basis functions has already been encountered implicitly in the discussion of patchwise local refinement in Chapter 3. The control point relationship, (12.37), is utilized in (3.65) to ensure that the solution on the interface between two patches is in the space of functions that can be represented on the coarser patch. The relationship between basis functions, (12.40), is used in (3.75) to ensure that the weighting functions used on the interface are only those emanating from the same coarse space. See Chapter 3.

Note

1. The term "blossom" has also been used for an independent development of these concepts by Ramshaw, 1987b. The equivalence between blossoms and polar forms was clarified in Ramshaw, 1989.

13

State-of-the-Art and Future Directions

This is our assessment of the state-of-the-art and promising future research directions.

13.1 State-of-the-art

The current status of isogeometric analysis is summarized as follows:

- A number of single and multiple patch NURBS-based parametric models have been developed and analyzed. Application areas include linear and nonlinear structures, laminar and turbulent flows, and fluid–structure interaction.
- A new projection technique has been developed for handling incompressibility for higher-order NURBS discretizations. Applications to linear and nonlinear problems in solid mechanics have proved successful. This is the first time a coherent strategy for higher-order elements has been developed. The accuracy of stresses is particularly noteworthy.
- Basic mesh refinement schemes have been investigated, namely, h-, p- and k-refinement, corresponding to, respectively, traditional mesh refinement, C^0 order elevation, and C^{p-1} order elevation. It has been shown that NURBS-based isogeometric analysis preserves geometry at all levels of refinement and that detailed features can be retained without excessive mesh refinement, in contrast with traditional finite element analysis. A constraint equation approach has been developed to transition between NURBS patches involving different levels of refinement.
- Superior accuracy to traditional finite element analysis has been demonstrated in all cases, and indications of significantly increased robustness in vibration and time-harmonic wave propagation analysis have been noted. A duality principle relating dispersion error analysis on an infinite domain and frequency analysis on a finite domain has been established. Superior accuracy of NURBS over finite elements has been established for turbulent flows.
- A mathematical theory of h-refinement has been developed. Mathematical investigations utilizing "n-widths" are under way to quantify spline-based approximations compared with traditional finite elements procedures. See Appendix 5.A and Evans *et al.*, 2009.

Isogeometric Analysis: Toward Integration of CAD and FEA   by J. A. Cottrell, T. J. R. Hughes, Y. Bazilevs
© 2009, John Wiley & Sons, Ltd

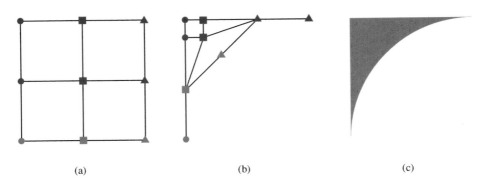

(a)   (b)   (c)

Figure 13.1 Control meshes can be severely distorted and still result in useful physical meshes, as in the case of this single element quadratic NURBS patch. (a) The control mesh in the index space. (b) The control mesh in the physical space. Note that some regions of the control elements have zero area. (c) The physical mesh. Two of the three sides are straight, while proper selection of the weights has resulted in a third side that is a circular arc. Such an element has been used in Lipton *et al.*, 2009 to model fillets in geometries containing reentrant corners. Despite the amount of distortion of the NURBS patch required to generate such a shape, isogeometric analysis with this fillet element resulted in accurate stresses *and* full convergence rates. See Lipton *et al.*, 2009.

- So far, isogeometric analysis has been applied to solid parts and simple thin-walled structures, and the extension to more complex stiffened thin-walled structures seems apparent. The challenge is to apply it to very complex modules and assemblages.
- It has been shown (Lipton *et al.*, 2009) that valid, higher-order NURBS geometries can be developed from control meshes that include various degrees of degeneration, including shapes such as tetrahedra, wedges, and pyramids, and non-convex elements. The control mesh can even include elements that are turned inside out and the physical mesh remains valid. Patch tests are passed in these situations. It is believed that this will loosen restrictions on the generation of three-dimensional meshes compared with traditional finite elements. An example of a "fillet element" resulting from such degeneration is shown in Figure 13.1.
- A T-spline analysis capability has been developed and initial structural analysis calculations have proved successful.
- An interface to LS-Dyna (Livermore Software Technology Corporation, 2007) is under development. It accommodates input of NURBS surface (bivariate) and volume (trivariate) models. NURBS surface models can be converted directly to Reissner–Mindlin shell elements without first generating a mesh (see Benson *et al.*, 2009). Successful initial calculations of shells and solids have been performed. To the best of the authors' knowledge, this is the first time a CAD file has been used directly in a commercial finite element code to perform a structural analysis without any intermediate steps of geometry clean-up or mesh generation. We believe that this is the beginning of a new trend in removing the barriers that exist between design and analysis.
- Successful calculations of solutions of the Cahn–Hilliard equation have been attained with higher-order NURBS bases. This sets the stage for further applications to phase-field models of physical interest.

13.2 Future directions

Here are some of the topics that we think should be pursued.

- **Local unstructured refinement algorithms.** Isogeometric analysis provides finite element codes with a precise geometry that may be refined without communication with the CAD file from which it was generated. However, efficient local unstructured refinement algorithms, suitable for analysis applications, do not yet exist. This is an extremely important research focus because isogeometric models created by designers may capture the geometry accurately but will most likely be too coarse for analysis. Consequently, it will be necessary to perform adaptive refinement in the analysis code in order to achieve the required accuracy for the application under consideration. There are several approaches that might be pursued. Here are three:
  - **Hierarchical refinement.** The idea is to treat knot spans as elements and to perform *h*-refinement by subdividing the knot spans. This approach is geometrically intuitive and similar to procedures used in finite elements. However, in the context of splines, the full smoothness of the original basis will need to be retained in the refinement. This is essential for the efficiency, accuracy and robustness of the refinement scheme. The added basis functions will be assigned "hierarchical" degrees-of-freedom, and a natural multilevel algorithmic architecture ensues. This has several benefits. For example, the control points for the original geometric model are retained at all levels of refinement, providing a concise geometric parameterization for design optimization. Additionally, hierarchically defined bases result in improved condition numbers, of critical importance to the efficiency of iterative solvers. Element-based approaches with splines are facilitated by conversion to a Bézier basis on knot spans. The Bézier basis is also frequently used in geometry applications because there are many efficient algorithms that have been developed for manipulating it. Octree data structures are also anticipated to be useful in developing element-based spline refinement schemes.
  - **Function subdivision.** Geometers tend to think in terms of basis functions rather than elements, despite the fact that knot spans provide a rather natural definition of elements. In smooth particle hydrodynamics (SPH, see Gingold and Monaghan, 1977), and the various meshless methods, the fundamental object is the basis function rather than the element. In the function subdivision approach, a basis function is decomposed into a sum of refined basis functions of the same class and the original function is then replaced by the new basis functions. The data structure for this procedure seems to be entirely different than for the element-based approach described above. The two approaches need to be compared as to their suitability for engineering analysis applications.
  - **Partition of unity method.** In the partition of unity method, new basis functions are added, but all previous basis functions are retained. Each of the basis functions is then divided by the sum of all the basis functions in order to reestablish the partition of unity property. Even if the original basis consists of only polynomials, the refined basis will consist of rational functions. In the case of splines it is typical to start with a rational basis composed of NURBS or T-splines. One of the key issues that must be carefully researched is the ability of the refinement scheme to avoid linear dependencies.
- **Dynamic structural applications.** So far the isogeometric approach has been applied primarily to linear and nonlinear static structural applications, structural eigenvalue problems,

and time-harmonic wave propagation. It needs to be tested on linear and nonlinear dynamic structural applications. If a consistent mass matrix would be employed, the full accuracy of NURBS would be attained. It is noted that ad hoc "row-sum" lumped mass matrices developed in previous work on structural vibrations only achieved second-order accuracy, that is, they did not maintain the full rate-of-convergence of consistent mass. Lumped mass approaches dominate certain areas of transient analysis, such as crash and blast analysis, metal forming, and wave propagation. It would be very desirable to develop higher-order accurate lumped mass matrices. One promising approach to achieving this end is through the construction of dual bases. These produce diagonal mass matrices and should retain full rate-of-coverage if implemented in a consistent, Petrov–Galerkin, weighted-residual format. The challenge to success here is the numerical stability of the internal force calculation and the complexity and lack of smoothness of dual basis functions. If these obstacles can be overcome, it may open the way toward higher-order accurate explicit procedures, which would be of enormous practical value.

- **Quadrature.** Isogeometric analysis has been shown to be more accurate than traditional finite element analysis per degree-of-freedom. So far, sufficiently accurate Gaussian quadrature has been utilized on knot spans, which engenders considerable overhead compared with higher-order C^0 elements. The reason for this is, roughly speaking, knot locations correspond to nodes in finite element analysis and Gaussian quadrature rules can be used over multiple knot spans within finite elements because basis functions are C^∞ there. Of course, for isogeometric Bézier elements (see Section 5.2) the cost of quadrature is identical to standard C^0 Lagrange elements.

  Initial attempts have been made to develop efficient quadrature rules for splines. See Hughes *et al.*, 2008b. The new rules take account of the precise level of smoothness across knots and improve upon the brute force approach of using Gaussian rules between knots. These rules have been developed for uniform and non-uniform knot spacing and are determined numerically in all but the simplest cases. They need to be generalized for local refinement in multiple dimensions. There may be opportunity to further improve the situation by utilizing additional considerations.

  It is important to keep the issue of quadrature in perspective. The cost of analysis typically does not scale linearly with the number of quadrature points except in special cases, such as explicit dynamic analysis (Livermore Software Technology Corporation, 2007). Quadrature is also highly parallelizable, whereas equation solving is typically only partially parallelizable. Equation solving also usually scales with a power greater than 1 of the number of elements and equivalently the number of quadrature points, and, consequently, dominates overall analysis cost, especially for large three-dimensional problems.

- **Collocation.** The smoothness of higher-order NURBS basis functions permits the construction of variational methods in which the strong form of the residual may be used (*i.e.*, the form in which integration-by-parts is not used). This offers the possibility of developing collocation methods in which the residual is evaluated at a number of collocation points equal to the number of control points in the model. This number is much less than the number of quadrature points necessary in a typical Galerkin formulation. The proper location of collocation points is intimately linked to the issue of efficient quadrature and thus this topic builds upon the previous one. The challenges that need to be addressed in the development of collocation schemes are maintaining the numerical stability of the inertial and internal forces with a small number of evaluation points and developing a methodology that identifies the optimal location of collocation points. Collocation techniques are preferred over Galerkin methods when the number of quadrature points dominates solution cost. The possibility of efficient higher-order collocation procedures seems unique to smooth spline bases.

- **Contact problems with friction.** One of the drawbacks of traditional finite elements is the faceted approximation of smooth boundaries. Problems manifest themselves in several application areas, such as flows about smooth hydrodynamic configurations, shape optimization of ships requiring curvature continuity, and imperfection-sensitive thin shell buckling analysis. An application that cannot be dealt with by faceted finite elements is sliding contact between solid bodies. The lack of smoothness necessitates "fixes" even in very simple situations. In some procedures the kinematics are projected onto smooth B-spline surfaces necessitating inserting these surfaces between contacting bodies and anticipating where contact may occur in the problem set-up. These techniques complicate code architecture and are not applicable to complex engineering designs. It would seem that smooth NURBS bases would be much better suited for sliding contact and may eliminate the need for special purpose procedures. The challenge to be met here would be to develop algorithms to efficiently locate contact regions and adaptively generate compatible refinements of the contact surface.

- **Shells without rotational degrees-of-freedom.** The smooth basis functions of NURBS presents the opportunity to develop thin shell elements without rotational degrees-of-freedom. In traditional finite element analysis, based on C^0 shape functions, rotational degrees-of-freedom are required to maintain compatibility across element interfaces. In the analysis of smooth shell structures, these are unnecessary with NURBS. However, to maintain proper continuity at locations where shells intersect at finite angles, and to enforce rotation (*i.e.*, slope) boundary conditions, either Lagrange multiplier or discontinuous Galerkin procedures are required. Eliminating rotations, especially in large-deformation applications, is an enormous simplification. This approach eliminates shear locking *ab initio*, results in half the number of degrees-of-freedom of traditional shell analysis, and eliminates complex parameterizations of rotational degrees-of-freedom in large-deformation analysis. (Finite rotations are not vectorial, they form a multiplicative matrix group, and require Euler angles or some other cumbersome parameterization.)

- **Curved NURBS beam element.** In order to develop a frame structural analysis capability and a compatible beam for a Reissner–Mindlin shell formulation, a curved Timoshenko beam element needs to be developed. This will be useful for the analysis of stiffened shell structures. Development of a curved beam element without rotational degrees-of-freedom, for use with the previously described shell formulation, would also be welcome.

- **Geometrical model development.** The key to eliminating the CAD/CAE bottleneck is to create parameterized geometries in the design phase. The recent development of T-spline surfaces illustrates the possibilities. Trimmed NURBS surfaces can be made into untrimmed T-splines, from which untrimmed NURBS can be generated (see Sederberg *et al.*, 2008). Either the untrimmed NURBS or T-spline surface files can be directly transferred to isogeometric analysis codes and shell structural analysis can be performed without the necessity of all the usual geometric clean-up, defeaturing, and mesh generation. This has already been demonstrated for some simple "obstacle course" shell calculations using the isogeometric interface to the LS-Dyna code. This is the first and most significant step toward automating the "design to analysis" cycle. The most significant challenge facing isogeometric analysis is developing three-dimensional spline parameterizations from surfaces. This is a problem of geometry generation. The most promising starting points seem to be based on the assumption of untrimmed T-spline or NURBS surfaces containing a volume. This would be applicable to solid parts and also internal flow geometries. The case of external flow geometries appears to be easier. Ideally, it would be desirable to retain the surface parameterization in the process. However, this probably could be relaxed for many practical applications. There are several promising directions to be researched. Different procedures may be better suited for specific application areas. Here are some:

- Develop the three-dimensional analogue to T. Sederberg's surface "sewing" algorithm, that is, view the solid as being trimmed by the T-spline or NURBS surface and create an untrimmed volumetric T-spline or NURBS. This could follow along the lines of existing octree based mesh generation procedures, the difference compared with traditional finite elements being the increased ability of higher-order NURBS to maintain geometric validity in the face of degenerated element shapes, non-convex elements, etc.
- If we do not insist on maintaining the surface parameterization and are content with approximating the surface geometry to a predefined precision, then existing hexahedral mesh generation technology may be given a new life because of the ability of higher-order NURBS to generate valid geometries from quite pathological hexahedral control meshes.
- The work by Y. He and D. Gu on generating conformal meshes on surfaces, Ricci flows, and polycube spline decompositions provides another direction to pursue (Li *et al.*, 2007; Gu and Yau, 2008).

- **Boundary Integral Methods.** Surface models composed of NURBS, T-splines, or subdivision surfaces could be used directly in an isogeometric boundary integral formulation (see Cervera and Trevelyan, 2005a, 2005b). This would appear to be a straightforward but practically important development. There are a number of problem classes where the boundary integral method is a viable analysis choice. The advantage is of course that no volume discretization is required.
- **Phase-field modeling.** The smoothness of NURBS and T-splines and their geometric flexibility make them ideal candidates for phase-field modeling, which invariably entails higher-order spatial differential operators. Smooth basis functions are necessary for developing simple Galerkin discretizations of the phase-field equations. Initial experiences with the Cahn–Hilliard equation, perhaps the most utilized phase-field model, have been very good. The approach developed is able to calculate early-time dynamics and late-time equilibrium solutions accurately and efficiently. Adaptive time-stepping is essential in calculating equilibria because the dynamics frequently varies over many orders of magnitude in an analysis. In addition, it has been determined how to desensitize calculations from mesh dependence. This opens the way to calculating topologically correct phase-field solutions on coarser meshes, of importance in practical engineering applications. The resolution of sharp interfaces is also remarkably crisp with NURBS basis functions. Current and future efforts should be devoted to the development of a phase-field model for water/water-vapor two-phase flows (the Navier–Stokes–Korteweg equations, Korteweg, 1901, and generalizations such as in Jamet *et al.*, 2001) and air/water/water-vapor three-phase flows. The last two theories may be useful in representing cavitation phenomena and the last theory may be applicable to water mists used to fight fires. Other applications of phase-field models are topology optimization and crack propagation. There are many others.
- **Shape optimization.** A key advantage of isogeometric analysis is the potential ability of integrating CAD, FEA and shape optimization. The control variables of the geometry provide a concise parameterization that can be used as design variables. Once an optimal design has

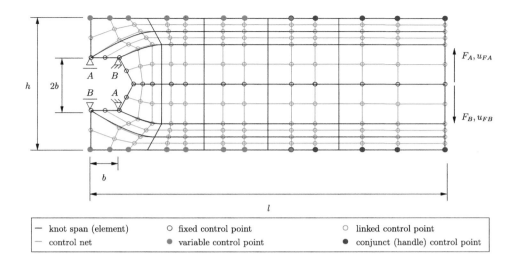

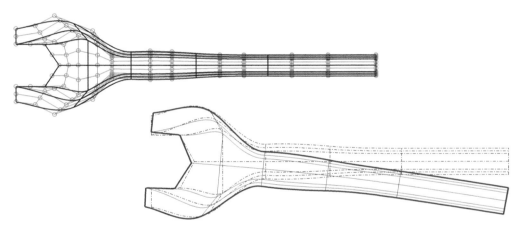

Figure 13.2   Isogeometric shape optimization applied to the open spanner problem. The initial design is shown (top) along with its control net. The optimal design is shown with control points (middle) and deformed (bottom). From Wall *et al.*, 2008.

been obtained, the design can be returned to the CAD system directly because it will already be in the "language" of the system, namely, NURBS, T-splines, etc. A recent study has initiated this pursuit (see Figure 13.2, from Wall *et al.*, 2008).

- **Isogeometric meshless methods.** There is an enormous interest in meshless methods, but so far the relationship to geometry has been almost universally ignored. This was identified as the major shortcoming of meshless technology by Sakurai, 2006. Meshless isogeometric methods could be developed utilizing the concept of PB-splines (*i.e.*, point-based splines), see Sederberg *et al.*, 2003. See Figure 13.3.

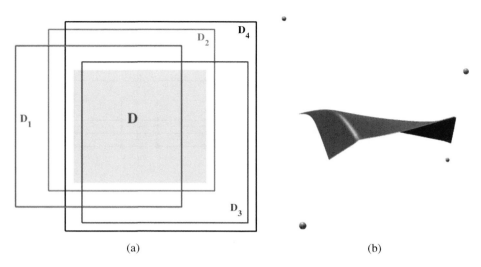

(a)   (b)

Figure 13.3   PB-splines. (a) The supports, $\mathbf{D}_\alpha$, and the parameter space for the spline, $\mathbf{D}$. (b) The corresponding biquadratic PB-spline with its four control points.

- **Toolkits for analysis model construction.** During the more than fifty years of development of finite element technology, many software tools have been developed to facilitate analysis model construction. There are numerous commercial, industrial, and academic software packages available. Tool sets for spline-based analysis are currently non-existent. This is an important future endeavor. One of the deliverables of the EXCITING project in Europe is an isogeometric analysis toolkit. The organization responsible for its development is the Institut National de Recherche en Informatique et Automatique (INRIA) in France.
- **Triangular and tetrahedral NURBS.** There are NURBS constructs for triangular and tetrahedral patches (see Lai and Schumaker, 2007). These shapes, however, have not been widely utilized in design, with the possible exception of the Loop subdivision surface method (Loop, 1987) which is based on triangles. As noted previously, however, subdivision surfaces have not yet found extensive use in engineering design applications.
- **Convex constraints.** If the solution of an analysis problem is required to reside in a convex set, the constraint can be applied to the control variables and the convex hull property will then ensure that the solution will satisfy the constraint in pointwise fashion. This is in contrast with classical higher-order finite elements.
- **Analysis with trimmed objects.** Trimmed NURBS surfaces are a ubiquitous feature of design. See Figure 13.4. On the one hand, it is possible to replace trimmed NURBS with T-splines (Sederberg *et al.*, 2003, 2004). However, this may not always be desirable and so there needs to be analysis methodology that can accommodate trimming. In fact, there is considerable legacy methodology in computational analysis that is applicable. Early examples are described as "fictitious domain methods" (see, *e.g.*, Saulev, 1962, 1963). More recently Glowinski *et al.*, 1999 have developed procedures for the dynamics of flows with spherical particles. Of late there has been increased interest in so-called "embedded boundary methods" and "immersed boundary methods" (see, *e.g.*, Roma *et al.*, 1999; Helzel *et al.*, 2005). An issue of the journal *Computer Methods in Applied Mechanics and Engineering* has

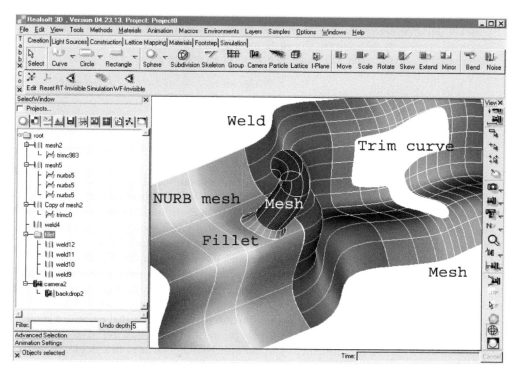

Figure 13.4 Screenshot from Realsoft 3D (Realsoft Graphics, 2008). Current CAD technology frequently relies on trimmed NURBS surfaces to describe complex geometries, particularly when two parametric surfaces intersect. Trimmed geometries present a host of difficulties for meshing and analysis.

been devoted to research developments on this topic (Fogelson *et al.*, 2008). Of particular note is the work of Parvizian *et al.*, 2007 in the context of *p*-methods. In this work, higher-order accuracy was attained and it would seem that extension to NURBS would be straightforward. Mention may also be made of Hollig, 2003. In all, there seem to be many promising procedures to deal with trimmed NURBS efficiently.

- **T-splines.** The major deficiency of NURBS is topological in that gaps and overlaps at intersections of surfaces cannot be avoided, complicating mesh generation. See Table 13.1. Another deficiency is that they utilize a tensor product structure making the representation of detailed local features inefficient. T-splines are a recently developed generalization of NURBS technology. T-splines correct the deficiencies of NURBS in that they permit local refinement and coarsening, and a solution to the gap/overlap problem. In CAD as in FEA, there is always some approximation somewhere. However, it needs to be controlled to the degree of accuracy determined by the application. T-splines is a technology satisfying this requirement. Commercial T-spline plug-ins have been introduced for Maya and Rhino, two NURBS-based design systems (see T-Splines, Inc., 2008a, 2008b).

A NURBS surface is defined using a set of control points, which lie, topologically, in a rectangular grid. This means that a large percentage of NURBS control points are superfluous

Table 13.1 Finite element analysis models the topology of a domain in an accurate fashion, though the geometry itself is only approximate. Computer aided design accurately represents the geometry, though the topology has historically been incorrect. Isogeometric analysis attempts to model both the topology and the geometry accurately

	Topology	Geometry
Finite element analysis	✓	✗
Computer aided design	✗	✓
Isogeometric analysis	✓	✓

in that they contain no significant geometric information, but merely are needed to satisfy topological constraints. In many cases 80% or more of NURBS control points are superfluous for design purposes. By contrast, a T-spline control mesh is allowed to have partial rows of control points. A partial row of control points terminates in a T-junction, hence the name T-splines. T-spline models typically require only 20% of the control points compared to NURBS models. For a designer, fewer control points means faster modeling time.

Refinement, the process of adding new control points to a control mesh without changing the surface, is an important basic operation used by designers. A limitation of NURBS is that refinement requires the insertion of an entire row of control points. T-junctions enable T-splines to be locally refined. Another limitation of NURBS is that because a single NURBS surface must have a rectangular topology, most objects must be modeled using several NURBS surfaces. It is difficult to join multiple NURBS surfaces in a single, smooth, watertight model, especially if corners of valence other than four are introduced. These are referred to as "extraordinary points." T-junctions make it possible to merge together several NURBS surfaces into a gap-free T-spline.

Another serious problem inherent in NURBS is that it is mathematically impossible for a trimmed NURBS to accurately represent the intersection of two NURBS surfaces without introducing gaps in the model. A reason for this is that a generic curve of intersection between two bicubic patches is degree 324 (Sederberg et al., 1984), whereas the degree of the image of a conventional trimming curve is only 18. An NSF-sponsored workshop (MSRI, 1999) identified the unavoidable gaps in trimmed NURBS as the most pressing unresolved problem in the field of CAD. This problem is a major cause of the incompatibility between CAD and analysis software (Kasik et al., 2005), which in 1999 was estimated to cost the U.S. automotive industry alone over $1 billion annually (Brunnermeier and Martin, 1999). The existing approach to identifying and resolving such problems is to employ "healing" software, which does not fix the problem, but only reduces the size of gaps. The problem is a significant ingredient in the design–analysis bottleneck because a CAD model must be closed in order to generate an analysis-suitable geometry and mesh. T-splines provide a way to close the gap and solve many of the problems with NURBS that have vexed the CAD community for three decades. T-splines are also forward and backward compatible with NURBS. Every NURBS is a special case of a T-spline (i.e., a T-spline with no T-junctions or extraordinary points) and every T-spline can be converted into one or more NURBS surfaces by performing repeated local refinement to eliminate all T-junctions. Compatibility is crucial for commercialization, especially in a mature industry like CAD.

Appendix A: Connectivity Arrays

Let us make the discussions of Chapters 3 and 4 regarding the various connectivity arrays a bit more concrete by considering a specific example. We take the simple two-dimensional mesh shown in Figure 2.16 of Chapter 2. With $p = q = 2$ and knot vectors $\Xi = \{0, 0, 0, 0.5, 1, 1, 1\}$ and $\mathcal{H} = \{0, 0, 0, 1, 1, 1\}$, it is not difficult to deduce that we will have two elements of non-zero area. Recalling our tensor product structure, it is also evident that we will have twelve biquadratic functions requiring some type of numbering convention. During assembly, however, we will only be looking at local entities and local numbering conventions. The purpose of the connectivity arrays is to maintain simple bookkeeping procedures that will relate these local and global schemes.

Before proceeding, a comment about notation is in order. In this appendix, when we consider the trivariate case with parametric directions ξ, η, and ζ, we will associate with them polynomial orders p, q, and r, respectively. The number of basis functions in each of these parametric directions is given by n, m, and l. Though potentially confusing, this reverse alphabetical ordering of these last three variable names is consistent with the standard practice of using n for the univariate case, and it allows us to avoid the potentially more confusing practice of using the letter o as a variable. With this convention in place and open knot vectors assumed, in the ξ-direction we have $\Xi = \{\xi_1, \ldots, \xi_{n+p+1}\}$, in the η-direction we have $\mathcal{H} = \{\eta_1, \ldots, \eta_{m+q+1}\}$, and in the ζ-direction we have $\mathcal{Z} = \{\zeta_1, \ldots, \zeta_{l+r+1}\}$.

A.1 The INC array

For higher-dimensional NURBS objects, it is very convenient to introduce the concept of **NURBS coordinates**. Examining the index space view in Figure A.1, the NURBS coordinates of any vertex in the mesh are simply the *indices* of the knots that define it. For example, the vertex created by the intersection of the knot lines corresponding to ξ_3 and η_2 has NURBS coordinates $(3, 2)$. Note that this is the vertex at which the support of the blue function begins. In fact, this is how we will most frequently use NURBS coordinates: to identify the knots at which the support of a function begins.

Isogeometric Analysis: Toward Integration of CAD and FEA   by J. A. Cottrell, T. J. R. Hughes, Y. Bazilevs
© 2009, John Wiley & Sons, Ltd

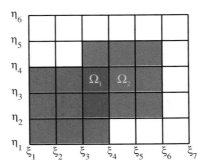

Figure A.1 The index space view of the mesh from Figure 2.16 in Chapter 2. The support of $\tilde{N}_{1,1;2,2}(\xi, \eta)$ is shown in red, while the support of $\tilde{N}_{3,2;2,2}(\xi, \eta)$ is in blue. The region in which they overlap is purple. Examination of the knot *values* indicates that only two of the knot spans correspond to elements with positive measure in the parameter space (*i.e.*, those denoted Ω_1 and Ω_2).

This leads us to a natural scheme for the global numbering of basis functions. If there are n functions in the ξ-direction and m functions in the η-direction, then define

$$A = n(j - 1) + i \tag{A.1}$$

such that the global bivariate function $\tilde{N}_A(\xi, \eta)$ is the tensor product of univariate functions $N_i(\xi)$ and $M_j(\eta)$. We define the INC ("NURBS coordinates") array such that given a global basis function number and a parametric direction, it returns the index of the one-dimensional basis function in the specified direction that was used to build the global function. Because the support of any one-dimensional NURBS function $N_i(\xi)$ is $[\xi_i, \xi_{i+p+1}]$, we can also interpret the INC array as relating the global basis function number and the specified parametric direction with the index of the knot in the appropriate knot vector at which the support of the function begins. Thus, with $\tilde{N}_A(\xi, \eta) = N_i(\xi)M_j(\eta)$ we have

$$i = \text{INC}(A, 1) \quad \text{and} \quad j = \text{INC}(A, 2). \tag{A.2}$$

Turning our attention to Figure A.1 and noting that $n = 4$, $p = 2$, $m = 3$, and $q = 2$, we have the INC array given in Table A.1. Thus we see that the red function is $\tilde{N}_1(\xi, \eta)$ and has NURBS

Table A.1 The INC array corresponding to the mesh in Figures A.1 and A.3. INC consumes a global basis function number and a parametric direction number and returns the corresponding NURBS coordinate

INC	A (global function number)											
	1	2	3	4	5	6	7	8	9	10	11	12
1 (ξ-coordinate)	1	2	3	4	1	2	3	4	1	2	3	4
2 (η-coordinate)	1	1	1	1	2	2	2	2	3	3	3	3

coordinates $(1, 1)$, while the blue function is $\tilde{N}_7(\xi, \eta)$ with NURBS coordinates $(3, 2)$. The NURBS coordinates are required by many routines, such as basis function evaluation, that explicitly utilize the knot vectors. They can be useful in many other settings as well, depending upon the data structures chosen.

A.2 The IEN array

The concept of NURBS coordinates provides us with an easy way to determine which functions have support in a given element. First, let us assign element numbers. Knowing that we are using open knot vectors, the number of elements in the ξ-direction is $n - p$; similarly, in the η-direction we have $m - q$ elements (note that due to the possibility of repeated *internal* knots, some of these elements may have zero measure in the parametric domain; this scheme does not, however, apply element numbers to the knot spans that are known *a priori* to have zero measure due to the use of open knot vectors). Consider an element $\Omega^e = [\xi_i, \xi_{i+1}] \times [\eta_j, \eta_{j+1}]$, where $p + 1 \leq i \leq n$ and $q + 1 \leq j \leq m$. A natural numbering scheme is to assign the element number

$$e = (j - q - 1)(n - p) + (i - p). \tag{A.3}$$

Thus, the "lower, left-hand corner" of element e has NURBS coordinates (i, j). See Figure A.2.

From our examination of univariate spline functions in Chapter 2, we know exactly which functions have support in element e, namely, any function of the form $N_\alpha(\xi)M_\beta(\eta)$ for integers α and β such that $i - p \leq \alpha \leq i$ and $j - q \leq \beta \leq j$. Thus, the total number of local basis functions is $n_{en} = (p + 1)(q + 1)$. Let us assign local function number 1 to the function with NURBS coordinates (i, j). We then assign the remaining local numbers, working backwards in ξ first, followed by η. Thus, with A as in (A.1), the global numbers of the first $p + 1$ local

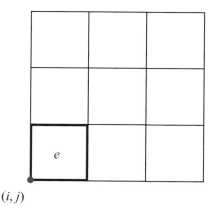

(i, j)

Figure A.2   Element number e and NURBS coordinates (i, j). NURBS coordinates are the indices where a basis function $\tilde{N}_A(\xi, \eta) = N_i(\xi)M_j(\eta)$ begins in the index space. The support of this basis function is shown, assuming $p = q = 2$.

functions are $A, A - 1, \ldots, A - p$. The function N_{A-p-1} does not have support in the element, so we move a row in the η-direction and continue numbering with $A - n, A - n - 1, \ldots, A - n - p$. Again, we must move to the next row and continue with $A - 2n, \ldots, A - 2n - p$. This continues until we reach our last set of function numbers, $A - qn, \ldots, A - qn - p$, at which point we are finished.

The IEN ("element nodes") array connects these global function numbers to their local ordering on the element. In finite elements, global basis function numbers are identified with global node numbers, and local basis function numbers are identified with local node numbers. It is for this reason that the IEN array is referred to as the "element nodes" array. Even though this designation no longer applies in the present case, we retain the name. Given the element number, e, and the local basis function number, b, the corresponding global basis function number, B, is given by

$$B = \text{IEN}(b, e).$$ (A.4)

Thus, if $A = \text{IEN}(1, e)$ as in the previous paragraph, then we have, for example, $A - 1 = \text{IEN}(2, e)$, $A - n = \text{IEN}(p + 2, e)$, and $A - qn - p = \text{IEN}((p + 1)(q + 1), e)$.

The IEN array corresponding to the mesh in Figures A.1 and A.3 is shown in Table A.2. Observe that the blue function, $\tilde{N}_7$, which has support in both elements, has local number $a = 4$ on element $e = 1$ and also local number $a = 5$ on element $e = 2$. That is, $\text{IEN}(4, 1) = \text{IEN}(5, 2) = 7$. The red function, $\tilde{N}_1$, has support in only the first element. The only entry corresponding to it is $\text{IEN}(9, 1) = 1$.

Let us consider the trivariate case. With a knowledge of just the polynomial orders and the number of univariate functions in each parametric direction, we can set up IEN and INC by implementing the pseudocode in Algorithm 7.

Table A.2 The IEN array corresponding to the mesh in Figures A.1 and A.3. IEN consumes a local basis function number and an element number and returns the corresponding global basis function number

IEN	\multicolumn{9}{c}{a (local basis function number)}								
	1	2	3	4	5	6	7	8	9
$e = \{1$	11	10	9	7	6	5	3	2	1
$\quad\ \ 2$	12	11	10	8	7	6	4	3	2

Algorithm 7: Building the INC and IEN arrays.

Data: The polynomial orders (p, q, and r) and number of univariate basis functions (n, m, and l) for each of the parametric directions (ξ, η, and ζ, respectively) must be included as inputs.

Result: We will construct the INC and IEN arrays. The total number of elements, n_{el}, the total number of global basis functions, n_{np}, and the number of local basis functions, n_{en}, will also be defined.

```
// Global variable definitions and initializations:
nel = (n-p)*(m-q)*(l-r);                           // number of elements
nnp = n*m*l;                             // number of global basis functions
nen = (p+1)*(q+1)*(r+1);                  // number of local basis functions
INN[nnp][3] = 0;                                  // NURBS coordinates array
IEN[nen][nel] = 0;                                     // connectivity array

// Local variable initializations:
e, A, B, b, i, j, k, iloc, jloc, kloc                     // should all be
                                                       // initialized to zero
```

```
for k = 1 to l do
    for j = 1 to m do
        for i = 1 to n do
            A = A + 1;                  // increment global function number
            INN[A][1] = i;
            INN[A][2] = j;                       // assign NURBS coordinates
            INN[A][3] = k;
            if i, j, and k ≥ (p+1), (q+1), and (r+1), respectively then
                e = e+1;                        // increment element number
                for kloc = 0 to r do
                    for jloc = 0 to q do
                        for iloc = 0 to p do
                            B = A - kloc*n*m - jloc*n - iloc;
                                                   // global function number
                            b = kloc*(p+1)*(q+1) + jloc*(p+1)+ iloc + 1;
                                                    // local function number
                            IEN[b][e] = B;         // assign connectivity
                        end
                    end
                end
            end
        end
    end
end
```

A.3 The ID array

A.3.1 The scalar case

Now that we have numbered all of the basis functions used to construct our geometry, established a local numbering convention, and collected the connectivity information relating the two points of view, we need to turn our attention to the specific requirements of analysis. First, let us consider the case of a scalar solution field. Recall that in Chapter 3 we assumed a numbering of the global functions such that each function with support on the Dirichlet boundary had a higher index than any without support on that boundary. This was convenient for the exposition of finite element concepts, but we have no reason to expect it to be compatible with the numbering system proposed in the previous section. In general, we have one equation corresponding to each function that does not have support on the Dirichlet boundary. This assumes Dirichlet boundary conditions are satisfied *strongly* (see Section 3.4 in Chapter 3 for elaboration). We must construct a mapping between the global index of those functions, and an equation number between 1 and n_{eq}, the total number of equations (which, in the scalar case, is less than or equal to the total number of functions). This information is stored in the ID ("destination") array.

The ID array itself will depend on the specifics of the boundary conditions. Referring to Figure A.3, assume that we have Dirichlet data prescribed along the edge from $(3, 1.5)$ to $(3, 5)$ in the physical space. We can tell from Figure A.1 that any function N_A such that $INC(A, 1) = 4$ is going to have support on that edge, and thus will *not* have an equation number corresponding to it. Though there are many conventions we might adopt, we simply assign equation numbers in ascending order, assigning 0 to any function with support on the Dirichlet boundary.[1] Thus, we arrive at the ID array shown in Table A.3.

A.3.2 The vector case

When the unknowns are vector-valued, we must expand ID to consume not only a global function number, A, but a degree-of-freedom number, i, as well. Again, there are several ways that we might do this, but one straightforward approach traverses an outer loop through the function numbers and an inner loop through the degrees-of-freedom, assigning equation

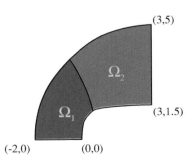

Figure A.3 The physical space view of the mesh from Figure 2.16 in Chapter 2. Both $\tilde{N}_{1,1;2,2}(\xi, \eta)$ and $\tilde{N}_{3,2;2,2}(\xi, \eta)$ have support in Ω_1, while only $\tilde{N}_{3,2;2,2}(\xi, \eta)$ has support in Ω_2. In the diagram, the coordinates of points in physical space are enclosed in parentheses.

Table A.3 The ID array corresponding to the mesh in Figures A.1 and A.3 and a scalar solution field. In the scalar case, ID consumes a global basis function number and returns the corresponding equation number. Functions with equation number 0 are associated with the lifting g^h (*i.e.*, strongly enforced Dirichlet boundary conditions) and do not correspond to active degrees-of-freedom

	A (global function number)											
ID	1	2	3	4	5	6	7	8	9	10	11	12
P (equation number)	1	2	3	0	4	5	6	0	7	8	9	0

numbers sequentially, with any constrained (i, A) pair being mapped to 0. Thus, the ID array relates equation number P to the corresponding degree-of-freedom, i, and global function number A by way of

$$P = \text{ID}(i, A). \tag{A.5}$$

Each component of the solution field may involve Dirichlet data prescribed on different portions of the domain, but this does not create a conflict as long as the information is stored properly in the ID array. For example, assume we wish to solve a problem of two-dimensional linear elasticity on the domain from Figure A.3. We might wish to prescribe displacements in the x-direction on the edge from $(3, 1.5)$ to $(3, 5)$, while prescribing displacements in the y-direction on the edge from $(-2, 0)$ to $(0, 0)$. This would result in the ID array given in Table A.4.

A.4 The LM array

The final connectivity array that we will consider is just a composition of the previous two. The most common form of the LM ("location matrix") array consumes a degree-of-freedom number, i, a local basis function number, a, and an element number, e, and it returns a global equation number,

$$P = \text{LM}(i, a, e) \tag{A.6}$$

Table A.4 The ID array corresponding to the mesh in Figures A.1 and A.3 and a vector-valued solution field. In the vectorial case, ID consumes a global basis function number and a degree-of-freedom number and returns the corresponding equation number. Combinations (i, A) with equation number 0 are associated with the lifting g_i^h (*i.e.*, strongly enforced Dirichlet boundary conditions) and do not correspond to active degrees-of-freedom

		A (global function number)											
ID		1	2	3	4	5	6	7	8	9	10	11	12
$i =$	1	1	2	4	0	7	8	10	0	13	14	16	0
	2	0	3	5	6	0	9	11	12	0	15	17	18

Table A.5 The LM array corresponding to the mesh in Figures A.1 and A.3 and a vector-valued solution field. In the vectorial case, LM consumes a degree-of-freedom number, a local basis function number, and an element number, and returns the corresponding equation number. Combinations (i, a, e) with equation number 0 are associated with the lifting g_i^h (*i.e.*, strongly enforced Dirichlet boundary conditions) and do not correspond to active degrees-of-freedom

LM		e (element number) 1	2
$a = 1$	$i = 1$	16	0
	$i = 2$	17	18
$a = 2$	$i = 1$	14	16
	$i = 2$	15	17
$a = 3$	$i = 1$	13	14
	$i = 2$	0	15
$a = 4$	$i = 1$	10	0
	$i = 2$	11	12
$a = 5$	$i = 1$	8	10
	$i = 2$	9	11
$a = 6$	$i = 1$	7	8
	$i = 2$	0	9
$a = 7$	$i = 1$	4	0
	$i = 2$	5	6
$a = 8$	$i = 1$	2	4
	$i = 2$	4	5
$a = 9$	$i = 1$	1	2
	$i = 2$	0	3

such that

$$\text{LM}(i, a, e) = \text{ID}(i, \text{IEN}(a, e)). \tag{A.7}$$

With this definition, we can build the LM array shown in Table A.5 from the data in Tables A.2 and A.4.

Alternatively, we can define LM as a two-dimensional array. To do this, define the number of degrees-of-freedom in each control variable to be n_{ed}. For example, in linear elasticity we usually have $n_{ed} = d$, where d is the number of spatial dimensions. Thus, define the local equation number, p, corresponding to degree-of-freedom number i and local basis function number a to be

$$p = n_{ed}(a - 1) + i,\qquad(\text{A.8})$$

where, clearly, we must be careful not to confuse this use of the symbol "p" here with its previous usage as a polynomial order. With this definition, we can implement the two-dimensional variant of LM such that

$$\text{LM}(p, e) = \text{LM}(i, a, e).\qquad(\text{A.9})$$

The data structures and arrays necessary for implementing isogeometric analysis are almost the same as for finite elements (see Hughes, 2000). Only the INC array does not have a direct counterpart in standard implementations of finite elements.

Note

1. This convention is appropriate for strongly enforced Dirichlet boundary data. If Dirichlet data are implemented weakly, all degrees-of-freedom are active and receive (positive) equation numbers. In fluid mechanics, we have come to prefer weakly enforced Dirichlet boundary conditions. For elaboration see Bazilevs and Hughes, 2007; Bazilevs *et al.*, 2007b, 2008b.

References

Abedi, P., Patracovici, B., and Haber, R.B. (2005). A spacetime discontinuous Galerkin method for linearized elastodynamics with element-wise momentum balance. *Computer Methods in Applied Mechanics and Engineering*, 193:1997–2018.

Akkerman, I., Bazilevs, Y., Calo, V., Hughes, T.J.R., and Hulshoff, S. (2008). The role of continuity in residual-based variational multiscale modeling of turbulence. *Computational Mechanics*, 41:371–378.

Argyris, J., Fried, I., and Scharpf, D.W. (1968). The TUBA family of plate elements for the matrix displacement method. *The Aeronautical Journal of the Royal Aeronautical Society*, 72:701–709.

Argyris, J.H. and Kelsey, S. (1960). *Energy Theorems and Structural Analysis*. Butterworths, London.

Arnold, D.N., Brezzi, F., Cockburn, B., and Marini, L.D. (2002). Unified analysis of Discontinuous Galerkin methods for elliptic problems. *SIAM Journal of Numerical Analysis*, 39:1749–1779.

Auricchio, F., Beirao de Veiga, L., Lovadina, C., Reali, A., and Sangalli, G. (2007). A fully "locking-free" isogeometric approach for plane linear elasticity problems: A stream function formulation. *Computer Methods in Applied Mechanics and Engineering*, 197:160–172.

Bajaj, C., Chen, J., and Xu, G. (1995). Modeling with cubic A-patches. *ACM Transactions on Graphics*, 14:103–133.

Bajaj, C., Schaefer, S., Warren, J., and Xu, G.L. (2002). A subdivision scheme for hexahedral meshes. *Visual Computer*, 18:343–356.

Barenblatt, G. I. (1979). *Similarity, Self-Similarity, and Intermediate Asymptotics*. Plenum Press.

Barth, T. J. (1998). Numerical methods for gasdynamic systems on unstructured meshes. In Kröner, D., Ohlberger, M., and Rohde, C., editors, *An Introduction to Recent Developments in Theory and Numerics for Conservation Laws*, volume 5 of *Lecture Notes in Computational Science and Engineering*, pages 195–285. Springer-Verlag.

Bazilevs, Y. and Hughes, T.J.R. (2007). Weak imposition of Dirichlet boundary conditions in fluid mechanics. *Computers and Fluids*, 36:12–26.

Bazilevs, Y. and Hughes, T.J.R. (2008). NURBS-based isogeometric analysis for the computation of flows about rotating components. *Computational Mechanics*, 43:143–150.

Bazilevs, Y., Beirao de Veiga, L., Cottrell, J.A., Hughes, T.J.R., and Sangalli, G. (2006a). Isogeometric analysis: approximation, stability and error estimates for *h*-refined meshes. *Mathematical Models and Methods in Applied Sciences*, 16:1031–1090.

Bazilevs, Y., Calo, V.M., Zhang, Y., and Hughes, T.J.R. (2006b). Isogeometric fluid–structure interaction analysis with applications to arterial blood flow. *Computational Mechanics*, 38:310–322.

Bazilevs, Y., Calo, V.M., Cottrell, J.A., Hughes, T.J.R., Reali, A., and Scovazzi, G. (2007a). Variational multiscale residual-based turbulence modeling for large eddy simulation of incompressible flows. *Computer Methods in Applied Mechanics and Engineering*, 197:173–201.

Bazilevs, Y., Michler, C., Calo, V.M., and Hughes, T.J.R. (2007b). Weak Dirichlet boundary conditions for wall-bounded turbulent flows. *Computer Methods in Applied Mechanics and Engineering*, 196:4853–4862.

Bazilevs, Y., Calo, V., Hughes, T.J.R., and Zhang, Y. (2008a). Isogeometric fluid–structure interaction: Theory, algorithms, and computations. *Computational Mechanics*, 43:3–37.

Bazilevs, Y., Michler, C., Calo, V.M., and Hughes, T.J.R. (2008b). Isogeometric variational multiscale modeling of wall-bounded turbulent flows with weakly enforced boundary conditions on unstretched meshes. *Computer Methods in Applied Mechanics and Engineering*. Published online. doi:10.1016/j.cma.2008.11.020.

Bazilevs, Y., Calo, V.M., Cottrell, J.A., Evans, J., Hughes, T.J.R., Lipton, S., Scott, M.A., and Sederberg, T.W. (2009). Isogeometric analysis using T-splines. *Computer Methods in Applied Mechanics and Engineering*, in press.

Bell, K. (1969). A refined triangular plate bending element. *International Journal of Numerical Methods in Engineering*, 1:101–122.

Belytschko, T. and Tsay, C.S. (1983). A stabilization procedure for the quadrilateral plate element with one-point quadrature. *International Journal of Numerical Methods in Engineering*, 19:405–419.

Belytschko, T., Lin, J.I., and Tsay, C.S. (1984). Explicit algorithms for nonlinear dynamics of shells. *Computer Methods in Applied Mechanics and Engineering*, 42:225–251.

Belytschko, T., Stolarski, H., Liu, W.K., Carpenter, N., and Ong, J.S.-J. (1985). Stress projection for membrane and shear locking in shell finite elements. *Computer Methods in Applied Mechanics and Engineering*, 51:221–258.

Belytschko, T., Gu, L., and Lu, Y.Y. (1994). Fracture and crack growth by element-free Galerkin methods. *Modelling and Simulation in Materials Science and Engineering*, 2:519–534.

Benson, D.J., Bazilevs, Y., Hsu, M.-C., and Hughes, T.J.R. (in press). Isogeometric shell analysis: The Reissner–Mindlin shell. *Computer Methods in Applied Mechanics and Engineering*. In review.

Bernstein, S. (1912). Démonstration du théorem de Weierstrass fondeé sur le calcul des probabilités. *Harkov Soobs. Matem ob-va*, 13:1–2.

Bezier, P. (1966). Définition numérique des courbes et surfâces I. *Automatisme*, XI:625–632.

Bezier, P. (1967). Définition numérique des courbes et surfâces II. *Automatisme*, XII:17–21.

Bezier, P. (1972). *Numerical Control: Mathematics and Applications*. Wiley. Translated from the French by A. R. Forrest.

Birkhoff, G. and Lynch, R.E. (1987). *Numerical Solution of Elliptic Problems*. Society for Industrial and Applied Mathematics.

Bischoff, M., Wall, W.A., Bletzinger, K.-U., and Ramm, E. (2004). Models and finite elements for thin-walled structures. In Stein, E., de Borst, R., and Hughes, T.J.R., editors, *Encyclopedia of Computational Mechanics, Vol. 2, Solids, Structures and Coupled Problems*, chapter 3. Wiley.

Bochev, P. B. and Gunzburger, M. D. (1998). Finite element methods of the least-squares type. *SIAM Review*, 40:789–837.

Brezzi, F. and Fortin, M. (1991). *Mixed and Hybrid Finite Element Methods*. Springer-Verlag.

Brezzi, F., Douglas, J., and Marini, L.D. (1985). Two families of mixed finite elements for 2nd order elliptic problems. *Numerische Mathematik*, 47:231–235.

Brillouin, L. (1953). *Wave Propagation in Periodic Structures*. Dover Publications, Inc.

Bronstein, A.M., Bronstein, M.M., and Kimmel, R. (2008). *Numerical Geometry of Non-rigid Shapes*. Springer-Verlag.

Brooks, A.N. and Hughes, T.J.R. (1982). Streamline upwind / Petrov–Galerkin formulations for convection dominated flows with particular emphasis on the incompressible Navier–Stokes equations. *Computer Methods in Applied Mechanics and Engineering*, 32:199–259.

Brunnermeier, S.B. and Martin, S.A. (1999). Interoperability cost analysis of the US automotive supply chain. National Institute of Standards and Technology. *NIST Planning Repor 99-1*.

Buehrle, R.D., Fleming, G.A., Pappa, R.S., and Grosveld, F.W. (2001). Finite element model development for aircraft fuselage structures. *Sound and Vibration*, 35:32–38.

Cahn, J.W. (1961). On spinodal decomposition. *Acta Metallurgica*, 9:795–801.

Cahn, J.W. and Hilliard, J.E. (1958). Free energy of a non-uniform system. I. Interfacial free energy. *Journal of Chemical Physics*, 28:258–267.

Canuto, C., Hussaini, M.Y., Quarteroni, A., and Zang, T.A. (1988). *Spectral Methods in Fluid Dynamics*. Springer-Verlag.

Catmull, E. and Clark, J. (1978). Recursively generated B-spline surfaces on arbitrary topological meshes. *Computer Aided Design*, 16:350–355.

Celniker, G. and Gossard, D. (1991). Deformable curve and surface finite elements for free-form shape design. *ACM Computer Graphics*, 25:157–266.

Cervera, E. and Trevelyan, J. (2005a). Evolutionary structural optimisation based on boundary representation of NURBS. Part i: 2d algorithms. *Computers and Structures*, 83:1902–1916.

Cervera, E. and Trevelyan, J. (2005b). Evolutionary structural optimisation based on boundary representation of NURBS. Part ii: 3d algorithms. *Computers and Structures*, 83:1917–1929.

Chaikin, G.M. (1974). An algorithm for high speed curve generation. *Computer Graphics and Image Processing*, 3:346–349.

Chavan, K.S., Lamichhane, B.P., and Wohlmuth, B.I. (2007). Locking-free finite element methods for linear and nonlinear elasticity in 2D and 3D. *Computer Methods in Applied Mechanics and Engineering*, 196:4075–4086.

Chung, J. and Hulbert, G.M. (1993). A time integration algorithm for structural dynamics with improved numerical dissipation: The generalized-α method. *Journal of Applied Mechanics*, 60:371–75.

Ciarlet, P.G. (1978). *The Finite Element Method for Elliptic Problems*. North-Holland.

Cirak, F. and Ortiz, M. (2001). Fully C^1-conforming subdivision elements for finite deformation thin shell analysis. *International Journal of Numerical Methods in Engineering*, 51:813–833.

Cirak, F., Ortiz, M., and Schröder, P. (2000). Subdivision surfaces: a new paradigm for thin shell analysis. *International Journal of Numerical Methods in Engineering*, 47:2039–2072.

Cirak, F., Scott, M.J., Antonsson, E.K., Ortiz, M., and Schröder, P. (2002). Integrated modeling, finite-element analysis, and engineering design for thin-shell structures using subdivision. *Computer-Aided Design*, 34:137–148.

Clough, R.W. (1960). The finite element method in plane stress analysis. In *Proceedings of the Second ASCE Conference on Electronic Computation*. Pittsburgh, PA.

Clough, R.W. and Tocher, J.L. (1965). Finite element stiffness matrices for analysis of plates in bending. In *Proc. Conference on Matrix Methods in Structural Mechanics*. Wright-Patterson A.F.B.

Cockburn, B. (2004). Discontinuous Galerkin methods for computational fluid dynamics. In Stein, E., de Borst, R., and Hughes, T.J.R., editors, *Encyclopedia of Computational Mechanics, Vol. 3, Computational Fluid Dynamics*, chapter 4. Wiley.

Codina, R., Principe, J., Guasch, O., and Badia, S. (2007). Time dependent subscales in the stabilized finite element approximation of incompressible flow problems. *Computer Methods in Applied Mechanics and Engineering*, 196:2413–2430.

Cohen, E., Lyche, T., and Riesenfeld, R. (1980). Discrete B-spline and subdivision techniques in computer aided geometric design and computer graphics. *Computer Graphics and Image Processing*, 14:87–111.

Cohen, E., Reisenfeld, R.F., and Elber, F. (2001). *Geometric Modeling with Splines: An Introduction*. A. K. Peters, Ltd.

Coons, S. A. (1967). Surfaces for computer-aided design of space forms. Project MAC-TR-41, MIT.

Cottrell, J.A., Reali, A., Bazilevs, Y., and Hughes, T.J.R. (2006). Isogeometric analysis of structural vibrations. *Computer Methods in Applied Mechanics and Engineering*, 195:5257–5296.

Cottrell, J.A., Hughes, T.J.R., and Reali, A. (2007). Studies of refinement and continuity in isogeometric analysis. *Computer Methods in Applied Mechanics and Engineering*, 196:4160–4183.

Couchman, L., Dey, S., and Barzow, T. (2003). *ATC Eigen-Analysis, STARS / ARPACK / NRL Solver*. Naval Research Laboratory, Englewood Cliffs, NJ.

Courant, R. (1943). Variational methods for solution of equilibrium and vibration. *Bulletin of the American Mathematical Society*, 49:1–43.

Cowper, G.R., Kosko, E., Lindberg, G.M., and Olson, M.D. (1968). Formulation of a new triangular plate bending element. *CASI Transaction – Canadian Aeronautics and Space Intstitute*, 1:86–90.

Cox, M.G. (1971). The numerical evaluation of B-splines. Technical report, National Physics Laboratory DNAC 4.

Cundall, P.A. and Strack, O.D.L. (1979). Discrete numerical-model for granular assemblies. *Geotechnique*, 29:47–65.

Curry, H.B. and Schoenberg, I.J. (1966). On Pólya frequency functions IV: The fundamental spline functions and their limits. *Journal d'Analyse Mathematique*, 17:71–107.

De Boor, C. (1972). On calculation with B-splines. *Journal of Approximation Theory*, 6:50–62.

De Boor, C. (1978). *A Practical Guide to Splines*. Springer-Verlag.

De Casteljau, P. (1959). Outillages méthodes calcul. Technical report, A. Citroën, Paris.

De Rham, G. (1956). Sur une courbe plane. *Journal of Pure and Applied Mathematics*, 35:25–42.

De Souza Neto, E.A., Peric, D., Dutko, M., and Owen, D.R.J. (1996). Design of simple low order finite elements for large strain analysis of nearly incompressible solids. *International Journal of Solids and Structures*, 33:3277–3296.

De Souza Neto, E.A., Pires, F.M. Andrade, and Owen, D.R.J. (2005). F-bar based linear triangles and tetrahedra for finite strain analysis of nearly incompressible solids. Part I: formulation and bench-marking. *International Journal for Numerical Methods in Engineering*, 62:353–383.

De Veubeke, B.F. (1968). An equilibrium model for plate bending. *International Journal of Solids and Structures*, 4:447–468.

Demkowicz, L. (2007). *Computing with hp-Adaptive Finite Elements, Vol 1: One and Two Dimensional Elliptic and Maxwell Problems*. Chapman & Hall/CRC.

Demkowicz, L., Kurtz, J., Pardo, D., Paszynski, M., Rachowicz, W., and Zdunek, A. (2008). *Computing with hp-Adaptive Finite Elements, Vol 2: Frontiers: Three Dimensional Elliptic and Maxwell Problems with Applications*. Chapman & Hall/CRC.

Dolbow, J. and Belytschko, T. (1999). Volumetric locking in the element free Galerkin method. *International Journal of Numerical Methods in Engineering*, 46:925–942.

Donea, J., Giuliani, S., and Halleux, J. P. (1982). An arbitrary Lagrangian–Eulerian finite element method for transient dynamics fluid–structure interactions. *Computer Methods in Applied Mechanics and Engineering*, 33:689–723.

Doo, D. and Sabin, M. (1978). Behavior of recursive division surfaces near extraordinary points. *Computer Aided Design*, 10:356–370.

Dorfel, M.R., Juttler, B., and Simeon, B. (2008). Adaptive isogeometric analysis by local h-refinement with t-Splines. *Computer Methods in Applied Mechanics and Engineering*. Published online. doi:10.1016/j.cma.2008.07.012.

Duarte, C.A. and Oden, J.T. (1996). An hp-adaptive method using clouds. *Computer Methods in Applied Mechanics and Engineering*, 139:237–262.

Dung, N.T. and Wells, G.N. (2008). Geometrically nonlinear formulation for thin shells without rotation degrees of freedom. *Computer Methods in Applied Mechanics and Engineering*, 197:2778–2788.

Elguedj, T., Bazilevs, Y., Calo, V.M., and Hughes, T.J.R. (2008). $\bar{B}$ and $\bar{F}$ projection methods for nearly incompressible linear and nonlinear elasticity and plasticity using higher-order NURBS elements. *Computer Methods in Applied Mechanics and Engineering*, 197:2732–2762.

Engel, G., Garikipati, K., Hughes, T.J.R., Larson, M.G., and Mazzei, L. (2002). Continuous/discontinuous finite element approximations of fourth-order elliptic problems in structural and continuum mechanics with applications to thin beams and plates, and strain gradient elasticity. *Computer Methods in Applied Mechanics and Engineering*, 191:3669–3750.

Ern, A. and Guermond, J.-L. (2004). *Theory and Practice of Finite Elements*. Springer-Verlag.

Eskilsson, C. and Sherwin, S. (2006). Spectral/hp discontinuous Galerkin methods for modelling 2D Boussinesq equations. *Journal of Computational Physics*, 212:566–589.

Evans, J.A., Bazilevs, Y., Babuška, I., and Hughes, T.J.R. (2009). n-widths, sup-infs, and optimality ratios for the k-version of the isogeometric finite element method. *Computer Methods in Applied Mechanics and Engineering*. Published online. doi:10.1016/j.cma.2009.01.021.

Farhat, C. and Geuzaine, P. (2004). Design and analysis of robust ALE time-integrators for the solution of unsteady flow problems on moving grids. *Computer Methods in Applied Mechanics and Engineering*, 193:4073–4095.

Farhat, C., Geuzaine, P., and Grandmont, C. (2001). The discrete geometric conservation law and the nonlinear stability of ALE schemes for the solution of flow problems on moving grids. *Journal of Computational Physics*, 174(2):669–694.

Farin, G. (1999a). *Curves and Surfaces for CAGD, A Practical Guide, Fifth Edition*. Morgan Kaufmann Publishers.

Farin, G.E. (1999b). *NURBS Curves and Surfaces: from Projective Geometry to Practical Use, Second Edition*. A. K. Peters, Ltd.

Figueroa, A., Vignon-Clementel, I.E., Jansen, K.E., Hughes, T.J.R., and Taylor, C.A. (2006). A coupled momentum method for modeling blood flow in three-dimensional deformable arteries. *Computer Methods in Applied Mechanics and Engineering*, 195:5685–5706.

Flanagan, D.P. and Belytschko, T. (1981). A uniform strain hexahedron and quadrilateral with orthogonal hourglass control. *International Journal of Numerical Methods in Engineering*, 17:679–706.

Flory, P.J. (1961). Thermodynamic relations for highly elastic materials. *Transactions of the Faraday Society*, 57:829–838.

Fogelson, A., Wang, X. S., and Liu, W.-K. (2008). Immersed boundary method and its extensions. *Computer Methods in Applied Mechanics and Engineering*, 197:2047–2372.

French, D.A. (1993). A space–time finite element method for the wave equation. *Computer Methods in Applied Mechanics and Engineering*, 107(1–2):145–157.

French, D.A. (1998). Discontinuous galerkin finite element methods for a forward–backward heat equation. *Applied Numerical Mathematics*, 28(1):37–44.

French, D.A. and Peterson, T.E. (1996). A continuous space–time finite element method for the wave equation. *Mathematics of Computation*, 65(214):491–506.

Gallagher, R.H., Padlog, J., and Bijlaard, P.P. (1962). Stress analysis of heated complex shapes. *American Rocket Society Journal*, 32:700–707.

Gee, M., Wall, W.A., and Ramm, E. (2005). Parallel multilevel solutions of nonlinear shell structures. *Computer Methods in Applied Mechanics and Engineering*, 194:2513–2533.

Gingold, R.A. and Monaghan, J.J. (1977). Smoothed particle hydrodynamics – theory and application to non-spherical stars. *Monthly Notices of the Royal Astronomical Society*, 181:375–389.

Glowinski, R., T.-W.Pan, Hesla, T.I., and Joseph, D.D. (1999). A distributed lagrange multiplier/fictitious domain method for particulate flows. *International Journal of Multiphase Flow*, 25:755–794.

Gomez, H., Calo, V.M., Bazilevs, Y., and Hughes, T.J.R. (2008). Isogeometric analysis of the Cahn-Hilliard phase-field model. *Computer Methods in Applied Mechanics and Engineering*, 197:4333–4352.

Gonzalez, D., Cueto, E., and Doblaré, M. (2008). Higher-order natural element methods: Toward an isogoemetric meshless method. *International Journal of Numerical Methods in Engineering*, 74:1928–1954.

Gordon, W. (1969). Spline-blended surface interpolation through curve networks. *Journal of Mathematics and Mechanics*, 18:931–952.

Gould, P. L. (1999). *Introduction to Linear Elasticity*. Springer-Verlag.

Gregory, J.A. (1983). *n*-sided surface patches. In Gregory, J.A., editor, *Mathematics of Surfaces*, pages 217–232. Clarendon Press.

Gresho, P. M. and Sani, R. L. (1998). *Incompressible Flow and the Finite Element Method*. Wiley.

Grosveld, F.W., Pritchard, J.I., Buehrle, R.D., and Pappa, R.S. (2002). Finite element modeling of the NASA Langley Aluminum Testbed Cylinder. *8th AIAA/CEAS Aeroacoustics Conference*, Breckenridge, CO. AIAA 2002-2418.

Gu, X.D. and Yau, S.-T. (2008). *Computational Conformal Geometry*. International Press.

Hansbo, P. and Hermansson, J. (2003). Nitsche's method for coupling non-matching meshes in fluid–structure vibration problems. *Computational Mechanics*, 32:134–139.

Hansbo, P., Hermansson, J., and Svedberg, T. (2004). Nitsche's method combined with space–time finite elements for ALE fluid–structure interaction problems. *Computer Methods in Applied Mechanics and Engineering*, 193:4195–4206.

Hauswirth, L., Pérez, J., Romon, A., and Ros, A. (2004). The periodic isoperimetric problem. *Transactions of the AMS*, 356:2025–2047.

Helzel, C., Berger, M.J., and Leveque, R.J. (2005). A high-resolution rotated grid method for conservation laws with embedded geometries. *SIAM Journal on Scientific Computing*, 26:785–809.

Hilber, H. and Hughes, T.J.R. (1978). Collocation, dissipation, and overshoot for time integration schemes in structural dynamics. *Earthquake Engineering and Structural Dynamics*, 6:99–117.

Hilber, H., Hughes, T.J.R., and Taylor, R. (1977). Improved numerical dissipation for time integration algorithms in structural dynamics. *Earthquake Engineering and Structural Dynamics*, 5:283–292.

Höllig, K. (2003). *Finite Element Methods with B-Splines*. Society of Industrial and Applied Mathematics.

Holzapfel, G.A. (2004). Computational biomechanics of soft biological tissue. In Stein, E., de Borst, R., and Hughes, T.J.R., editors, *Encyclopedia of Computational Mechanics, Vol. 2, Solids and Structures*, chapter 18. Wiley.

Huerta, A. and Fernandez-Mendez, S. (2001). Locking in the incompressible limit for the Element-Free Galerkin method. *International Journal of Numerical Methods in Engineering*, 51:1362–1383.

Hughes, T.J.R. (1977). Equivalence of finite elements for nearly incompressible elasticity. *Journal of Applied Mechanics, Transactions ASME*, 44(1):181–183.

Hughes, T.J.R. (1980). Generalization of selective integration procedure to anisotropic and nonlinear media. *International Journal for Numerical Methods in Engineering*, 15:1413–1418.

Hughes, T.J.R. (1995). Multiscale phenomena: Green's functions, the Dirichlet-to-Neumann formulation, subgrid scale models, bubbles and the origins of stabilized methods. *Computer Methods in Applied Mechanics and Engineering*, 127:387–401.

Hughes, T.J.R. (2000). *The Finite Element Method: Linear Static and Dynamic Finite Element Analysis*. Dover Publications.

Hughes, T.J.R. and Allik, H. (1969). Finite elements for compressible and incompressible continua. In *Proceedings of the Symposium on Civil Engineering*, pages 27–62, Vanderbilt University, Nashville, Tenn.

Hughes, T.J.R. and Franca, L.P. (1988). A mixed finite element formulation for Reissner–Mindlin plate theory: Uniform convergence of all higher order spaces. *Computer Methods in Applied Mechanics and Engineering*, 67:223–240.

Hughes, T.J.R. and Garikipati, K. (2004). On the continuious/discontinuous galerkin (cdg) formulation of Poisson-Kirchhoff plate theory. In Mathiesen, K.M., Kvamsdal, T., and Okstad, K.M., editors, *Computational Mechanics – Theory and Practice*, pages 29–35. CINME, Barcelona.

Hughes, T.J.R. and Hulbert, G.M. (1988). Space–time finite-element methods for elastodynamics formulations and error estimates. *Computer Methods in Applied Mechanics and Engineering*, 66:339–363.

Hughes, T.J.R. and Liu, W.K. (1981a). Nonlinear finite element analysis of shells: Part 1. three-dimensional shells. *Computer Methods in Applied Mechanics and Engineering*, 26:331–362.

Hughes, T.J.R. and Liu, W.K. (1981b). Nonlinear finite element analysis of shells: Part 2. two-dimensional shells. *Computer Methods in Applied Mechanics and Engineering*, 27:167–181.

Hughes, T.J.R. and Mallet, M. (1986). A new finite element formulation for fluid dynamics: III. The generalized streamline operator for multidimensional advective-diffusive systems. *Computer Methods in Applied Mechanics and Engineering*, 58:305–328.

Hughes, T.J.R. and Sangalli, G. (2007). Variational multiscale analysis: the fine-scale Green's function, projection, optimization, localization, and stabilized methods. *SIAM Journal of Numerical Analysis*, 45:539–557.

Hughes, T.J.R., Taylor, R.L., and Sackman, J.L. (1975). Finite element formulation and solution of contact-impact problems in continuum mechanics – III. SESM Report 75-3, Department of Civil Engineering, The University of California, Berkeley.

Hughes, T.J.R., Taylor, R.L., and Kanoknukulchai, W. (1977). Simple and efficient finite element for plate bending. *International Journal of Numerical Methods in Engineering*, 11:1529–1543.

Hughes, T.J.R., Cohen, M., and Haroun, M. (1978). Reduced and selective integration techniques in finite element analysis of plates. *Nuclear Engineering and Design*, 46:203–222.

Hughes, T.J.R., Pister, K.S., and Taylor, R.L. (1979). Implicit-explicit finite elements in non-linear transient analysis. *Computer Methods in Applied Mechanics and Engineering*, 17/18:159–182.

Hughes, T.J.R., Liu, W.K., and Zimmermann, T.K. (1981). Lagrangian–Eulerian finite element formulation for incompressible viscous flows. *Computer Methods in Applied Mechanics and Engineering*, 29:329–349.

Hughes, T.J.R., Feijóo., G., Mazzei, L., and Quincy, J.B. (1998). The variational multiscale method – A paradigm for computational mechanics. *Computer Methods in Applied Mechanics and Engineering*, 166:3–24.

Hughes, T.J.R., Scovazzi, G., and Franca, L.P. (2004). Multiscale and stabilized methods. In Stein, E., de Borst, R., and Hughes, T.J.R., editors, *Encyclopedia of Computational Mechanics, Vol. 3, Computational Fluid Dynamics*, chapter 2. Wiley.

Hughes, T.J.R., Cottrell, J.A., and Bazilevs, Y. (2005). Isogeometric analysis: CAD, finite elements, NURBS, exact geometry, and mesh refinement. *Computer Methods in Applied Mechanics and Engineering*, 194:4135–4195.

Hughes, T.J.R., Reali, A., and Sangalli, G. (2008a). Duality and unified analysis of discrete approximations in structural dynamics and wave propagation: Comparison of p-method finite elements with k-method NURBS. *Computer Methods in Applied Mechanics and Engineering*, 197:4104–4124.

Hughes, T.J.R., Reali, A., and Sangalli, G. (2008b). Efficient Quadrature for NURBS-based Isogeometric Analysis. *Computer Methods in Applied Mechanics and Engineering*. Published online. doi:10.1016/j.cma.2008.12.004.

Hulbert, G.M. and Hughes, T.J.R. (1990). Space–time finite-element methods for 2nd-order hyperbolic equations. *Computer Methods in Applied Mechanics and Engineering*, 84:327–348.

Humphrey, J.D. (2002). *Cardiovascular Solid Mechanics*. Springer-Verlag.

Irons, B.M. (1966). Engineering application of numerical integration in stiffness method. *Journal of the American Institute of Aeronautics and Astronautics*, 14:2035–2037.

Jamet, D., Lebaigue, O., Coutris, N., and Delhaye, J.M. (2001). The second gradient method for the direct numerical simulation of liquid-vapor flows with phase change. *Journal of Computational Physics*, 169:624–651.

Jansen, K.E., Whiting, C.H., and Hulbert, G.M. (1999). A generalized-α method for integrating the filtered Navier–Stokes equations with a stabilized finite element method. *Computer Methods in Applied Mechanics and Engineering*, 190:305–319.

Johnson, A.A. and Tezduyar, T.E. (1994). Mesh update strategies in parallel finite element computations of flow problems with moving boundaries and interfaces. *Computer Methods in Applied Mechanics and Engineering*, 119:73–94.

Johnson, C. (1987). *Numerical Solution of Partial Differential Equations by the Finite Element Method*. Cambridge University Press.

Johnson, R.W. (2005a). A B-spline collocation method for solving the incompressible Navier–Stokes equations using an ad hoc method: the Boundary Residual method. *Computers and Fluids*, 34:121–149.

Johnson, R.W. (2005b). Higher order B-spline collocation at the Greville abscissae. *Applied Numerical Mathematics*, 52:63–75.

Kagan, P., Fischer, A., and Bar-Yoseph, P. Z. (1998). New B-spline finite element approach for geometrical design and mechanical analysis. *International Journal of Numerical Methods in Engineering*, 41:435–458.

Kagan, P., Fischer, A., and Bar-Yoseph, P. Z. (2003). Mechanically based models: Adaptive refinement for B-spline finite element. *International Journal of Numerical Methods in Engineering*, 57:1145–1175.

Kasik, D.J., Buxton, W., and Ferguson, D.R. (2005). Ten CAD model challenges. *IEEE Computer Graphics and Applications*, 25:81–92.

Korteweg, D.J. (1901). Sur la forme que prennent les équations du mouvement des fluides si l'on tient compte des forces capillaires causées par des variations de densité considérables mais continues et sur la théorie de la capillarité dans l'hypothése d'une variation continue de la densité. *Arch. Néerl. Sci. Exactes Nat.*, 6:1.

Kuttler, U., Forster, C., and Wall, W.A. (2006). A solution for the incompressibility dilemma in partitioned fluid–structure interaction with pure Dirichlet fluid domains. *Computational Mechanics*, 38:417–429.

Kwok, W. Y., Moser, R.D., and Jiménez, J. (2001). A critical evaluation of the resolution properties of B-spline and compact finite difference methods. *Journal of Computational Physics*, 174:510–551.

Lai, M.J. and Schumaker, L.L. (2007). *Spline Functions on Triangulations*. Cambridge University Press.

Lane, J. and Riesenfeld, R. F. (1980). A theoretical development for the computer generation and display of piecewise polynomial surfaces. *IEEE Transactions on Pattern Analysis and Machine Intelligence*, 2:35–46.

Laursen, T.A. (2002). *Computational Contact and Impact Mechanics*. Springer-Verlag.

Lele, S.K. (1992). Compact finite difference schemes with spectral-like resolution. *Journal of Computational Physics*, 103:16–42.

LeTallec, P. and Mouro, J. (2001). Fluid structure interaction with large structural displacements. *Computer Methods in Applied Mechanics and Engineering*, 190:3039–3068.

Li, X., Guo, X., Wang, H., He, Y., Gu, X., and Qin, H. (2007). Harmonic volumetric mapping for solid modeling applications. In *SPM '07: Proceedings of the 2007 ACM Symposium on Solid and Physical Modeling*, pages 109–120, New York, NY. ACM.

Lim, C. (1999). A universal parameterization in B-spline curve and surface interpolation. *Computer Aided Geometric Design*, 16:407–422.

Lipton, S., Evans, J.A., Bazilevs, Y., Elguedj, T., and Hughes, T.J.R. (2009). Robustness of isogeometric structural discretizations under severe mesh distortion. *Computer Methods in Applied Mechanics and Engineering*. Published online. doi:10.1016/j.cma.2009.01.022.

Liu, W.K., Jun, S., and Zhang, Y.F. (1995). Reproducing kernel particle methods. *International Journal of Numerical Methods in Engineering*, 20:1081–1106.

Livermore Software Technology Corporation (2007). *LS-DYNA Keyword User's Manual*. Livermore Software Technology Corporation. http://www.lstc.com.

Loop, C.T. (1987). Smooth subdivision surfaces based on triangles. Master's thesis, University of Utah.

Loop, C.T. and DeRose, T.D. (1989). A multisided generalization of Bézier surfaces. *ACM Transactions on Graphics*, 8:204–234.

Maker, B.N. (1995). NIKE3D a non-linear, implicit, three-dimensional finite element code for solid and structural mechanics. Technical Report UCRL-MA-105268 Rev. 1, Lawrence Livermore National Laboratory, *University of California*, Livermore.

Malkus, D.S. and Hughes, T.J.R. (1978). Mixed finite element methods – reduced and selective integration techniques: A unification of concepts. *Computer Methods in Applied Mechanics and Engineering*, 15:63–81.

Meirovitch, L. (1967). *Analytical Methods in Vibrations*. The MacMillan Company.

Melenk, J.M. and Babuska, I. (1996). The partition of unity finite element method: Basic theory and applications. *Computer Methods in Applied Mechanics and Engineering*, 139:289–314.

Mikhlin, S.G. (1964). *Variational Methods in Mathematical Physics*. Pergamon.

Miranda, I., Ferencz, R.M., and Hughes, T.J.R. (1989). An improved implicit-explicit time integration method for structural dynamics. *Earthquake Engineering and Structural Dynamics*, 18:643–653.

Moin, P. (2001). *Fundamentals of Engineering Numerical Analysis*. Cambridge University Press.

Moser, R., Kim, J., and Mansour, R. (1999). DNS of turbulent channel flow up to Re=590. *Physics of Fluids*, 11:943–945.

MSRI (1999). Workshop on mathematical foundations of computer aided design. Mathematical Sciences Research Institute, Berkeley, CA.

Munjiza, A., Owen, D.R.J., and Bicanic, N (1995). A combined finite-discrete element method in transient dynamics of fracturing solids. *Engineering Computations*, 12:145–174.

Nagtegaal, J.C., Park, D.M., and Rice, J.R. (1974). On numerically accurate finite element solutions in the fully plastic range. *Computer Methods in Applied Mechanics and Engineering*, 4:153–177.

Natekar, D., Zhang, X. F., and Subbarayan, G. (2004). Constructive solid analysis: a hierarchical, geometry-based meshless analysis procedure for integrated design and analysis. *Computer-Aided Design*, 36(5):473–486.

Nayroles, B., Touzot, G., and Villon, P. (1992). Generalizing the finite element method: diffuse approximations and diffuse elements. *Computational Mechanics*, 10:307–318.

Nedelec, J.C. (1980). Mixed finite elements in ir3. *Numerische Mathematik*, 35:315–341.

Newmark, N.M. (1959). A Method of Computation for Structural Dynamics. *Journal of the Engineering Mechanics Division, ASCE*, pages 67–94.

Oñate, E., Idelsohn, S., Zienkiewicz, O.C., and Taylor, R.L. (1996). A finite point method in computational mechanics. applications to convective transport and fluid flow. *International Journal of Numerical Methods in Engineering*, 39:3839–3866.

Oñate, E. and Zarate, F. (2000). Rotation-free triangular plate and shell elements. *International Journal of Numerical Methods in Engineering*, 47:557–603.

Parvizian, J., Düster, A., and Rank, E. (2007). Finite cell method: h- and p-extension for embedded domain problems in solid mechanics. *Computational Mechanics*, 41:121–133.

Peters, J. and Reif, U. (2008). *Subdivision Surfaces*. Springer-Verlag.

Phaal, R. and Calladine, C. R. (1992). A simple class of finite-elements for plate and shell problems: 1. Elements for beams and thin flat plates. *International Journal of Numerical Methods in Engineering*, 35:955–977.

Piegl, L. and Tiller, W. (1997). *The NURBS Book (Monographs in Visual Communication), Second Edition*. Springer-Verlag.

Prenter, P.M. (1975). *Splines and Variational Methods*. Wiley.

Provatidis, C.G. (2009). Integration-free Coons macroelements for the solution of 2D Poisson problems. *International Journal of Numerical Methods in Engineering*, 77:536–557.

Ramshaw, L. (1987a). Blossoming: a connect-the-dots approach to splines. Technical report, Digital Systems Research Center, Palo Alto, CA.

Ramshaw, L. (1987b). Blossoms: a connect-the-dots approach to splines. Technical report, Digital Systems Research Center, Palo Alto, CA.

Ramshaw, L. (1989). Blossoms are polar forms. *Computer Aided Geometric Design*, 6:323–358.

Rank, E., Düster, A., Nübel, V., Preusch, K., and Bruhns, O. T. (2005). High order finite elements for shells. *Computer Methods in Applied Mechanics and Engineering*, 194 (21–24):2494–2512.

Raviart, P.A. and Thomas, J.M. (1977). Primal hybrid finite element methods for 2nd-order elliptic equations. *Mathematics of Computation*, 31:391–413.

Realsoft Graphics (2008). http://www.realsoft.com.

Riesenfeld, R.F. (1972). *Application of B-spline Approximation to Geometric Problems of Computer Aided Design*. PhD thesis, Syracuse University.

Rogers, D. F. (2001). *An Introduction to NURBS With Historical Perspective*. Academic Press.

Roma, A.M., Peskin, C.S., and Berger, M.J. (1999). An adaptive version of the immersed boundary method. *Journal of Computational Physics*, 153:509–534.

Saad, Y. (2003). *Iterative Methods for Sparse Linear Systems, Second Edition*. Society for Industrial and Applied Mathematics.

Saad, Y. and Schultz, M.H. (1986). GMRES: A generalized minimal residual algorithm for solving nonsymmetric linear systems. *SIAM Journal of Scientific and Statistical Computing*, 7:856–869.

Sabin, M.A. (1997). *Spline Finite Elements*. PhD thesis, Cambridge University.

Sakurai, H. (2006). Element-Free Methods vs. Mesh-Less CAE. *International Journal of Computational Methods*, 3:445–464.

Saulev, V.K. (1962). A method for automatization of the solution of boundary value problems on high performance computers. *Doklady Akademii Nauk SSSR*, 144:497–500.

Saulev, V.K. (1963). On solution of some boundary value problems on high performance computers by fictitious domain method. *Siberian Mathematics Journal*, 4:912–925.

Schoenberg, I. (1946). Contributions to the problem of approximation of equidistant data by analytic functions. *Quarterly of Applied Mathematics*, 4:45–99.

Schultz, M.H. (1973). *Spline Analysis*. Prentice-Hall.

Schumaker, L.L. (2007). *Spline Functions: Basic Theory, Third Edition*. Krieger.

Sederberg, T.W., Anderson, D.C., and Goldman, R.N. (1984). Implicit representation of parametric curves and surfaces. *Computer Vision, Graphics and Image Processing*, 28:72–84.

Sederberg, T.N., Zhengs, J.M., Bakenov, A., and Nasri, A. (2003). T-splines and T-NURCCSs. *ACM Transactions on Graphics*, 22 (3):477–484.

Sederberg, T.W., Cardon, D.L., Finnigan, G.T., North, N.S., Zheng, J., and Lyche, T. (2004). T-spline simplification and local refinement. *ACM Transactions on Graphics*, 23 (3):276–283.

Sederberg, T.W., Finnigan, G.T., Li, X., Lin, H., and Ipson, H. (2008). Watertight trimmed NURBS. In *SIGGRAPH '08: ACM SIGGRAPH 2008 papers*, pages 1–8, New York, NY. ACM.

Shakib, F., Hughes, T.J.R., and Johan, Z. (1991). A new finite element formulation for computational fluid dynamics: X. The compressible Euler and Navier-Stokes equations. *Computer Methods in Applied Mechanics and Engineering*, 89:141–219.

Simo, J.C. and Hughes, T.J.R. (1998). *Computational Inelasticity*. Springer-Verlag, New York.

Simo, J.C. and Taylor, R.L. (1991). Quasi-incompressible finite elasticity in principal stretches. Continuum basis and numerical algorithm. *Computer Methods in Applied Mechanics and Engineering*, 85:273–310.

Simo, J.C., Taylor, R.L., and Pister, K.S. (1985). Variational and projection methods for the volume contraint in finite deformation elasto-plasticity. *Computer Methods in Applied Mechanics and Engineering*, 51:177–208.

Stanley, G.M. (1985). *Continuum-based Shell Elements*. PhD thesis, Division of Applied Mechanics, Stanford University.

Stoll, C. (May, 2006). When slide rules ruled. *Scientific American*, pages 80–87.

Strang, G. and Fix, G.J. (1973). *An Analysis of the Finite Element Method*. Prentice-Hall.

Surana, K.S., Ahmadi, A.R., and Reddy, J.N. (2002). The k-version finite element method for self-adjoint operators in BVP. *International Journal of Computational Engineering Science*, 3:155–218.

Szabo, B. and Babuska, I. (1991). *Finite Element Analysis*. Wiley.

Szabo, B., Düster, A., and Rank, E. (2004). The p-version of the finite element method. In Stein, E., de Borst, R., and Hughes, T.J.R., editors, *Encyclopedia of Computational Mechanics, Vol. 1, Fundamentals*, chapter 5. Wiley.

T-Splines, Inc. (2008a). http://www.tsplines.com/maya/.

T-Splines, Inc. (2008b). http://www.tsplines.com/rhino/.

Taig, I.C. (1961). Structural analysis by the matrix displacement method. Technical report, English Electric Aviation.

Taylor, C.A., Hughes, T.J.R., and Zarins, C.K. (1998). Finite element modeling of three-dimensional pulsatile flow in the abdominal aorta: relevance to atherosclerosis. *Annals of Biomedical Engineering*, 26:975–987.

Taylor, C.A., Hughes, T.J.R., and Zarins, C.K. (1999). Effect of exercise on hemodynamic conditions in the abdominal aorta. *Journal of Vascular Surgery*, 29:1077–1089.

Taylor, R.L. (1972). On completeness of shape functions for finite element analysis. *International Journal of Numerical Methods in Engineering*, 4:17–22.

Terzopoulos, D. and Qin, H. (1994). Dynamic NURBS with geometric constraints for interactive sculpting. *ACM Transactions on Graphics*, 13:103–136.

Tezduyar, T.E. (2003). Computation of moving boundaries and interfaces and stabilization parameters. *International Journal of Numerical Methods in Fluids*, 43:555–575.

Tezduyar, T. E. and Sathe, S. (2007). Modelling of fluid–structure interactions with the space-time finite elements: Solution techniques. *International Journal of Numerical Methods in Fluids*, 54:855–900.

Tezduyar, T.E., Behr, M., Mittal, S., and Johnson, A.A. (1992). Computation of unsteady incompressible flows with the stabilized finite element methods – space-time formulations, iterative strategies and massively parallel implementations. In *New Methods in Transient Analysis*, PVP-Vol. 246/ AMD-Vol. 143, pages 7–24. ASME, New York.

Thurston, W.P. (1982). Three-dimensional manifolds, kleinian groups and hyperbolic geometry. *Bulletin of the American Mathematical Society (New Series)*, 6:357–381.

Thurston, W.P. (1997). *Three-Dimensional Geometry and Topology, Vol. 1*. Princeton University Press.

Timoshenko, S. and Woinowsky-Krieger, S. (1959). *Theory of Plates and Shells (Engineering Societies Monograph), Second Edition*. McGraw-Hill.

Turner, M.J., Clough, R.W., Martin, H.C., and Topp, L.J. (1956). Stiffness and deflection analysis of complex structures. *Journal of Aeronautical Sciences*, 23:805–823.

Versprille, K.J. (1975). *Computer-aided Design Applications of the Rational B-spline Approximation Form*. PhD thesis, Syracuse University.

Vichnevetsky, R. and Bowles, J.B. (1982). *Fourier Analysis of Numerical Approximations of Hyperbolic Equations*. Society of Industrial and Applied Mathematics.

Volpin, O., Bercovier, M., and Matskewich, T. (1999). A comparison of invariant energies for free-form surface construction. *Visual Computer*, 15:199–210.

Wahlbin, L.B. (1991). Local behavior in finite element methods. In Ciarlet, P.G. and Lions, J.L., editors, *Finite Element Methods (Part 1)*, volume 2 of *Handbook of Numerical Analysis*, pages 353–522. North-Holland.

Wall, W.A., Frenzel, M.A., and Cyron, C. (2008). Isogeometric structural shape optimization. *Computer Methods in Applied Mechanics and Engineering*, 197:2976–2988.

Warren, J. and Weimer, H. (2002). *Subdivision Methods for Geometric Design*. Morgan Kaufmann Publishers.

Wells, G.N. and Dung, N.T. (2007). A C^0 discontinuous Galerkin formulation for Kirchhoff plates. *Computer Methods in Applied Mechanics and Engineering*, 196:4985–5000.

Wells, G.N., Kuhl, E., and Garikipati, K. (2006). A discontinuous Galerkin method for the Cahn–Hilliard equation. *Journal of Computational Physics*, 218:1480–1498.

Wheeler, M.F. (1978). An elliptic collocation–finite element method with interior penalties. *SIAM Journal of Numerical Analysis*, 15:152–161.

Whitham, G.B. (1974). *Linear and Nonlinear Waves*. Wiley.

Wood, W.L., Bossak, M., and Zienkiewicz, O.C. (1980). An alpha modification of newmark's method. *International Journal of Numerical Methods in Engineering*, 15:1562–1566.

Wriggers, P. (2002). *Computational Contact Mechanics*. Wiley.

Wriggers, P. and Zavarise, G. (2007). A formulation for frictionless contact problems using a weak form introduced by Nitsche. *Computational Mechanics*. DOI Number 10.1007/s00466-007-0196-4.

Zhang, Y., Bazilevs, Y., Goswami, S., Bajaj, C., and Hughes, T.J.R. (2007). Patient-specific vascular NURBS modeling for isogeometric analysis of blood flow. *Computer Methods in Applied Mechanics and Engineering*, 196:2943–2959.

Zienkiewicz, O.C. and Cheung, Y.K. (1968). *The Finite Element Method in Structural and Continuum Mechanics*. McGraw-Hill.

Zienkiewicz, O.C., Irons, B.M., Campbell, J., and Scott, F.C. (1970). Three dimensional stress analysis. In *IUTAM Symposium on High Speed Computing in Elasticity*. Liège

Zienkiewicz, O.C., Taylor, R.L., and Too, J.M. (1971). Reduced integration technique in general analysis of plates and shells. *International Journal of Numerical Methods in Engineering*, 3:275–290.

Index

Printed and bound in the UK by
CPI Antony Rowe, Eastbourne

Printed and bound by CPI Group (UK) Ltd, Croydon, CR0 4YY

16/04/2025

14658405-0001